Wheat Landraces

Nusret Zencirci • Faheem S. Baloch
Ephrem Habyarimana • Gyuhwa Chung
Editors

Wheat Landraces

Springer

Editors
Nusret Zencirci
Biology Department
Bolu Abant İzzet Baysal University
Gölköy, Bolu, Turkey

Faheem S. Baloch
Agricultural Sciences and
Technology Faculty
Sivas Science and Technology University
Gölköy, Bolu, Turkey

Ephrem Habyarimana
Accelerated Crop Improvement Program
International Crops Research Institute
for the Semi-Arid Tropics
Patancheru, Telangana, India

Gyuhwa Chung
Department of Biotechnology
Chonnam National University
Yeosu, Korea (Republic of)

ISBN 978-3-030-77387-8 ISBN 978-3-030-77388-5 (eBook)
https://doi.org/10.1007/978-3-030-77388-5

This Springer imprint is published by the registered company Springer Nature Switzerland AG
The registered company address is: Gewerbestrasse 11, 6330 Cham, Switzerland

Foreword

Landraces are a repository of the gene pool that enriches and maintains biodiversity and sustainably stabilize the ecosystem to make it functional. Plant landraces as heterogeneous local adaptations of domesticated species provide genetic resources that meet current and future challenges for farming in stressful environments. These local ecotypes can show variable phenology and low to moderate edible yield but are often highly nutritious crops. The main contributions of landraces to plant breeding have been traits for more efficient nutrient uptake and utilization, as well as useful genes for adaptation to stressful environments such as cold, drought, salinity, and higher temperatures. A systematic landrace evaluation may define patterns of diversity, which will facilitate identifying alleles for enhancing yield and abiotic stress adaptation, thus raising the productivity and stability of staple crops in vulnerable environments. Farmers have been growing wheat landraces composed of traditional varieties for years, induced by natural and human selection, which, in return, adapt consequently to local ecological conditions and management practices

Turkey falls within Vavilov's center of origin and harbors large genetic diversity of several economically important crop species. Wild crop relatives, and their respective domesticated forms, in addition to a multitude of other crop species, have been cultivated for millennia in several parts of Turkey, especially in the northern part of the Fertile Crescent. Turkish farmers derived landraces from these crops; however, these landraces and the indigenous knowledge gained over many generations are being lost due to several anthropogenic and other factors. Though, landraces of wheat still play an important role in the livelihood of small-scale farmers in Turkey.

Wheat has been a staple crop in the Anatolian region since prehistoric times. Anatolia has been home to vast numbers of farming cultures, from the initial waves of Neolithic migrants to modern times. Early Indo-European cultures, Hittites, Hellenic, and Byzantine cultures, Romans, and Turkish cultures have successively occupied Anatolia since the beginnings of agriculture there. The diversity of wheat in Anatolia is large, as farmers have identified, multiplied, and preserved them for millennia. Farmers have been growing wheat landraces composed of traditional varieties through years of natural and human selection that are as a consequence adapted to local ecological conditions and management practices

The first thing to keep in mind is that loss of landraces is continuous and irreversible without any end. Landraces will inevitably continue to be replaced by genetically uniform cultivars. There may be ways of maintaining some of the landraces sustainably for long periods, but there seems to be no way of conserving all the landraces forever.

Landraces of wheat still play an important role in the livelihood of small-scale farmers in Turkey. Several factors including physical, climatic, socio-economic, market facilities, and pricing policies play a major role in the cultivation of landraces. It is estimated that cereal landraces cover almost 800,000 ha in Turkey. Market creation, development of benefit-sharing regimes, conservation of traditional knowledge related to landrace utilization, development of on-farm breeding facilities, growing landrace mixtures, amendment of the seed registration system and/or creation of a special registry system for landraces, and use of landraces in organic farming systems are suggested for the sustainability of crop landraces.

Çukurova University, Agricultural Faculty
Field Crops Department Adana, Turkey

Hakan Özkan

Contents

1 **Introduction** 1
Mehmet Örgeç, Çisem Nildem Doğan, Ferdi Ağıl, Günce Şahin, and Nusret Zencirci

2 **Wheat Landraces in Mesopotamia** 13
Fethiye Özberk and Irfan Özberk

3 **Conservation of Plant Genetic Resources** 35
Kürşad Özbek and Nusret Zencirci

4 **Characterisation of Genetic Diversity in Wheat Landraces** 59
Özlem Özbek

5 **Macro-Microelements in Wheat Landraces and Their Use in Breeding** 83
Ayten Salantur and Cuma Karaoğlu

6 **Nutritional and Technological Properties of Wheat Landraces** 93
Asuman Kaplan Evlice

7 **Total Economic Value of Wheat Landraces** 121
Sevinç Karabak and Mustafa Kan

8 **Chemical Contents of Wheat Landraces and Their Contribution to Human Health** 147
Cisem Nildem Keskin, Fatma Pehlivan Karakas, and Ferdi Ağıl

9 **Climate Change and Global Warming Effect(s) on Wheat Landraces: A General Approach** 169
Hakan Ulukan

10 Wheat Landraces Versus Resistance to Biotic and Abiotic Stresses 193
Awatef Ali Shlibak, Mehmet Örgeç, and Nusret Zencirci

11 Contribution of Landraces in Wheat Breeding 215
Muhammad Azhar Nadeem, Mehmet Zahit Yeken, Mehmet Tekin, Zemran Mustafa, Rüştü Hatipoğlu, Husnu Aktaş, Ahmad Alsaleh, Evren Cabi, Ephrem Habyarimana, Nusret Zencirci, Tolga Karaköy, Hakan Özkan, and Faheem Shehzad Baloch

Index 259

Chapter 1
Introduction

Mehmet Örgeç, Çisem Nildem Doğan, Ferdi Ağıl, Günce Şahin, and Nusret Zencirci

1.1 Origin and Evolution of Wheat

When humans are being turned to the agriculture-based society from the hunter-gatherer society, domesticating this cereal (wheat) became essential. Because of its easy harvest, high yield, and long-term storage, people unwittingly selected this crop and consequently its useful genes. Natural selection and hybridization between different species are also made and improved changes in wheat cultivars (Gustafson et al. 2009). Hybridization, drift, migration, and natural selection have impressed the generation of modern cultivars' genotype and as shown in their evolution by researches (Nevo et al. 2002).

Wheat species and the whole Triticeae tribe have been known to have n = 1x = 7 chromosome number since the 1900s. Einkorn, emmer, durum, rivet, Polish, Persian, spelt, bread, club, and Indian shot are some examples of cultivated wheats. Einkorn (*Triticum monococcum* ssp. *monococcum*) has diploid (2n = 2x = 14, AA) chromosomes; emmer (*Triticum dicoccum*) and durum (*Triticum durum*) have 2n = 4x = 28, BBAA; and spelt (*Triticum spelta*) and bread wheat (*Triticum aestivum* L.) have hexaploid (2n = 6x = 42, BBAADD) chromosomes. Chromosomes (1 to 7) in the diverse diploid genomes (B, A, and D) are suggested that they are related with wheat evolution (Gustafson et al. 2009).

One of the three genomes is A in wheat which is a part of the modern wheat evolution. It is known that *T. urartu Thumanjan ex Gandilyan* is the donor to A genome. Although there are some discussions about the donor of B genome, *Ae. speltoides Tausch* is strongly considered to be the origin. There is also a consensus about the donor of D which comes from *Ae. tauschii Coss*. Those with AA, BB, and

M. Örgeç · Ç. N. Doğan · F. Ağıl · G. Şahin · N. Zencirci (✉)
Department of Biology, Faculty of Science and Art, Bolu Abant Izzet Baysal University, Gölköy, Bolu, Turkey

N. Zencirci et al. (eds.), *Wheat Landraces*,
https://doi.org/10.1007/978-3-030-77388-5_1

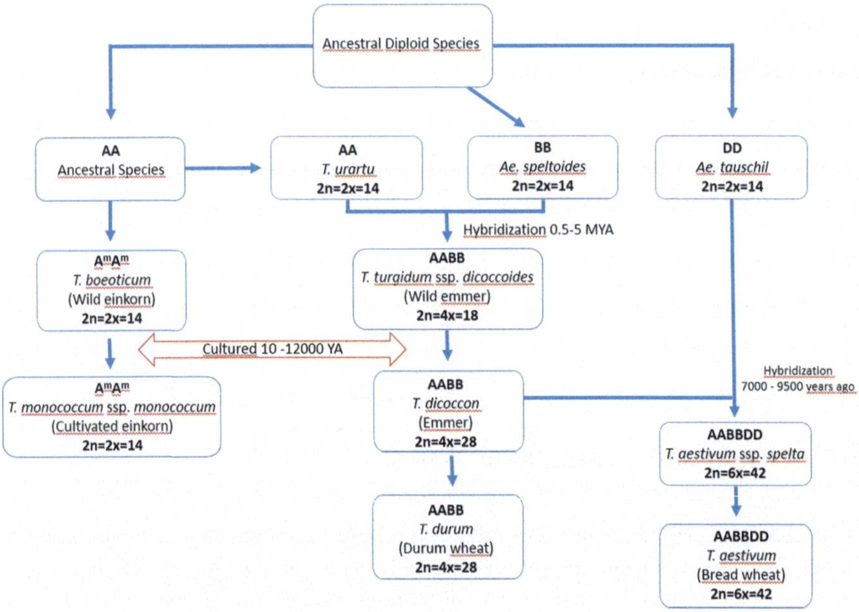

Fig. 1.1 Evolution of wheat. (Modified from Chantret et al. (2005))

DD chromosomes have 2n = 14, those with AABB chromosomes have 2n = 28, and those with AABBDD chromosomes have 2n = 42 chromosomes (Fig. 1.1).

There are two types of modern wheat cultivars: one of them is hexaploid bread wheat (*Triticum aestivum* L., 2n = 6x = 42) and the other is tetraploid hard or durum wheat (*T. turgidum* L. (Thell.)) (Gustafson et al. 2009). It is known that the modern cultivar of wild emmer came from the hybridization of *Ae. speltoides* (2n = 2x = 14) and *T. urartu* (2n = 2x = 14). In the process of evolution, first, the wild emmer (*T. dicoccoides Körn. ex Asch. & Graebn. Schweinf.*) transformed to emmer (*T. dicoccon*) and then to tetraploid wheats such as *T. turgidum* L., *T. polonicum* L., *T. carthlicum Nevski*, and *T. durum Desf.* Moreover, bread wheat emerged from the hybrid of *T. dicoccoides* and *Ae. Tauschii* (Özberk et al. 2016).

1.2 History of Wheat and Breeding Studies in Turkey

Southwest Asia is the place from where all wheat originated. First evidence of domestic wheats was in the Fertile Crescent, Central Asia, and southern China, around 10,000–12,000 years ago (Gustafson et al. 2009). Current evidence from genetics, botanical, and archaeology showed that the core of cereal agriculture was located in today's Southeastern Turkey and north Syria (Lev-Yadun et al. 2000). Einkorn and emmer, early domesticated wheat species, have been known to be cultivated around 10,000 BP (Harlan 1981). Site of Çatalhöyük in Turkey is a place that the first evidence had been found about the domesticated bread wheat around 7800 BP (Harlan 1975). Until 5000 BP, einkorn and emmer had been grown for about 3000 years, and according to the archeological findings, they suddenly almost vanished in Southeastern Turkey. Today, einkorn, emmer, and, of course, durum and bread wheat are popular again since they are important for diet as they are revalued in West Asia, North Africa, and Turkey as they were for the last 120,000 years (Braun 2011).

Because of its higher wheat production, Turkey is among the top ten wheat-producing countries in the world (Braun 1999). The total annual production of wheat in Turkey changes between 16 and 21 million tons (Özberk et al. 2005). The annual production of durum wheat reaches five million tons. Because of that, Turkey also became a leader of durum wheat production, especially among North African and West Asian countries (Özberk et al. 2005). Also, nearly one million ton of durum wheat grown in Turkey produced 750,000 tons of macaroni in 1998. Therefore, Turkey ranks fifth in wheat production (Özberk et al. 2005). The usage of wheat is varying from making bread to making pasta. That is why its consumption is very high worldwide, and the annual wheat consumption in Turkey per capita is 200 kg (Braun et al. 2001).

Talking about wheat landraces, one scientist deserves mentioning is Mirza Gökgöl. He is a leading plant scientist who collected various cultivated wheat samples all around Turkey between 1929 and 1955 and reported valuable information about Turkey's wheat landraces. In 1935, Gökgöl collected wheat samples and researched their genetic variations. At the end of his study, 18,000 types and 256 new varieties of wheat have been described (Karagöz and Özberk 2014; Zencirci et al. 2018). Wheats which were grown even in distinct areas of Turkey were reported in detail. According to him, Turkey does not need to introduce foreign wheat material, because it has very rich wheat varieties (Knüpffer et al. 2015).

Before the Turkish Republic was founded in 1923, researches about wheats were started in 1920. In 1925, the first Seed Improvement Station was founded in Eskişehir. The other stations followed Eskişehir were Adapazarı, Yeşilköy (Istanbul), Ankara, Samsun, Adana, and Antalya (Zencirci et al. 1996). Ak 702 wheat cultivar was released after the research in Eskişehir Research Station in 1931. Then, Sertak 52, Köse Melez 1713, and Kara Kılcık were released. Yayla 305 which was released in 1939 was selected by Emcet Yektay, another distinct Turkish wheat scientist. One advantage of Yayla 305 was its resistance to bunt; therefore, it was preferred to be cultivated by farmers. Besides these, Melez-13 was the first variety which was obtained from a crossbreeding program in 1944 (Braun et al. 2001). Köse 220–39 and Sivas 111–33 population selections were also cultivated in larger acreages for long years. At the same time, Manitoba cultivar from Canada to Central Anatolia, Mentana from Italy, and Florence from Australia were also introduced. Later, Kunduru 414/44 was improved by reselection, and Ankara 093/44 and Akova wheats were released after earlier crossing programs. Due to World War II, the wheat production and acreage sown decreased in Turkey. Because of this fall, people requested for an increase in wheat production. Despite the increase of wheat production after World War II, wheat production was still not enough for the increasing population in Turkey. Accelerating wheat breeding and exchanging of material were restarted in 1950. Local materials (with winter hardiness, good quality, and wide adaptation) were crossed with foreign materials (disease resistance and higher yield) to improve better adapted cultivars. Mexico established, with the support from Rockefeller Foundation, International Wheat and Maize Improvement Center (CIMMYT) in 1943. Then, some materials were sent to Turkey such as Sonora 64. These cultivars increased the wheat yield from 1.5 ton/hectare to 4 ton/hectare in Çukurova. Lerma Rojo 64, Penjamo 62, Sonora 63, Sonora 64, Mayo 64, and Super X were the other Mexican spring wheat cultivars which were introduced to Turkey. Also, Brevor, Scout, Gaines, Burt, Wanser, Gage, Warrior, Lancer, Duruchamp, and Nugaines cultivars were introduced from the USA and Bezostaja-1, Odeskaya-51, Harkovskaya, and Miranovskaya-808 from Russia in 1967 (Zencirci et al. 1996; Özberk et al. 2016).

The National Cool Season Cereals Research and Training Project was founded by the agreement between the Rockefeller Foundation and Turkish Government in 1967. The basic aims of this project were to develop techniques for the improvement of cultivars, enhance water-use efficiency in soil, decrease damages caused by insects and diseases, and educate people and seek new agricultural techniques and to search for economic wheat production. Moreover, some technical information, employees, germplasm, and education opportunities were provided by International Wheat and Maize Improvement Center (CIMMYT), Mexico, and Oregon State University, USA (Zencirci et al. 1996). With this project, the production of wheat nearly doubled by 1982. Therefore, the average yields have grown from 1.1 to 1.82 ton/ha (Özberk et al. 2005). Eskişehir Research Institute released Gerek-79 when the project was still alive. Some other cultivars of these early years of start were Çakmak-79, Gökgöl-79, Tunca-79, and Haymana-79. Gerek-79 was sown on more than 1.5 million hectares by 1996, and it stayed as the leading winter wheat cultivar for years (Braun et al. 2001).

International Winter Wheat Improvement Program (IWWIP) is a program which was initiated by the Government of Turkey, International Center for Agricultural Research in the Dry Areas (ICARDA), and CIMMYT in the mid-1980s. The main aim of this program is to develop winter and facultative wheat germplasm for Central and West Asia. IWWIP also encouraged the exchange of winter germplasm for the global breeding community. ICARDA was incorporated into a project in 1991 by integrating the facultative wheat breeding activities in Syria (iwwip.org). IWWIP developed winter and facultative wheats which were released in Argentina, Pakistan, Iran, Afghanistan, Turkey, and Tajikistan. Winter wheats (>157) are scattered around the world by IWWIP (Braun et al. 2001).

Recently, a total of 12 different local wheat varieties are relatively important and have been reportedly sown in Turkey mostly. These are Zerun, Ak, Kırmızı, Sarı, Karakılçık, Kirik, Siyez, Koca, Topbaş, Şahman, Üveyik, and Göderedi. The largest local wheat production has been in Southeastern Anatolia region. These populations can be considered to be used for different purposes. Some are especially preferred for bread making such as Zerun, Kırmızı Buğday, Kırik, etc. On the other hand, some are more important for bulgur making (Siyez, Şahman, Sarı Wheat etc.) (Kan et al. 2017).

1.3 Importance of Wheat Landraces

Wheat, one of the main foodstuffs in human nutrition, is among the most important food products in the world. Progressive increases in the population direct the development and call for production policies which aimed at ensuring food safety and security within the sustainability framework. Wheat is one of the most important sources of income in rural areas together with its nutrition and strategic importance (Karabak et al. 2012). Moreover, wheat has economic, social, cultural, historical, and even archeological value. Throughout the history, wheat has been deeply influenced and developed by many civilizations (Özberk et al. 2016).

In 2017, 771 million tons of wheat was produced worldwide. China, India, Russia, and the USA are among the leading wheat-producing countries in the world. Turkey is among the top ten wheat-producing countries worldwide. Although the amount of wheat produced increases regularly, its production area is gradually decreasing (FAO, Food and Agriculture Organization 2017; TUIK, Turkish Statistical Institute 2017). Local varieties are used as the basic genetic materials in wheat breeding programs across the world and in Turkey. "Particularly rapid increase in human population, technological changes, and infrastructure development" destroyed wheat production facilities rapidly (Frankel 1970). For this reason, the demand for basic wheat nutrients is increasing, including countries whose climates are not suitable for wheat growing. Similarly, climate change induced by global warming has started to affect, directly or indirectly, modern and traditional wheat production systems. Therefore, wheat production, which is expected still to be one of the main nutrients in the future, is expected to decrease. Then, it would become hard to meet the required amount of nutrients with the reduced production (Jaradat 2013).

With the green revolution, worldwide primitive landrace varieties have been replaced by bred cultural varieties, obviously resulting in a serious genetic erosion. Many of old wheat and barley varieties existed in Turkey, Iraq, Iran, and Pakistan have been replaced by these new cultivars. Genetic uniformity has increased with modern cultivars improved based on local wheat genetic resources. This genetic uniformity disrupted the heterogeneity of plant plasma in conventional agricultural systems and led to the emergence of epidemic diseases that attack these genetically uniform products. The reason why cultural crops are greatly damaged by diseases in the world is that there are very few limited resistance genes present in the available cultivars. On the other hand, the number of diseases increased and diversified. Starting from the 1950s, genetic natural wheat resources were rapidly destroyed (Altındal and Akgün 2015).

Wheat is a food source that is extensively produced in the world and consumed in different ways. In general, wheat used in bread making is also used to produce bulgur, pasta, biscuits, and flour (Özberk et al. 2016). Wheat is rich in micronutrients, including mineral substances, B vitamins, E vitamins, total phenol content, and antioxidant content, in the local varieties (Zhao et al. 2009; Cummins and Roberts-Thomson 2009). Together with these phytochemical and biological activities, wheat has also some medicinal effects on diseases (cardiovascular diseases, diabetes, and cancer). Some studies show a 21% lower risk of cardiovascular disease in the whole grain and bran fibers. In addition, 27% of daily cereal nutrition consumption decreased the risk of Type 2 diabetes. Due to its antioxidant activity, wheat prevents cancer diseases (Mozaffarian et al. 2003; Liu 2003; Şahin et al. 2017).

Turkey is a country so eligible for wheat farming in terms of both the farming culture and the environmental structures. When the statistical data by organizations of TUIK and FAO examination points out an annual production about 20 million tons of wheat in Turkey and it shows that Turkey has an income of approximately 5-7 billion dollars from wheat (FAOSTAT 2017; TUIK 2017). The improvement of new wheat species seen throughout the world is also observed in Turkey. This situation constitutes a threat to the existence of native landraces/species in Turkey. Moreover, highly efficient wheat production programs in Turkey have caused serious genetic erosion (Karagöz 2014a; b). This situation can be prevented by collecting local wheat varieties and preserving them in ex situ (outside their natural growing area) or in situ (in their natural growing area) regions while agricultural activities continue (Kan et al. 2017) (Fig. 1.2).

A local variety can be defined in many ways. Harlan (1995) describes the landraces as "well-matched populations – variable in equilibrium with both environment

Fig. 1.2 A view at Einkorn Wheat Field Day in Seben, Bolu, Turkey (2018)

and pathogens – and genetically dynamic." They are "local" when seed from that variety has been planted in the region for at least one farmer generation (Louette 2000). According to Biodiversity International, landrace is defined as follows: "A landrace of a seed-propagated crop is a variable population, which is identifiable and usually has a local name. It lacks 'formal' crop improvement, is characterized by non-specific adaptation to the environmental conditions of the area of the cultivation (tolerant to the biotic and abiotic stresses of that area) and is closely associated with the uses, knowledge, habits, dialects, and celebrations of people who developed and continued to grow it" (Karagöz 2014a; b).

Einkorn and emmer, ancient wheat species, are the first cultivated wheats. Einkorn wheat was found firstly in Karacadağ region, Turkey (Heun et al. 1997). Emmer wheat is the tetraploid ancestor of durum and bread wheat (Emebiri et al. 2008). Among crop types, diversity in Turkish wheat has always attracted greater interest since the beginning of the twentieth century. In the first quarter of the twentieth century, the leading Turkish scholar Mirza Gökgöl has collected local wheat varieties and evaluated them for basic features. As a result of this analysis, Gökgöl showed that almost all wheat varieties existed in Turkey, and the region provided an endless treasure for the breeders (Gökgöl 1935; Gökgöl 1955; Zencirci et al. 2018). Karagöz (1996) reported that farmed einkorn wheat was grown by farmers in fields with small amounts and no irrigation facilities in the north of Turkey such as Sinop, Kastamonu, Bursa, and Bolu provinces. These varieties, which are generally used as animal feed, are also used in bulgur, bread, and food. Although the varieties produced are lower yielding than modern varieties, they have higher nutritional values. Lower yield is due to the fact that the local wheat varieties are better adapted to the negative production conditions (biotic and abiotic) than modern varieties (Tan 2002). Russian scientist N. I. Vavilov identified eight rich centers for both wild relatives and old local varieties of crops around the world. Turkey is taking part in two of these centers (Vavilov 1951).

1.4 Advantages and Disadvantages

Wheat landraces are traditional crop varieties developed by farmers over the years by natural or artificial human selection and adapted to local growing conditions and management practices (Zeven 1999). Several factors such as physical, climatic, and socioeconomic conditions, market facilities, etc. play an important role in cultivation of landraces (Karagöz 2014a, b). Modern wheat species may not be enough to fill social, such as traditional food making, and economic conditions as wheat landraces do. Landraces can provide reliable sustenance and sustainable food source to local communities such as bulgur, macaroni, and fodder for their animals.

Landraces have higher level of nutrients such as copper, iron, magnesium, manganese, phosphorus, selenium, and zinc more than those in modern wheat cultivars (Jaradat 2013). Due to the high concentration of tocols, carotenoids, and lutein in these landrace varieties, they are more protective against chronic diseases such as cancer and diabetes. Because of these features, landraces have been considered as a healthy food; therefore, people consume those wheat varieties more (Azeez et al. 2018). With these kinds of intentions, people increase values of landraces. With all these reasons, wheat landraces like other crop landraces have become so important resources in breeding programs.

In traditional agricultural systems, farmers generate and conserve new varieties, most of which are landraces. Those farmers are only consumers of these landrace products selected, saved, and recycled by themselves. This individual knowledge process leads to a very dynamic genetic landrace structure (Özbek 2014). Reliable yield level, another important reason for retention of landraces, is because of resistance of landraces to marginal stress conditions.

Wheat landraces adapt better to changing climate conditions and stressed environments than modern cultivars (Jaradat 2013), and they provide useful genetic traits (Azeez et al. 2018). While landraces arisen through natural or human selection by years (Dotlağıl et al. 2010), most modern cultivars have been improved by professional breeders. Therefore, genetic base has become narrower in wheat or in other crops as well. In a wheat improvement program, scientist needs to take advantages of new genetic diversity resource (Dotlağıl et al. 2010). In that context, wheat landraces are important resources to improve the genetic base of modern wheat cultivars by providing valuable breeding characteristics by their more comprehensive genetic base.

Landraces also have tolerance to other abiotic stresses and the resulting good yield and developmental stability. Previous studies about einkorn and bread wheat under cold and drought stress applications shown that einkorn wheat landraces performed better than those stresses according to plant development parameters such as germination rate, germination power, and leaf length when compared to bread wheat (Aslan et al. 2016; Aktaş et al. 2017). The development of new varieties from landrace populations is an applicable strategy to increase landrace yield and yield stability and to endure expected future climate change conditions (Witcombe et al. 1996).

The genetic structure of wheat landraces is an evolutionary output to survival especially under arid and semiarid conditions. The effects of natural and human selections have led to structure of genotypes representing different combinations of traits, such as growth habit; cold, heat, or drought tolerance; early growth vigor time to heading and maturity; and quality traits. As a result, wheat landraces have become complex, variable, and diverse populations in equilibrium with both biotic and abiotic stresses in their environment.

On the other hand, wheat landraces have some disadvantages. Wheat landraces have been largely relegated by high-yielding cultivars in many developing countries and rarely cultivated in developed countries because of their low yield potential (Azeez et al. 2018). The other disadvantage of wheat landrace varieties is traditionally lower market price, though it has changed these days, and limited selling strategies (FAO 2015). Due to their good flavor, taste, and cultivation in local regions, they have higher prices and are sold at niche markets and luxury stores these days.

These two issues also decrease consumption of wheat landraces at a large scale by society.

Agriculture has been a great milestone at evolution of human society of lifestyle. Crops with useful features such as easy harvest, high yield, long-term storage, and easy transport have been important to humans. There are nine different crops that are cultivated in today's world, and wheat is one of these important crops for human consumption.

Evolution and diversity of wheat have been spread widely by not only natural selection but also conventional/modern breeding techniques which are made by human. Scientists usually focus on improvement of wheat cultivars on the purpose of increased resistance against biotic and abiotic stress conditions. To have wheat cultivars with important traits such as higher yield potential in rigid areas, resistant to biotic and abiotic stress, and healthy product content is important than even before because of the rapid increase in the human population. Wheat landraces have advantages according to their high amount of nutrient, protein, tocols, carotenoids, etc. contents when compared to modern wheat cultivars. Due to these valuable contents, wheat landraces are one of the important ingredients for healthy food.

Consumption of wheat landraces was limited at local areas because of increased use of modern wheat cultivars worldwide. Because of this reason, wheat landrace varieties conserve their natural traits, and some important ones have become more dominant because of the artificial selection by farmers, but mostly there has been continuation of their evolution according to their own natural environment. This was highly important for wheat landraces to conserve their genetic pool and their genetic diversity wider than modern wheat cultivars. The important genetic traits of wheat landraces can serve as a potential tool for wheat improvement programs including breeding, genetic engineering, and genetic transformation.

Health concerns, feeding the increasing human population, and development of new varieties will be some of the big problems scientists will face in the near future. Therefore, focusing on landrace cultivars, informing the people about important values of landrace for the future, and increasing experimental studies by scientists will bring big opportunities to overcome these problems.

References

Aktaş H, Aslan D, Ordu B et al (2017) Evaluation of bread and einkorn wheat under *in vitro* drought stress. J Ani Plant Sci 27(6)

Altındal D, Akgün İ (2015) Plant genetic resources and the status of cereals. Adnan Menderes University Journal of the Faculty of Agriculture 12(1):147–153

Aslan D, Ordu B, Zencirci N (2016) Einkorn wheat (*Triticum monococcum* spp. *monococcum*) tolerates cold stress better than bread wheat during germination. Tarla Bitkileri Merkez Araştırma Enstitüsü Dergisi 25(2):182–192. [In Turkish]

Azeez AM, Adubi OA, Durodola FA (2018) Landraces and crop genetic improvement. Rediscovery of Landraces as a Resource for the Future:1–19

Braun HJ (1999) Prospects of Turkey's wheat industry, breeding and bio-technology. Paper presented at the Orta Anadoluda Hububat Tarımının Sorunları ve Çözüm Yolları Sempozyumu, Konya,Turkey, 8–11 June

Braun HJ, Zencirci N, Altay F et al (2001) Turkish wheat Pool. In: Bonjean AP and Angus WJ editors. The world wheat book: a history of wheat breeding, intercept ltd. & Lavosier publishing Inc. Paris France. P: 851–879

Chantret N, Salse J, Sabot F et al (2005) Molecular basis of evolutionary events that shaped the hardness locus in diploid and polyploid wheat species (Triticum and Aegilops). Plant Cell 17:1033–1045

Cummins AG, Roberts-Thomson IC (2009) Prevalence of celiac disease in the AsiaPacific region. J Gastroenterol Hepatol:1347–1351. https://doi.org/10.1111/j.1440-1746.2009.05932.x

Dotlağıl L, Hermuth J, Stehno Z et al (2010) How can wheat landraces con-tribute to present breeding? Czech J Genet Plant Breed 46:70–74

Emebiri LC, Ogbonnaya F, Moody D (2008) Population structure in wild emmer wheat (*Triticum dicoccoides*) based on EST-SSR

FAO Food and Agriculture Organization (2017) http://www.fao.org/faostat/en/#data/QC

Frankel OH (1970) Genetic conservation in perspective. In: Frankel OH, Bennett E (eds) Genetic resources in plants-their exploration and conservation, pp 469–489

Gökgöl M (1935) Turkey's wheats, tom I (in Turkish) Istanbul ministry of agriculture. Istanbul-Yeşilköy Seed Breeding Station Publication 7:436

Gökgöl M (1955) Classification key for wheats (in Turkish). Ministry of Agriculture, Publication and Communication Directorate, No 716:172

Gustafson P, Raskina O, Ma XF et al (2009) Wheat evolution, domestication, and improvement. In: Carver BF (ed) Wheat science and trade, 1st edn. Kansas, US, Wiley-Blackwell, pp 5–30

Harlan JR (1975) Crops and man. American society of agronomy, crop science society, WI. USA

Harlan JR (1981) The early history of wheat: earliest traces to the sack of Rome. Wheat science-today and tomorrow. Cambridge

Harlan JR (1995) Crops & man. American society of agronomy, Inc, crop science society of America, Inc., Madison, Wisconsin, USA, p 284

Heun M, Schäfer-Pregl R, Klawan D et al (1997) Site of einkorn wheat domestication identified by DNA fingerprinting. Science 278(5341):1312–1314

Jaradat AA (2013) Wheat landraces: a mini review. Emir J Food Agric 25(1):20–29

Kan M, Kucukcongar M, Morgounov A et al (2017) The general situation of wheat landrace populations and factors affecting production decisions of wheat land-race producers in Turkey. Journal of Agricultural Faculty of Gaziosmanpasa University 34(2):54–64

Karabak S, Taşçı R, Özkan N et al (2012) Adoption and economic analyses of Wheat Varieties in Ankara. Paper presented at 10th book of agricultural economics congress. Konya 2:694–702

Karagöz A (2014a) Wheat landraces of Turkey. Emir J food Agric 26(2):149

Karagöz A (2014b) Wheat landraces of Turkey. Emir J Food Agric 26(2):149–156

Karagöz A, Özberk İ (2014) Wheat genetic resources and exploitation in breeding in Turkey. Paper presented at international Mesopotamia agriculture congress, Diyarbakır, 22-25 sept. p:244–251

Knüpffer H, Zencirci N, Yılmaz H et al (2015) Mirza Gökgöl – a pioneer of wheat genetic resources from Turkey. Paper presented at international conference dedicated to the 85th anniversary of the Azerbaijan state agricultural university (Ganja), at Ganja, p: 227–231

Lev-Yadun S, Gopher A, Abbo S (2000) The cradle of agriculture. Science 288:1602–1603

Liu S (2003) Whole-grain foods, dietary fiber, and type 2 diabetes: searching for a kernel of truth. Am J Clin Nutr 77:527–529

Louette D (2000) Traditional management of seed and genetic diversity: what is a landrace? In: Brush SB (ed) Genes in the field. On farm conservation of crop diversity, IPGRI, IDRC. Lewis Publisher, pp 109–142

Mozaffarian D, Kumanyika SK, Lemaitre R et al (2003) Cereal, fruit, and vegetable fiber intake and the risk of cardiovascular disease in elderly individuals. JAMA 289(13):1659–1666

Nevo E, Korol AB, Beiles A et al (2002) Population genetics, genetic re-sources, and genome organization of wheat's progenitor, *Triticum dicoccoides*. In: Nevo E, Korol AB, Beiles A, Fahima T (eds) Evolution of wild emmer and wheat improvement. Springer, Berlin, p 364
Özbek Ö (2014) Turkish wheat landraces: population structure and function. Emir J Food Agric 26(2):137–148
Özberk İ, Atay S, Altay F et al (2016) Wheat atlas of Turkey (in Turkish). WWF-Turkey. Büyük Postane cad. No:19 Garanti Han Kat:5 Bahçekapı 34420 – İstanbul
Özberk İ, Özberk F, Atlı A et al (2005) Durum wheat in Turkey; yesterday, today and tomorrow. In: Royo C, Fonzo ND (eds) Durum wheat breeding: current approaches and future strategies, Vol. 1 and 2, 1st edn, USA, pp 981–1005
Şahin Y, Yıldırım A, Yücesan B et al (2017) Phytochemical content and antioxidant activity of einkorn (*Triticum monococcum* ssp. *monococcum*), bread (*Triticum aestivum* L.), and durum (*Triticum durum* Desf.) wheat. Prog Nutr 19(4):450–459
Tan A (2002) In situ on-farm conservation of landraces grown in North-Western transitional zone of Turkey (in Turkish). Final report. Tubitak-Togtag-2347. Tübitak, Ankara
TUIK, Turkish Statistical Institute (2017) https://www.tuik.gov.tr
Vavilov NI (1951) The origin, variation, immunity and breeding of cultivated plants. LWW 72(6):482
Witcombe JR, Joshi A, Joshi KD et al (1996) Farmer participatory crop improvement. I. Varietal selection and breeding methods and their impact on biodiversity. Exp Agric 32:445–460
Zencirci N, Kınacı E, Atlı A et al (1996) Wheat research in Turkey. In: Braun HJ, Altay F, Kronstad WE, Beniwal SPS, McNab A (eds) Wheat: prospects for global improvement. Kluwer Academic Publishers, Dordrecht, The Nederland, pp 11–15
Zencirci N, Yılmaz H, Garaybayova N (2018) Mirza (Hacızade) Gökgöl (1897–1981): the great explorer of wheat genetic resources in Turkey. Genet Resour Crop Evol 65(3):693–711
Zeven AC (1999) The traditional inexplicable replacement of seed and seed ware of landraces and cultivars: a review. Euphytica 110:181–191
Zhao FJ, Su YH, Dunham SJ (2009) Variation in mineral micronutrient concentrations in grain of wheat lines of diverse origin. J Cereal Sci 49(2):290–295

Chapter 2
Wheat Landraces in Mesopotamia

Fethiye Özberk and Irfan Özberk

2.1 Historical Background

Fertile Crescent and Southeastern Turkey are known as the centre of civilization surrounded by arid and semiarid lands in western Asia. The term Fertile Crescent was first used by James Henry Breasted in 1938 (Braidwood 1972). Fertile Crescent became home to wild wheats and traditional varieties and other crops of the modern world (Diamond 2002). Vavilov (1926), in his *Phyto-geographical Bases of Plant Breeding* book, showed the eastern Mediterranean region as the origin of wheat. Plant domestication from this region over thousands of years has also resulted in the development of enormous diversity. Progressive adaptation to a wide range of environments and responding to various selection pressures including biotic, abiotic, and human intervention have resulted in characteristic intraspecific diversity and differentiations represented by many landraces with specific histories and ecogeographical origins (Teshome et al. 2001). The information gathered from several excavations suggests that the agriculture started to evolve in Anatolia almost 10,000 years ago. Anatolia hosted many civilizations in the past and was the pathway between Asia and Europe in the history (Harlan 1995; van Zeist and de Roller 1995; Karagöz et al. 2010). Recent excavations in Göbeklitepe of Sanliurfa province have a potential to shed light on the periods prior to known date of agriculture (Bird 1999). For more than two decades, the use of molecular markers has been providing new information on genetic diversity of crop plants in relation to wild relatives, centers of domestication, time frame of the domestication process, and specific alleles supporting domesticated traits. The connection between molecular markers and domestication geography took root in the paper by Heun et al. (1997), who found that, on the basis of AFLP (amplified fragment length polymorphism)

F. Özberk · I. Özberk (✉)
Harran University, Agricultural Faculty, Field Crops Department, Şanlıurfa, Turkey
e-mail: ozberki@harran.edu.tr

N. Zencirci et al. (eds.), *Wheat Landraces*,
https://doi.org/10.1007/978-3-030-77388-5_2

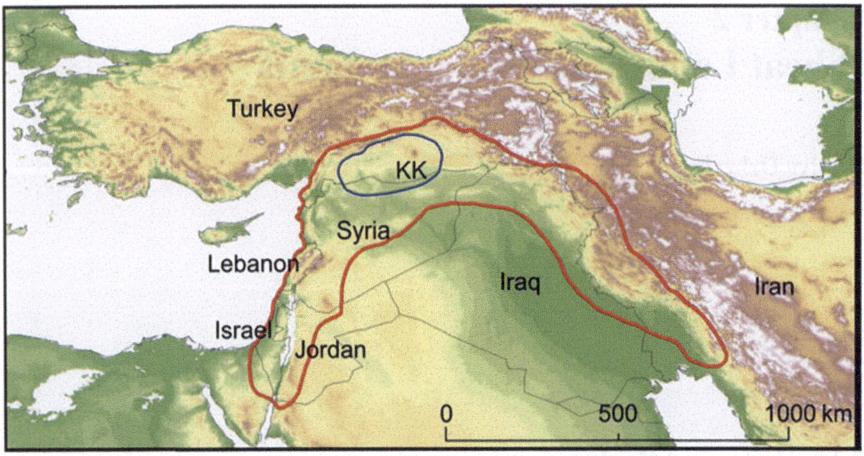

Fig. 2.1 Fertile Crescent and "core area" of plant domestication within the Fertile Crescent. The Fertile Crescent is indicated with a red line and the "core area" is shown with a blue line. KK Karacadağ mountain range in Southeastern Turkey

markers, the closest wild relatives of domesticated einkorn (*Triticum monococcum* ssp. *monococcum*, diploid) occur in a very restricted area within the Karacadağ mountain range in Southeastern Turkey (Fig. 2.1). From that, they concluded, not unreasonably, that this represents the site where humans first domesticated einkorn. Important contributions using different molecular markers for other species followed: einkorn (Kilian et al. 2007) and emmer (Ozkan et al. 2002, 2005; Mori et al. 2003; Luo et al. 2007).

Archaeological evidence documents the occurrence of plant remains at different excavation sites, in different stratigraphic layers that are analyzed and radiocarbon dated (Hillman 2000), from which a generally consistent picture emerges indicating that western agriculture originated in the Fertile Crescent after the last Ice Age, in aceramic Pre-Pottery Neolithic (PPN) from about 12,000 to 9500 years ago (Zohary and Hopf 2000; Nesbitt 2002; Salamini et al. 2002). It is now widely held that Fertile Crescent agriculture originated in a "core area" in Southeastern Turkey to northern Syria (Fig. 2.1), where the distribution of wild forms existed (Fig. 2.2).

Several issues concerning geography and domestication of wild emmer wheat were recently reviewed by Özkan et al. (2010). The authors considered published molecular and archaeological data and reanalyzed the data of Özkan et al. (2005). Wild emmer was probably domesticated in Southeastern Turkey (Ozkan et al. 2002, 2005; Mori et al. 2003; Luo et al. 2007; Jaradat 2013). A reconsideration of the domestication geography of tetraploid wheats has been considered by Ozkan et al. (2005) and by Luo et al. (2007). Phylogenetic analysis indicates that two different races of *T. dicoccoides* exist, the western one, colonizing Israel, Syria, Lebanon, and Jordan, and the central-eastern one, which has been frequently sampled in Turkey and rarely in Iraq and Iran. It is the central-eastern race that has played the role of the progenitor of the domesticated germplasm. This is supported by the results from

Fig. 2.2 Wild einkorn, wild emmer, and *Aegilops* species in their natural habitat within the Karacadağ mountain range. (Picture taken by Hakan Özkan in early July 2004)

the collections of Ozkan et al. (2002), Mori et al. (2003), and Luo et al. (2007). A disagreement is nevertheless appearing at the local geographical scale: the chloroplast DNA data indicate the Kartal mountains at the western border of the "core area" (Abbo et al. 2006), while AFLP fingerprinting points to the Karacadağ range as the putative site of tetraploid wheat domestication. From this area, emmer expanded across Asia, Europe, and Africa (Dubcovsky and Dvorak 2007). Southwestern expansion of domesticated emmer generated sympatry with the southern populations of *T. dicoccoides* and the rise of a secondary diversity center (Luo et al. 2007).

Durum wheat (*T. turgidum* ssp. *durum*) has been of great historical significance, because it provided a range of subspecies that were cultivated widely across the globe for thousands of years (Feuillet et al. 2007). Durum wheat spread out from the Fertile Crescent and through southern Europe, reaching North Africa around 7000 BC (Feldman 2001). It came into cultivation originally in the Damascus basin in southern Syria about 9800 BC (Zohary and Hopf 2000). A second route of migration occurred through North Africa during the Middle Ages (Moragues et al. 2006). Geographical expansion of durum wheat was intimately associated with human migrations. It is cultivated mainly in the marginal areas of Mediterranean region, Southern Europe, and North Africa, while more recently, it has started to expand to Southern Asia (Baloch et al. 2017).

Two of the most important traits in the evolution of wheat and other cultivated grasses constitute the domestication syndrome. These were (1) an increase in grain size, which was associated with successful germination and growth of seedlings in the cultivated fields; (2) the development of non-shattering seed, which prevented natural seed dispersal and allowed the humans to harvest and collect the seed in an optimal timing (Jaradat 2013); and (3) improvement of free threshing ability and non-brittle rachis. Additional modifications taking place during domestication and subsequent breeding concerned kernel row type (more rows in the domesticated species), plant height, grain hardness, tillering, seed dormancy, photoperiod, vernalization, and heading date (Salamini et al. 2002; Kilian et al. 2009).

2.2 Landraces and Common Characteristics in Mesopotamia

Landrace is a dynamic population of a cultivated plant that has historical origin and distinct identity and lacks formal crop improvement, as well as often being genetically diverse, locally adapted, and associated with traditional farming systems (Camacho Villa et al. 2006). A wheat landrace is composed of genetically heterogeneous populations comprising breeding lines and hybrid segregates which have evolved over many generations in contrasting environments and different local farming systems (Allard 1990). During the course of evolution, there have been many different landraces that existed adapting to various arid and semiarid conditions (Jaradat 1992; Brown 2000). As distinct plant populations, landraces were named and maintained by traditional farmers to meet their social, economic, cultural, and environmental needs. They were alternately called farmers' varieties or folk varieties (Belay et al. 1995). In landrace nomination, morphological structure (spike characteristics), grain color and hardness, location being grown, growth habit, fitness for end use, and the name of seed holder were referred. The genetic structure of wheat landraces is an evolutionary approach to survival and performance (Brown 2000), especially under arid and semiarid growing conditions (Jaradat 1992).The combined effect of natural and human selection has led to architecture of genotypes representing different combinations of traits, such as growth habit, cold and drought tolerance, early growth vigor, time to heading and maturity, grain filling duration, and quality traits (Masood et al. 2005). As compared to modern varieties, landraces, with relatively higher biomass, may not invest in larger root biomass, but rather in increased partitioning of root mass to deeper soil profiles, increased ability to extract moisture from those depths, and higher transpiration efficiency. In addition, their increased concentration of soluble carbohydrates in the stem shortly after anthesis ensures adequate translocation of assimilates to the developing grains. Therefore, early maturity, with some yield penalty, is a valuable trait that can be derived from wheat landraces to combat the typically encountered terminal drought in rain-fed wheat-growing areas (Ayed et al. 2010). Facultative growth habit is a unique characteristic of wheat landraces; it provides flexibility of sowing either in the fall as a winter crop or after the failure of the crop to overwinter, again in the spring. Under the growing conditions with limited nitrogen availability, wheat landraces and old varieties with a taller growth habit and lower harvest index absorb and translocate more nitrogen into the grain than modern varieties (Genç et al. 2005), presumably due to the greater pre-anthesis uptake and increased buffering capacity in genotypes with high vegetative biomass. Therefore, appropriately selected landraces with well-developed root system could be a source of variation for nutrient uptake and the improvement of seed quality (Jaradat 2013). Wheat landraces of Fertile Crescent are usually characterized by tall plants, long coleoptiles, and early vigor, which are very important for early ground cover and weed suppression (Jaradat et al. 1996; Moragues et al. 2006; Rawashdeh and Rawashdeh 2011). A wheat landrace is far from being a stable, distinct, and uniform unit; its diversity is linked to the diversity of the material sown in its immediate geographical area and to the level and frequency of seed exchange among farmers (Moragues et al. 2006).

2.3 Wheat Landraces Characteristics and Researches in Southeastern Turkey

Southeastern Anatolia is known to be the durum wheat belt of the country (Ozberk et al. 2005b). Therefore, there are more durum wheat landraces than bread wheat. Undesirable characteristics of durum wheat landraces in the area are defined by tallness, lodging type, being less responsive to chemical fertilizers, being adapted to low fertile soils, limited yield, and susceptibility to major foliar diseases such as rusts. However, their desirable characteristics are broad adaptation ability to stressed environments, high grain protein content, suitability to local dishes, and palatable straw for animal feeding. In the early twentieth century, durum and bread wheat landraces grown in Turkey were called "Ak Buğdaylar" and "Sarı Buğdaylar," respectively. Turkish farmers cultivated their landraces widely until the second half of the twentieth century. After World War II, a program was started in Turkey through an agreement with Rockefeller Foundation. Although it was a modest start in agriculture research, mechanization, and use of fertilizers and chemicals, it resulted in unexpected consequences. Among several plant groups in the country, wheat program had the greatest impact. It did not take long for the new varieties to replace the landraces. The heritage began to be demolished after so-called high-yielding "Mexican wheats" were introduced to the country. The acreage of the landraces grown in Turkey went down to about 0.55 mil. ha. (Karagöz 2014).

The first collection was completed at the first quarter of the twentieth century. Pioneering Turkish scientist Mirza Gökgöl collected 2120 wheat landraces from all over Turkey and evaluated them for basic characteristics. The name of the book who wrote is *Türkiye Buğdayları*. Gökgöl identified about 18,000 types of wheat, and among them, he identified 256 new varieties (Gökgöl 1939). Gökgöl detected the following landraces: Abuzer, Beyaz, Devedişi, Geore, Humrik, İskenderi, Karakılçık, Kırmızı, Kışlık beyaz, Kışlık büyükbaş, Komoy-karakılçık, Memeli, Pırçıklı sorgül, Ruto = Köse, Sorgül, Yazlık, Yazlık beyaz, Yazlık kırmızı and Yusufi in Diyarbakir and Ak Şami, Berzinnar, Bişeri (beşiri), Beyaz, Beyazsert, Beyaz topbaş, Beyaz yumuşak, Beyaziye, Birecik, Bozova, Havrani, İskenderiye (İskenderi), Karakılçık, Kendehari beyaz, Kendehari kırmızı, Kırmızı buğday, Kırmızı havran, Kırmızı kara,

Kırmızı menceki, Kırmızı mısri, Menceki, Mestişani, Mısri, Niseyri (Nuseyri), Samsai, Siri seyhan, Şami, Ufak daneli, Yazlık kendehari, Yerli karakılçık, and Yusufi in Sanliurfa, respectively. Major durum wheat landraces of the region grown until the 1960s were Bağacak, Sorgül, Beyaziye, Menceki, İskenderi, Mısri, and Havrani. Aşure is a unique bread wheat landrace grown in the Elazığ province and its neighborhood (Aktaş et al. 2017). But only a few of them are still grown by remote mountainous villagers of the region for home use. In the same period as Gökgöl, well-known Russian scientist Zhukovsky conducted three collecting missions to Turkey during 1925–1927. Zhukovsky was encouraged by Vavilov, and his missions were supported by The Botany Society of the Soviet Union (Zhukovsky 1927). During 3 years in Turkey, Zhukovsky collected around 10,000 samples of cereals, forages, and vegetables. The material was an enormous contribution to plant varieties in the Soviet Union (Zhukovsky 1951).

Another landrace collection was completed by Harlan during 1948–1949 with the contribution of the Agronomy Department of the University of Ankara, the Toprak Ofisi of the Ministry of Trade, and the Plant Breeding Stations of the Office of the Director General of Agriculture. The collection includes 2121 wheat accessions (incl. *T. monococcum* ssp. *monococcum*) and 55 wild wheat relatives. These populations were analyzed for botanical and agronomic composition, providing an unusual opportunity for studies on the behavior of botanical varieties in mixed populations under diverse climatic conditions. The wheats in Turkey were represented by remarkable diversity and great varietal wealth (Harlan 1950). It was proved that one of Turkish landraces contributed to American wheat production. "Türkiye Kırmızısı" (i.e., Turkish Red) bread wheat landrace was distributed to the German origin Menno's living in the Crimea under Ottoman authority in the mid-seventeenth century. Crimea was captured by Russians in 1783. They lived under Russian authority until 1870 with some difficulties in worship. They migrated to Kansas between 1870 and 1875 and took some wheat seeds with themselves. They grew the Türkiye Kırmızısı landrace in that area, and this became the base of American wheat history (Quisenberry and Reitz 1974; Braun et al. 2001). Hakkari originated "Horonek" (spring type durum wheat landrace) was found to be better than Russian varieties for its earliness; resistance to heat, fusarium, and Hessian fly; and high-yielding ability (Zhukovsky 1951). PI 2121 code number material collected by Harlan in the 1950s from Şemdinli township of Hakkari province was conserved in gene bank years in the USA. After the occurrence of rust epidemics, it was reinspected and found to be resistant to 51 races of various diseases, and the USA earned millions of dollars from this accession (Qualset et al. 1997). A comprehensive review of the history, characteristics, and use of wheat landrace in Turkey has been recently published by Karagöz (2014). There are about 22.000 Turkish wheat landraces in ex situ conservation worldwide (http://www.genesys-pgr.org).

Damania et al. (1996) evaluated the collection of 2420 accessions derived from single-spike population samples of durum wheat landraces collected in 1984 from 172 sites in 28 provinces in Turkey. They found diversity in these accessions for number of days to heading, maturity, and grain filling day as well as for plant height, peduncle length, and number of spikelet per spike, spike length, awn length, and kernel weight. As a result of the canonical analysis, significant correlation existed among provinces, meaning temperatures, altitude, latitude, and length of growing seasons. Eight distinct groups of provinces were identified by cluster analysis. They concluded that accessions could be utilized in crop improvement programs targeted at either favorable or stressed environments. Several other regional or local collection missions were fulfilled (Karagöz 1996; Qualset et al. 1997; Tan 2002; Karagöz and Zencirci 2005; Akçura and Topal 2006; Giuliani et al. 2009).

The last survey was carried out in 65 provinces of Turkey between 2009 and 2014 (Kan et al. 2015, 2016; Morgounov et al. 2016a; b). As a result of the survey, 162 local wheat landrace names were detected. The wheat landraces were ordered from the highest to the lowest frequency. In Turkey, the most common ten wheat landraces according to the frequency were shown as follows: (1) Ak Buğday (durum/bread wheat), (2) Sarı Buğday (durum/bread wheat), (3) Kırmızı Buğday (bread wheat), (4) Karakılçık (durum/bread wheat), (5) Zerun (bread wheat), (6) Kırik (bread wheat), (7) Koca Buğday (durum/bread wheat), (8) Siyez Buğdayı, (9) Topbaş (durum/bread wheat), (10) Üveyik Buğdayı (durum wheat). Moreover, some new landraces such as Alibayır (durum wheat), Kel buğday, and Boz buğday (bread wheat) were collected in Gaziantep and Kilis provinces of southeastern Anatolia.

In all characterization studies given above for wheat landraces, morphotype approach referring to highly heritable traits were employed. Cluster, ordination, and Shannon-Weaver indexes were mostly referred for classifications. Yıldırım et al. (2011) tried to assess genetic diversity among Turkish durum wheat landraces by microsatellites and found high genetic variability. Baloch et al. (2017) assessed the genetic diversity of central Fertile Crescent durum wheat landraces including Turkish and Syrian through whole genome DArTseq and SNP analysis and characterized 91 landraces employing 39,568 DArTseq and 20.661 SNP markers. They found that grouping pattern was not associated with the geographical distribution of durum wheat due to the mixing of Turkish and Syrian landraces by clustering based on near joining analysis, principal coordinate, and Bayesian model. Genç et al. (1993) compared durum wheat landraces of SE Anatolia with modern varieties under various drought levels and found that landraces were promising for biological yield and 1000 kernel weights, and they fell behind modern varieties for harvest index and number of kernel per spike. Özberk et al. (2005b) assessed yield and yield components employing 34 durum wheat landraces of SE Anatolia and found that there were positive correlations between numbers of kernel per spike, kernel weight per spike, and grain yield, whereas there was a negative correlation between plant height and grain yields. They concluded that similar yield components affected the grain yield in landraces like modern varieties. Koç (1993) detected the flag leaf photosynthesis ratio of southeastern Anatolian durum wheat landraces under increasing light intensity increased faster than that of modern varieties. Oran et al. (1971) carried out a rust survey in Sanliurfa, Diyarbakir, and Mardin provinces in 1969 and found that Beyaziye, Topbaş, Havrani, Akbaşak, Topik, Köse (23%), Kırmızı buğday, Aşure, Menceki, and Sorgül landraces were affected less than those of modern varieties such as Burt (40%), Wanser (39%), Floransa (39%), Herkovskaya (37%), and Ciete-Cerros (34%). Bez_I was found to be resistant to rusts. Mayo-64, Nadodores, Sonora-68, and Pitik-62 were the less affected modern varieties. Some of quality characteristics of SE Anatolia's durum wheat landraces such as Bağacak, Beyaziye, İskenderi, Sorgül, Karakılçık, Beyaz buğday, Ağ buğday, Bintepe, Havrani, Çalıbasan, Hacı Halil, and Akçakale were studied, and the results were compared to those of Zenith and Kyle. Taking into account protein quantity and quality, of the landraces, Havrani, and Çalıbasan were promising. With respect to yellow pigmentation and the oxidative enzymes, landraces Havrani, Hacı Halil, and Sorgül had great potential (Sayaslan et al. 2012). Micronutrient contents such as Fe, Zn, and Mn of 86 landraces belonging to various regions of Turkey were inspected and were found to be higher than those of modern varieties (Akcura and Kokten 2017).

There has been none of the variety released through the selection from landraces in the Southeastern Anatolia. Durum wheat cultivars of Urfa-2005 and Özberk released in 2005 have possessed some landrace parents such as Yerli and Akbaşak 073–44 in their pedigrees. But the landrace material belonging to this area was extensively employed by Ankara, Eskişehir, and Trakya Agricultural Research Institutes.

2.4 Wheat Landraces, Characteristics, and Researches in Syria, Iraq, Jordan, Lebanon, and Palestine

Wheat is the major crop in Syria with an annual production of 2.5 million tons (Shoaib and Arabi 2006). Farm-level surveys showed low spatial diversity of wheat where only a few dominant varieties occupied a large proportion of wheat area (Bishaw et al. 2015). The five top wheat varieties were ACSAD 65, Cham I, Cham 3, Lahan, and Cham 6, and they occupied 81% of the wheat area and were grown by 78% of the sampled farmers. Surviving durum wheat landraces in remote areas were Bayadi, Hamari, Haurani, and Swadi. Salamony was the only bread wheat landrace grown in the study area. Other durum wheat landraces are Jiduri, Zedi, Gharbi, Joulani, Jabali, Siklawi, and Yabroudi. Durum landrace variety Bayadi has drought and heat tolerance, disease resistance, and excellent pasta making quality. Bayadi is adapted to drought and poor soils by its higher tillering capacity. Sweidi, Sheirieh, and Shihani are other durum wheat landraces (Elings 1993). A 38 Syrian durum wheat landraces from diverse collection, as well, were evaluated for agronomic performance under arid conditions over two seasons at four locations (Elings 1993). In most cases, maximum yield was achieved by landraces. The population effect was significant for straw and grain yield. Nitrogen application was effective if moisture availability was the major growth-limiting factor. Hauran landrace provided high grain yield over diverse environments. Genetic diversity of Syrian wheat cultivars including six durum wheat landraces (Jazeera 17, Joury, Hamary, Haurany Ayoubeeh, and Bayady) and one bread wheat landrace (Salamony) were studied by amplified fragment length polymorphism (AFLP) markers (Shoaib and Arabi 2006). Cluster analysis with the entire AFLP data divided all cultivars into two major groups. The first group consisted of bread wheat cultivars and the second group durum wheat cultivars and durum wheat landraces, respectively. Narrow genetic diversity was detected among cultivars. In Syria, bread wheat showed lower average diversity and weighed diversity than durum wheat (Bishaw et al. 2015). Variance components analysis showed significant variations for 17 different agronomic traits such as plant height, yield components, and grain yield among wheat varieties and landraces. Principal component analysis and cluster analysis based on agronomic traits grouped modern varieties and landraces into separate clusters. Disease status of wheat landraces was studied seldom in Syria. Mamluk et al. (1990) carried out a survey between 1984 and 1988 and found that Haurani, Shyhani, and Sawadi were susceptible to major foliar diseases. Haurani was susceptible to yellow rust (60MS), leaf rust (60 MS), stem rust (45MS), and *Septoria* infection (9 vertical disease development score). Shyhani was susceptible to common bunt (33%) and Sawadi to flag smut (8%).

The Mediterranean climate in Jordan is characterized by dry hot summers with regional variation in temperature and mild wet winters with extreme variability in annual rainfall (Black 2009). Jordanian durum wheat landraces can be grouped into Ajloun and Karak landraces (Rawashdeh et al. 2007, 2010). Ajloun landraces are Sakneh, Srabeese, Soug, Sefsafeh, Fakreh, Koufranjeh, and Wadi Ajloun. Karak landraces are Raba, Qaser, Alouseh, Mazar, and Mu'tah. Louseh, Noorseh, and Hourani are the other durum wheat landraces grown in Jordan. Genetic diversity of Jordanian durum wheat landraces was studied by Al-Ajlouni and Jaradat (1994), and 250 landrace genotypes were assessed for 13 agronomic traits and the data was evaluated through principal component analysis and multivariate analysis. Jordanian wheat landraces were rich in source of genetic variation and could be used for improvement programs. Genetic diversity of 164 landrace accessions was also inspected for 26 agronomic traits by Abdel_Ghani (1999). Eighty-three percent of durum wheat landrace variation was attributed to Hourani landraces, and 15.2% was identified as Safra Ma'an landrace. Phenotypic diversity of Jordanian durum wheat landraces was further assessed by Rawashdeh et al. (2007). Fourteen morphological and agronomic traits were scored, and phenotypic similarity index was estimated and cluster analysis was performed. There was huge genetic variation among landraces. Genetic variation in Ajloun was much larger than that of Karak.

Wheat acreage of Iraq varies between 1.5 and 2.0 million hectares between 1990 and 1995 (Abbas 2001). National wheat production meets about 30% of total consumption, and the rest 3.5 million tons of wheat is met by import which costs about 2.3 billion US dollars per year. There have been very few references accessible about Iraqi wheat landraces. CIMMYT (International Wheat and Corn Improvement Center) has published Iraqi cultivars, but they were not identified as bread wheat and durum wheat. Saberbeg is known to be one of the important bread wheat landraces grown in Iraq. It is known to be resistant to drought with good bread making quality. Contrarily, it is susceptible to rust (Al-Sheick 2007). Durum wheat landrace of Omrabi that originated from northern Iraq. There is no clear information if some other bread wheat cultivars such as Al-Kaed, Al-Khair, Al-Nour, Al-Melad, Al-Neda, Al-Hashemia, Intsar, Iratom, Al-ize, Latifia, Rabia, Salı, Tahadi, Tamuz, Telafer, Al-Zehra, Abu-Gharip, and Serdar are landraces. Sixteen Iraqi wheat varieties were assessed for genetic diversity through ISSR, SRAP, and RAPD markers (Al-Kaab et al. 2016). The degree of genetic diversity, polymorphism information content, and resolving power was estimated. SRAP markers were found to be suitable for characterization of Iraqi wheat varieties. Among the cereal crops, wheat ranks first with a total cultivation area of 52.800 ha producing 116.200 tons of wheat in Lebanon. Most of the consumption is covered by import, which is 3.5 times greater than local production. Major durum wheat landraces are Hourani, Bekaii, Douchani, and Nabeljemal and bread wheat landrace is Salamouni. Landraces are still grown in Bekaa and Akkar plains (Chalak et al. 2010). Wheat landraces of White Debeya, Lakheesh, and Anbar are known to be Palestinian originated landraces (Fig. 2.3).

2.5 Current Situation of Landraces in Mesopotamia

Until the 1950s, wheat landraces constituted as dynamic and indispensable component of the overall agricultural biodiversity, especially in Mediterranean Basin, which has been largely valued as a source of agronomic, physiological, and genetic traits that can be used to improve new cultivars (Jaradat 2014). The green revolution, which occurred throughout from the 1940s to the 1960s, led to the development of high-yielding, disease-resistant wheat varieties with dwarfing genes, erect type, and highly responsive to inputs (Lopes et al. 2015). During the 1970s and 1980s and in many developing countries in Fertile Crescent, wheat landraces have been displaced by newly developed high yielding cultivars. These landraces have never been considered with sympathy by the developed countries due to their low-yielding ability and susceptibility to diseases compared with high-yielding varieties under high external input farming systems. An estimated 75% of the genetic

Fig. 2.3 Grain and plant status of durum wheat landrace of Noorseh in Jordan

diversity of crop plants was lost in the last century (FAO 1998; Hammer et al. 1999). Wheat landraces in the Fertile Crescent were no exception (Bettencourt 2011; WIEWS, 2012). Many farmers abandoned the age-old tradition of maintenance selection and their landraces became vulnerable to genetic erosion (Hammer et al. 1999; Harlan 1975, 1977). Genetic erosion refers to the permanent reduction in richness or evenness of common localized alleles over time in a defined area.

Stability of the planted area of local varieties can be an indicator of status and change in agrobiodiversity. Households were surveyed on the areas of local varieties planted between 2000 and 2004 in Syria, Jordan, Lebanon, and Palestine (Mazid et al. 2013). In Jordan, 6% of farmers abandoned some local wheat varieties; similarly, in Syria and Palestine, only 4% and 10% of farmers abandoned some local wheat varieties, respectively. In Lebanon, however, 14–19% of households abandoned wheat (Mazid et al. 2013).

The results of a countrywide wheat landraces collection mission carried out by Kan et al. (2015) in Turkey were compared to those of Gökgöl's 1935 and 1939 collection missions (Morgounov et al. 2016a; b). For a combination of 17 provinces and districts, the coverage of Gökgöl's surveys was matched by the latter study. It was found that name of the landraces changed over time almost entirely. Gökgöl mentioned generic names of 93 landraces, whereas the current collection from the same provinces and districts mentioned only 81 names. As the agronomy evolved, climate change adaptive requirements for wheat production changed; the landraces evolved and acquired new names reflecting their proper ties, use, and origin. Nearly 40% of wheat is used to be planted in the spring in the 1930s, whereas 95% of wheat is used to be planted in fall at present. In the 1920s, bread and durum wheat landraces occupied almost equal acreage: 39.9 and 41.5%, respectively, with an additional 14.2% of area devoted for club wheat. Currently, bread wheat landraces dominate the acreage with 52.9%. The acreage of durum wheat landraces have slightly decreased to 38.0%. The club wheat acreage has nearly disappeared with 2.2%. The loss of genetic diversity can be estimated by the frequency of rare species. The number of morphotypes listed by Gökgöl for the provinces and districts under study was 213 in the 1920s. Only 63 morphotypes were detected in the same provinces at present. The loss of genetic diversity was 70%.

Depending on the region, up to 80% of the farmers have tried modern cultivars, and most of them kept growing them along with landraces. The proportion of area growing wheat landraces to total wheat area in farmers' fields varied from 45 to 55% in the central Black Sea region and up to 98% in the southern coastal region. Farmers have access to modern cultivars but still keep their landraces. The main reason for maintaining landraces is satisfaction with the landraces' performance. While, on average, only 25 and 30% (bread wheat and durum wheat growers, respectively) of the farmers rated yield of the landraces as good; 83% of the respondents for bread wheat and 93% for durum wheat were happy with the grain quality and its suitability for homemade products (Fig. 2.4). The other highest ranked traits for bread wheat and durum wheat, respectively, were straw yield (74 and 80%) and straw quality (70 and 76%), cold tolerance (78 and 82%), and drought tolerance (71

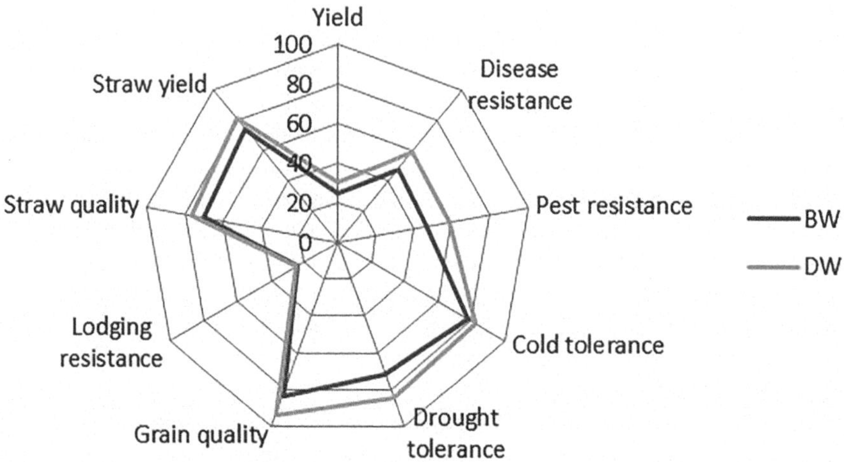

Fig. 2.4 Percentage of farmers' ratings of different traits of bread wheat (BW) and durum wheat (DW) landraces as good based on a survey of 1026 households in Turkey from 2009 to 2014

and 84%). For most of these traits, durum wheat landraces were rated slightly higher than bread wheat landraces (figure below) (Morgounov et al. 2016a; b).

Wheat grain in the rural areas is used for two main purposes: bread, including typical loaves and thin types, and bulgur or cracked wheat, in which wheat is cooked in water and processed. Bread and durum wheats, respectively, are normally used for these two products. Based on the survey of the farmers in the regions growing primarily bread wheat (Aegean, central Anatolia, northeastern Anatolia, and central-eastern Anatolia), its grain is mainly used for bread (64.3 to 83% of farmers). Of the four regions dominated by durum wheat, grain in the southern coastal and eastern Mediterranean regions is mainly used for bulgur (55.5 and 87.1%, respectively). The durum grain in the central Black Sea and southeastern Anatolia regions is used for both bulgur and bread (61.1 and 83.3%, respectively). Generally, the farmers were quite flexible in dual use of their grain for bread, bulgur, and other homemade products. Most of the club or compact wheat is used for dual purposes. Hulled einkorn wheat is used for bulgur in the Bolu region and for animal feed elsewhere. Emmer wheat is almost entirely used for animal feed. Durum wheat farmers in central Anatolia region were 100% satisfied with the grain, mostly using it for bulgur. In the southeastern Anatolia and central-eastern Anatolia regions, the durum farmers also gave very high ratings to the quality of their landraces, using them for dual purposes (bread and bulgur) (Morgounov et al. 2016a, b).

There are two potential bottlenecks in wheat diversity. The first relates to the recent origin of common wheat (nearly 8000 years ago; Cox 1997) and the presumption that there are relatively few tetraploid and diploid progenitor crosses. Hence, only a portion of diversity of *T. dicoccoides* and *Aegilops squarrosa* exists in common wheat. The second bottleneck relates to founder lines for local populations where the breeding programs often rely on a relatively limited number of parent lines in developing germplasm pools. The second bottleneck in bread wheat is believed to have reduced the population size by 6% (Cavanagh et al. 2013). The latter also points out to explain the value of germplasm exchange and the use of landraces. Allelic variation of genes can be recovered by going back to landraces.

As it was mentioned earlier, wheat landraces of the Fertile Crescent, typically characterized by tall plants, long coleoptiles, and early vigor, provide early ground cover which is vital for weed suppression in early growing stage of weed (Jaradat 2014). They have better quality attributes than high-yielding modern cultivars under organic and low-input farming systems (Koc et al. 2000; Moragues et al. 2006; Rawashdeh and Rawashdeh 2011). Grain yield, food and bread quality, marketability, grain color, and grain size were among the most important traits for which wheat landraces have been developed in the Fertile Crescent (Blum et al. 1987; Motzo et al. 2004). A combination of multiple genotypes within a wheat landrace and various levels of agronomic traits contributes to the yield stabilization under stressed environments (Kiaer et al. 2009). In addition, these landraces may be of special importance for the uptake of nitrogen in nitrogen-limited environments due to their late maturity and their high nutrient use efficiency and ability to translocate more nitrogen into the grain than modern varieties. Mediterranean wheat landraces retained a wide range of genetic diversity for gluten composition that has been mostly lost in modern cultivars (Dencic et al. 2000; Hamdi et al. 2010; Mohammadi and Amri 2013). Wheat landraces have better nutritional value than modern high-yielding varieties, a perception supported by chemical analysis and quality testing (Motzo et al. 2004).

Although there is presence of regional differences, the general breeding aims of cultivating durum wheat in Turkey is to obtain varieties that have high yield; yellow semolina color; gluten quality; resistance to lodging; tolerance to cold, heat, and drought; and tolerance to rust diseases (Özberk et al. 2010). In modern era of durum wheat breeding in Turkey, a variety development studies were initiated through the line selection from widely grown landraces. Therefore, Sarı Buğday 710 in 1931, Akbaşak 073/44 and Kunduru 414/44 in 1944, Fata'S′ 185/1 in 1961–1963, and Kunduru 1149 in 1967 were improved (Özberk et al. 2016). Apart from molecular genetic studies, many morphological, physiological, and quality characterization studies were carried out employing durum wheat landraces. Many beneficial traits were detected and tried to be exploited in modern breeding programs (Genç et al. 1993; Koç 1993; Barutçular et al. 1993; Alp and ve Kün 1999; Sönmez et al. 1999; Altınbaş and Tosun 2002; Özberk ve al. 2005a, b; Alp 2005; Alp and Aktaş 2005; Alp and Aktaş 2005; Kara and Akman 2007; Serpen et al. 2008; Köksel et al. 2008; Kütük et al. 2008; Öztürk et al. 2008; Gümüş et al. 2008; Alp and ve Sağır 2009; Koyuncu 2009; Sayaslan et al. 2012; Akçura 2009; Zencirci and Karagoz 2005). Molecular genetic studies mainly based on characterizations employing some morphological, physiological, and technological characteristics of landraces were assessed (Yıldırım et al. 2011; Baloch et al. 2017).

2.6 Future Perspectives

The basic prerequisite for sustainable conservation of wheat landraces is to have an enough amount of profit to compete with modern cultivars. Unless being profitable for grain production or products derived from any landrace, none of the landraces can be sustainable. Several agronomic and socioeconomic studies in the Fertile Crescent (Brush 1995; Dencic et al. 2000; Jaradat 2006) indicated that farmers' selection for desirable agronomic traits was a major force that shaped the population dynamic of wheat and other cereal landraces; therefore, continued and sustained on-farm conservation and sustainable utilization of these landraces ensure their continued evolution and contribution to sustainable local food systems (Rijal 2010; Mazid et al. 2013). On-farm landrace conservation requires continuation of the farmer-induced selection processes on how these landraces have been developed and how their genetic structures have been shaped. Historically and during the long history of wheat landrace emergence and evolution, farmers practiced seed replacement and renewal (Zeven 1999; Aderajew and Berg 2006). However, some farmers tended to keep their own seed for more than 20 years, presumably due to its food quality attributes (Rijal 2010; Galiè 2013). Participatory plant breeding and collaboration of breeders and local farmers (Fasoula 2004; Galiè 2013) allow farmers to access the improved landrace seed. Traditionally, farmers pass on the indigenous knowledge of seed management and its qualities, along with landrace seed itself, from generation to generation and among farmers even over relatively long (>100 Km) distances (Mazid et al. 2013). The unfortunate large-scale loss of local wheat landraces along with the local knowledge about their various qualities and uses can be attributed to the gradual disappearance of small farmers and local food cultures in the large part of the Fertile Crescent. Documentation and acknowledgment of indigenous knowledge are critical to retain the innumerable benefits of centuries of keen observation on the functioning of agroecosystems, on characteristics of various crop plants, and on their multiple uses. Climate change is expected to differentially affect components of complex biological interactions in modern and traditional wheat production systems. Wheat yield and quality will be affected by climate change directly or indirectly through diseases. Wheat landraces and their populations in and outside their centers of diversity that might respond to climate change will determine their continued productivity, utility, and survival. Nonbreeding approaches to create demand for landrace products to promote on-farm dynamic conservation and sustainable utilization of wheat landraces include the following: (1) raising public awareness regarding current and future value of landraces, (2) diversity fairs to allow for the exchange of landrace materials associated with indigenous knowledge, (3) visits among farmers in various localities to share the seed and experience, (4) contests for choice of highest diversity holding farmer, (5) recipe development and niche market creation for landrace products (Jaradat 2013), (6) growing mixtures for similar phenotypes to meet more local dish demands, (7) amendments in seed certification system allowing landraces to have diversity within the predetermined ranges, and (8) expanding organic farming practices employing more landraces (Karagöz 2014).

Durum and bread wheat landraces have been largely replaced, in their centers of diversity by monoculture of pure genotypes. This genetic erosion resulted in a significant loss of valuable genetic diversity for quality traits and resistance or tolerance to biotic and abiotic stresses. Diversity of wheat landrace populations, when structured to build spatial and temporal heterogeneity into cropping systems, enhances resilience to abiotic and biotic stresses (Bonman et al. 2007). New genes and alleles from landraces can be introgressed into modern varieties by hybridization. However, the importance of widening genetic diversity requires several actions in addition to hybridizations within breeding programs. These include monitoring genetic diversity and increasing rare alleles using landraces. Findings of new allelic variations promote phenotypic characterization of landraces for adaptation to climate change and facilitating information sharing. These strategies can be summarized as follows: genetic diversity can be determined by the use of different data types such as pedigree, morphological and biochemical markers, and DNA molecular markers (Beaumont et al. 1998; Li et al. 2010; Vinu et al. 2013). The first step in identifying genetic diversity patterns in a given population is to estimate the genetic similarity among genotypes. There are several algorithms developed to estimate genetic similarity as described by references (Beaumont et al. 1998; Mohammadi and Prasanna 2003; Kosman and Leonard 200; Aremu 2011). Principal component analysis (PCA), principal coordinate analysis (PCoorA), and multidimensional scaling (MDS) are three ordination methods used for genetic similarity among genotypes. Finding new allelic variation for known functional genes among landraces is quite important in modern breeding programs. Allele mining is a research field aimed at identifying allelic variation of relevant traits within genetic resource collection. For identified genes of known function and basic DNA sequence, genetic resource collections may be screened for allelic variation (Bhullar et al. 2010) using different molecular technologies (Kumar et al. 2010) (Lopes et al. 2015). Landraces with increased biomass and total photosynthesis and thousand kernel weight have potentially new allelic variation that should be exploited in wheat breeding. Strategies to retain genetic diversity existing in the landraces are available. The first one is measuring genetic diversity to build core collections. Therefore, genetic diversity can be preserved till later generations in a breeding program. The second is addressing the allelic variation for key traits in breeding programs with well-retained genetic diversity. The third one is retaining phenotypic variation and related genetic association for specific traits through large-scale and precision phenotyping coupled with GWAS for identification of new markers. Finally, data sharing among stakeholders may greatly benefit wheat improvement for adaptation to global warming (Lopes et al. 2015).

References

Abbas AI (2001) Focus on seed programs. The seed industry in Iraq. IPA agricultural research center, P.O.Box 39094. Abu Ghraib, Baghdad

Abbo S, Gopher A, Peleg Z, Saranga,Y, Fahima T, Salamani, F, Lev-Yadun S (2006) The ripples of "the big (agricultural) bang": the spread of early wheat cultivation. Genome 49:861–863

Abdel-Ghani, A.H.; Duwayri, M.; Kafawin, O (1999) Phenotypic diversity among wheat landraces from Jordan: morphological and developmental traits. Agris Vol 18(2):31–38

Aderajew, H. and T. Berg (2006) Selectors and non-selectors: agricultural and socio-economic implications of on-farm seed selection in Ethiopia [2006]. Noticiario de Recursos Fitogeneticos (IPGRI/FAO), Vol:145:1–10

Akçura M (2009) Genetic variability and interrelationship among grain yield and some quality traits in Turkish winter durum wheat landraces. Turk J Agric 33:547–556

Akcura M, Kokten K (2017) Variations in grain mineral concentrations of Turkish wheat landraces germplasm. Qual Assur Saf Crops Foods, 2017 9(2):153–159

Akçura M, Topal A (2006) Phenotypic diversity of winter bread wheat landraces in Turkey. J Crop Res (Turkey) 3:8–16

Al-Kaab DH, Hamdalla MS, Dweikat IM, Al-Saedi N (2016) Estimation of the degree of diversity for some Iraqi wheat varieties through ISSR,SRAP and RAPD markers. Agronomy-Horticulture –Faculty Publicaitons:884

Allard RW (1990) The genetics of host-pathogen coevolution: implications for genetic resource conservation. J Hered 81:1–6

Alp A (2005) Güneydoğu Anadolu Bölgesi Sulu Koşullarına Uygun Bazı Makarnalık Buğday Çeşitlerinin Verim ve Kalitelerinin Bölge Yerel Buğday Çeşitleriyle Karşılaştırılması. Türkiye VI. Tarla Bitkileri Kongresi, 5–9 Eylül 2005. Antalya, Cilt 2:707–712. (Turkish)

Alp A, ve Aktaş H (2005) Güneydoğu Anadolu Bölgesindeki Buğdaygil Genetik kaynaklarının Toplanması, Karakterizasyonu ve ÖnDeğerlendirmesi. GAP IV. Tarla Bit. Kongresi 21-23 Eylül, 2005. Şanlıurfa, Cilt 1:763–768

Alp A, ve Kün E (1999) Güneydoğu Anadolu Bölgesi Yerel Makarnalık Buğday Çeşitlerinin Tarımsal ve Kalite Özellikleri Üzeri Araştırmalar. Türkiye III. Tarla Bit. Kongresi. Çukurova Üniv, Adana (Turkish)

Alp A, ve Sağır A (2009) The evalution of durum wheat land races for resistance to yellow rust in the SE Anatolia region, Turkey. J Food Agri Environ 7(1):171–175

Al-Sheick SAW (2007) Country report to the FAO international technical conference on plant genetic resources, Baghdad

Altınbaş M, Ve Tosun M (2002) Makarnalık Buğday (T. durum Desf.) İle Yabani Tetraploid Buğday (T. dicoccoides Körn.) Melezlerinin BazıAgronomik ve Kalite Özellikleri ve Aralarındaki İlişkiler. Anadolu J AARI 12:51–64. (Turkish)

Aktaş H, Baloch FS (2017) Allelic variations of glutenin subunits and their association with quality traits in bread wheat genotypes. Turk J Agric For 41:127–134

Aremu CO (2011) Exploring statistical tools in measuring genetic diversity for crop improvement. In: Caliskan M (ed) Genetic diversity in plants. InTech. https://doi.org/10.5772/34950

Ayed S, Karmous C, Trifa Y, Salma-Ayed O (2010) Phenotypic diversity of Tunisian durum wheat landraces. Afr Crop Sci J 18:35–42

Baloch FS, Alsaleh A, Shahid MQ, Çiftçi VE,E. SaÂenz De Miera L, Aasim M, et al (2017) A whole genome DArTseq and SNP analysis for genetic Diversity assessment in Durum heat

Barutçular C, Koç M, Genç İ (1993) Bazı yerel ve ıslah edilmiş makarnalık buğday çeşitlerinde bayrak yaprak stoma direncinin tane dolum dönemindeki seyri. Makarnalık buğday ve mamulleri sempozyumu, 30 Kasım-3 Aralık, Ankara, 467–485

Beaumont MA, Ibrahim KM, Boursot P, Bruford MW (1998) Measuring genetic distance. In: Karp A, Isaac PG, Ingram DS (eds) Molecular tools for screening biodiversity. Springer, Dordrecht, The Netherlands, pp 315–327

Belay G, Tesemma T, Bechere E, Mitiku D (1995) Natural and human selection for purple grain tetraploid wheats in the Ethiopian highlands. Gen Res Crop Evol 42:387–391
Bettencourt E (2011) Sources of information on existing germplasm collections. Crop GeneBank Knowledge Base. Available from http://cropgenebank.sgrp.cgiar.org/index.php?
Bhullar NK, Zhang Z, Wicker T, Keller B (2010) Wheat gene bank accessions as a source of new alleles of the powdery mildew resistance gene Pm3: a large scale allele mining project. BMC Plant Biol 10:88
Bird M (1999) TIME. December 6, 1999 154(23)
Bishaw Z, Struik PC, vanGastel G, A.J. (2015) Wheat and barley seed system in Syria: how diverse are wheat and barley varieties and landraces from farmer's field? Inter J Plant Product 9(1):117–150
Black E (2009) The climate change on daily precipitation statistics in Jordan and Israel. Atmos Sci Lett 10(3):192–200
Blum AB, Sinmena G, Mayer J (1987) The grain quality of landraces of wheat as compared with modern cultivars. Plant breeding 99:226–233
Bonman JM, Beckelman HE, Jin Y, Gironella A (2007) Geographic
Braidwood RJ (1972) Prehistoric investigations in southwestern Asia. Proc Am Philos Soc 116:310–320
Braun H-J, Zencirci N, Altay F, Atli A, Avci M, Eser V, Kambertay M, Payne TS (2001) Turkish wheat pool. In: Bonjean AP, Angus WJ (eds) The world wheat book. A history of wheat breeding, Paris, pp 852–879
Brown AHD (2000) The genetic structure of crop landraces and the challenge to conserve them in situ on farms. In: Brush SB (ed) Genes in the field: on-farm conservation of crop diversity. Lewis Publishers, Boca Raton, FL, USA, pp 29–48
Brush SP (1995) In situ conservation of landraces in centers of crop diversity. Crop Sci 35:346–354
Camacho Villa TC, Maxted N, Scholten M, Ford-Lloyd B (2006) Defining and identifying crop landraces. Plant Genet Res 3:373–384
Cavanagh CR, Chao S, Wang S et al (2013) Genome-wide comparative diversity uncovers multiple targets of selection for improvement in hexaploid wheat landraces and cultivars. Proceedings of the National Academy of Sciences, USA 110:8057–8062
Chalak L, Noun J, El Haj S, Rizk H, Raghed Assi R, Attieh J, Maalouf F, Antoun MA, Sabra N (2010) Current status of agro-biodiversity in Lebanon and future challenges. Gene conserve 10(39):23–41
Cox TS (1997) Depending the wheat gene pool. J.Croğ Prod.1:1-25
Damania AB, Pecetti L, Qualset CQ, Humeid BO (1996) Diversity and geographic distribution of adaptive traits in Triticum turgidum L. (durum group) wheat landraces from Turkey. Genet Resour Crop Evolut 43:409–422
Dencic, S., R. Kastori, B. Kobiljski and B. Duggan. 2000. Evaluation of grain yield and its components in wheat cultivars and landraces under near optimal and drought conditions. Conservation of Plant Genetic Resources. In: Servinski, J. and Faberova, I (Eds)pp:29–33
Diamond J (2002) Evolution, consequences and future of plant and distribution of stem rust resistance in wheat landraces. Crop Sci 47:1955–1963
Dubcovsky J, Dvorak J (2007) Genome plasticity a key factor in the success of polyploid wheat under domestication. Science 316:1862–1866
Elings A (1993) Durum wheat landraces from Syria. Euphytica 70:85–96
FAO(1998) The state of the world's plant genetic resources for food and ageiculture, Food and Agriculture Organization of the United nations Rome,p 510
Fasoula,D.A.(2004) Accurate whole-plant phenotyping: An important component for successful marker assisted selection (MAS) In: J.Vollman H.Grousgruber and P.Ruckenbauer (Eds) pp 203-206. Genetic variation for plant breeding (17 th EUCARPIA General Congress)
Feuillet C, Langridge P, Waugh R (2007) Cereal breeding takes a walk on the wild side. Trends Genetic 24:24-32

Feldman M (2001) Origin of cultivated wheat. In: Bonjean AP, Angus WJ (eds) The world wheat book, A history of wheat Breeding. Lavoisier Publishing, Paris, pp 3–57
Galie A (2013) Governance of seed and food security through participatory plant breeding: Empirical evidence and gender analysis from Syria Natural Resources Forum 37:31–42
Genç İ, Koç M, Ve Barutçular C (1993) Bazı Yerel ve Islah Edilmiş Makarnalık Buğday Çeşitlerinde Biyolojik Verim ve Tane Veriminin Tane Dolum Dönemi Kurak Koşullarında Etkilenişi. Makarnalık Buğday ve Mamulleri Sempozyumu, 30 Kasım - 3 Aralık, Ankara, 443–459 (Turkish)
Genç Y, Humphries JM, Lyons GH, Graham RD (2005) Exploring genotypic variation in plant nutrient accumulation to alleviate micronutrient deficiency in populations. J.Trace Elements Med Biol 18:319–324
Giuliani A, Karagöz A, Zencirci N (2009) Emmer (Triticum dicoc¬con) production and market potential in marginal mountainous areas of Turkey. Mt Res Dev 29:220–229. https://doi.org/10.1659/mrd.00016
Gökgöl M (1935) Türkiye Buğdayları. Tarım Bakanlığı Yeşilköy Tohum Yegritim Enstitüsü Yayın no:7 İstanbul
Gökgöl M (1939) Turkish wheat, Istanbul
Gümüş S,Basman A,Karagöz A, Koksel H.(2008).Noodle qualty of ancient wheat flours. In: Koksel HUygun U, Basman A (Eds) Proceedings of Bhosphorus 2008ICC Int Conf. April 24–26, p 229
Hamdi W, Bellil I, Branlard G, Khelifi D (2010) Genetic Variation and Geographical Diversity for Seed Storage Proteins of Seventeen Durum Wheat Populations Collected in Algeria. Notulae Botanicae Horti Agrobotanici Cluj-Napoca, 38(2):22–32. https://doi.org/10.15835/nbha3824738
Hammer K, Diederichsen A, Spahillar M (1999) Basic studies toward strategies for conservation of plant genetic resources, In: J.Proc. Tech. meeting on the methodology of the FAO World information and Early Warning Systems on Plant genetic resources FAO,Rome
Harlan JR (1950) Collection of crop plants in Turkey. Agron J 42:258–259
Harlan JR (1975) Our vanishing genetic resources. Science 188:618–621
Harlan JR (1977) The origins of cereal agriculture in the Old World. In: Reed CA (ed) Origins of agriculture. Moulton Publ, The Hague, Netherlands, pp 357–383
Harlan JR (1995) Crops & Man. American Society of Agronomy, Inc. Crop Science Society of America, Inc., Madison, Wisconsin, USA, p 284
Heun M, Schafer-Pregl R, Klawan D, Castagna R, Accerbi M, Borghi B, Salamini F (1997) Site of einkorn wheat domestication identified by DNA fingerprinting. Science 278:1312–1314
Hillman G (2000) Plant food economy of Abu Hureyra. In: Moore A, Hillman G, Legge T (eds) Village on the Euphrates, from foraging to farming at Abu Hureyra. Oxford University Press, New York, pp 372–392
Jaradat AA (1992) Breeding potential of durum wheat landraces from Jordan I. Phenotypic diversity. Hereditas 11(6):301–304
Jaradat AA (2006) Phenotypic divergence in the meta-population of the Hourani durum wheat landrace. J. Food Agric Env 4:186–191
Jaradat AA (2013) Wheat landraces: a mini review. Emir J Food Agric 25(1):20–29
Jaradat AA (2014) The vanishing landraces of the Fertile Crescent. Emir J Food Agric 26(2):203–207
Jaradat AA, Ajlouni MM (1994) Breeding potential of durum wheat land races from JordanIII. Rate and duration of grain fill. Herbarium publications. Paper 12:74–77
Jaradat AA, Ajlouni MM, Duwayri MA (1996) Genetic resources of cereals and their wild relatives in Jordan:Revisited,In:A.A Jaradat (Ed)pp 55-75. Plant Genetic Resource of Jordan, IPGR, Rergional Office for West Asia and North Africa, Aleppo, Syria
Kan M, Küçükçongar M, Keser M, Morgunov A, Muminjanov H, Özdemir F, Qualset C (2015) Wheat landraces in farmers' fields in Turkey: National survey, collection, and conservation, 2009–2015. FAO, Ankara, Turkey

Kan M, Küçükçongar M, Morgunov A, Keser M, Özdemir F, Muminjanov H, Qualset C (2016) Wheat landraces production on farm level in Turkey: Who is growing where? Pak J Agric Sci 53:159–169

Kara B, ve Akman Z (2007) Yerel Buğday Ekotiplerinde Özellikler Arası İlişkiler ve Path Analizi. Süleyman Demirel Üniversitesi Fen Bilimleri Ens. Dergisi 11(3):219–224

Karagöz A (1996) Agronomic practices and socioeconomic aspects of emmer and einkorn cultivation in Turkey. In: Padulosi S, Hammer K, Heller J (eds) Proc. First Int. Workshop on Hulled Wheats. International Plant Genetic Resources Institute, Rome, Italy, Castelvecchio Pascoli, Tuscany, Italy. 21–22 July 1995, pp 172–177

Karagöz A (2014) Wheat Landraces of Turkey. Emir J Food Agric 26(2):149–156. http://www.ejfa.info/

Karagöz A, Zencirci N (2005) Variation in wheat (Triticum spp.) landraces from different altitudes of three regions of Turkey. Genet. Resour. Crop Evol 52:775–785. https://doi.org/10.1007/s10722-004-3556-3

Karagöz A, Zencirci N, Tan A, Taşkın T, Köksel H, Sürek M, Toker C, Özbek K (2010) Conservation and utilization of genetic resources (in Turkish). In: Vth technical congress of the agriculture, 17–21 Jan., 2010. Chamber of Agricultural Engineers, Ankara, pp 155–177

Kiær LP, Skovgard IM, Østergård H (2009) Grain yield increase in cereal variety mixtures: A meta analysis of field trials. Field Crop Res. 51:665–671

Kilian B, Ozkan H, Pozzi C, Salamini F (2009) Domestication of the Triticeae in the fertile crescent. In: Feuillet C, Meuhlbauer J (eds) Genetics and genomics of the Triticeae. Plant genetics and genomics: crops and models 7. Springer, New York, pp 81–119

Kilian B, Ozkan H, Walther A, Kohl J, Dagan T, Salamini F, Martin W (2007) Molecular diversity at 18 loci in 321 wild and 92 domesticate lines reveal no reduction of nucleotide diversity during Triticum monococcum (einkorn) domestication: Implications for the origin of agriculture. Mol Biol Evol 24:2657–2668

Koç M (1993) Bazı Yerel ve Islah Edilmiş Makarnalık Buğday Çeşitlerinde Bayrak Yaprak Fotosentez Hızı Üzerinde Araştırmalar. Makarnalık Buğday ve Mamulleri Sempozyumu, 30 Kasım - 3 Aralık, Ankara, 460–466 (Turkish)

Köksel H, Çelik S, Karagöz A, Ng PKW (2008) Partial characterization of starch in flours of ancient wheat and wild progenitor accessions. Ital J Food Sci 1(20):101–109

Koyuncu M (2009) Yerel Durum Buğday Çeşitlerinin Makarnalık Kalitelerini Etkileyen Önemli Parametreler Bakımından Taranması. YL.Tezi. Gazi Osman Paşa Üniv. Fen Bil. Enst. Gıda Müh. A.B.D. Tokat 49 s

Koç, M., Barutcular C., Zencirci N. (2000). Grain protein and grain yield of durum wheats from south-eastern Anatolia, Turkey. Aust. J. of Agric. Res. 51: 665-671

Kumar GR., Sakthivel K, Sundaram RM, Neeraja CN, Balachandran SM, Rani NS., Viraktamth BC, Madhav MS (2010). Allele mining in crops: prospects and potentials. Biotechnol Adv. 28:451-461

Kütük D, Masatcıoğlu MT, Öztürk S, Karagöz A, Köksel H (2008). Cracker making quality of ancient wheat flours. In Kooöksel H, Uygun U, Basman A (eds) proceeding of Bhosphorus 2008 ICC Int Conf. April 24-26, p214

Lopes MS, El-Basyoni I, Baenziger PS, Singh S, Royo C, Ozbek K, Aktas H, Ozer E, Ozdemir F, Ban AAT, Vikram P (2015) Exploiting genetic diversity from landraces in wheat breeding for adaptation to climate change. J Exper Bot 66(12):3477–3486

Luo MC, Yang ZL, You FM, Kawahara T, Waines JG, Dvorak J (2007) The structure of wild and domesticated emmer wheat populations, gene flow between them, and the site of emmer domestication. Theor Appl Genet 114:947–959

Mamluk OF, Al-Ahmed M, Makki MA (1990) Current status of wheat diseases in Syria. Phytopathologia Mediterranea 29(3):143–150

Masood MS, Javaid A, Rabbani MA, Anwar R (2005) Phenotypic diversity and trait association in bread wheat (Triticum aestivum L.) landraces from Baluchistan. Pakistan Pak J Bot 37:949–957

Mazid A, Shideed K, Amri A (2013) Assessment of on-farm conservation of dryland agrobiodiversity and its impact on rural livelihoods in the Fertile Crescent. Renew Agric Food Sys 29(4):366–377

Mohammadi R, Amri A (2013) Phenotypic diversity and relationships among a worldwide durum wheat (Triticum turgidum L. var. durum) germplasm collection under rainfed conditions in Iran. Crops Pasture Sci 64:87–99

Mohammadi SA, Prasanna BM (2003) Analysis of genetic diversity in crop plants. Salient statistical tools. Rev Interpret:1235–1248

Moragues M, Garcia del Moral LF, Moralejo M, Royo C (2006a) Yield formation strategies of durum wheat landraces with distinct pattern of dispersal within the Mfeiterranean basin. I. Yield components. Field Crop Res 95:194–205

Morgounov A, Keser M, Kan M, Küçükçongar M, Özdemir F, Gummadov N, Muminjanov H, Zuev E, Qualset CO (2016b) Wheat landraces currently grown in Turkey: Distribution, diversity and use. Crop Sci 56:3112–3124

Morgounov A, Keser M, Kan M et al (2016a) Wheat landraces currently grown in Turkey: distribution, diversity, and use. Crop Sci 56:3112–3124

Moragues M, García del Moral,LF, Moralejo M, Royo C (2006) Yield formation strategies of durum wheat landraces with distinct pattern of dispersal within the Mediterranean basin. Field Crops Res 95 (2–3):182–193

Mori N, Ishii T, Ishido T, Hirosawa S, Watatani H, Kawahara T, Nesbitt M, Belay G, Takumi S, Ogihara Y, Nakamura C (2003) Origin of domesticated emmer and common wheat inferred from chloroplast DNA fingerprinting. In: 10th international wheat genetics symposium, Paestum, pp 25–28

Motzo R, Fois S, Giunta F (2004) Relationship between grain yield and quality of durum wheats from different eras of breeding. Euphytica 140:147–154

Nesbitt M (2002) When and where did domesticated cereals first occur in southwest Asia? In: Cappers R, Bottema S (eds) The dawn of farming in the Near East. Ex Oriente, Berlin, pp 113–132

Oran YK, Parlak Y, Yılmazdemir Y.(19971). Doğu ve Güneydoğu Anadolu Bölgelerinde ekilen yerli ve yabancı menşeli buğday çeşitlerinin 1969 yılında pasa yakalanma durumları. Bitki koruma bülteni Cilt 11. No: 4 s 203-217

Özberk İ, Atay S, Altay F, Cabi E, Ozkan H, Atlı A (2016) Türkiyenin buğday atlası.WWF-Türkiye (Doğal hayatı koruma vakfı), İstanbul, Eylül, 2016. (Turkish)

Özberk I, Atlı A, Pfeiffer W, Özberk F, Coşkun Y (2005b) The effect of sunn pest (Eurygaster integriceps) damage on durum wheat: impact in the market place. Crop Protection 24(3):267–274

Özberk İ, Özberk F, Atlı A, Cetin L, Aydemir T, Keklikci Z, Önal MA, Braun HJ (2005a) Durum wheat in Turkey; yesterday, today and tomorrow. Durum wheat breeding: current aproaches and future strategies. In: Royo C, Nachit MN, Difonzo N, Araus JL, Pfeiffer WH, Slafer GA (eds) Chapter: 33. The Howard Press İnc., USA, p 1049

Özberk İ, Zencirci N, Özkan H, Özberk F, ve Eser V (2010) Dünden Bugüne Makarnalık Buğday Islahı ve Geleceğe Bakış. Makarnalık Buğday ve Mamulleri Konferansı, 17–18 Mayıs, 2010, 43–66 (Turkish)

Ozkan H, Brandolini A, Pozzi C, Effgen S, Wunder J, Salamini F (2005) A reconsideration of the domestication geography of tetraploid wheats. Theor Appl Genet 110:1052–1060

Ozkan H, Brandolini A, Schaefer-Pregl R, Salamini F (2002) AFLP analysis of a collection of tetraploid wheats indicates the origin of emmer and hard wheat domestication in southeast Turkey. Mol Biol Evol 19:1797–1801

Ozkan H, Willcox G, Graner A, Salamini F, Kilian B (2010) Geographic distribution and domestication of wild emmer wheat (Triticum dicoccoides). Genetic Resources and Crop Evolution 58(1):11–53

Öztürk S, Kahraman K, Karagöz A, Köksel H (2008) Utilization of Ancient Wheat Flours in Cookie Baking. In: Köksel H, Uygun U, Başman A (eds) Proceedings of Bosphorus 2008 ICC Int. Conf. April 24–26, p 216

Qualset CO, Damania AB, Zanatta CA, Brush SB (1997) Locally based crop plant conservation. In: Maxted N, Ford-Lloyd BV, Hawkes JG (eds) Plant genetic resources conservation: The in-situ approach. Chapman & Hall, London, UK, pp 160–175
Quinsberry KS, Reitz LP (1974) Türk buğdayı: Bir imparatorluğun temel taşı. Tarımsal tarih cilt 48 No: 1 Orta Batıda Çiftçilik, 1840-1900 a Symposiumo Ocak 1974 s: 98–110
Rawashdeh I, Rawashdeh NQ (2011) Exploring genetic diversity in Jordanian wheat landraces collected from different agroecological regions using random amplified polymorphic DNA analysis. Int J Agric Biol 13:325–331
Rawashdeh MI, Rawashdeh NQ, Haddad NI, Amri A (2010) Genetic diversity analysis among Jordanian durum wheat landraces, improved cultivars and wild types (T. dicoccoides) using RAPD Technique. Jordan J Agric Sci 6(4):548–556
Rawashdeh, N.K., Haddad, N.I., Al-Ajlouni, M.M. et al. Phenotypic diversity of durum wheat (Triticum durum Desf.) from Jordan. Genet Resour Crop Evol 54, 129–138 (2007). https://doi.org/10.1007/s10722-005-2636-3
Rijal DK (2010) Role of food tradition in conserving crop landraces on-farm. J. Agric. Ruckenbauer (Eds.), pp. 203–206. Genetic variation for Plant Breeding (17th EUCARPIA
Salamini F, Ozkan H, Brandolini A, Schafer-Pregl R, Martin W (2002) Genetics and geography of wild cereal domestication in the Near East. Nat Rev Genet 3:429–441
Sayaslan A, Koyuncu M, Yıldırım A, Güleç TE, Sönmezoğlu ÖA, Kandemir N (2012) Some quality characteristics of selected durum wheat (T.durum) landraces. Tur J Agri Forstry 36:749–756
Serpen A, Gökmen V, Karagöz A, Köksel H (2008) Phytochemical Quantificaiton and total antioxidant capacities of emmer (T. dicocconShrank) and Einkorn (T. monococcum L.) wheat landraces. J Agric Food Chemistry 56:7285–7292
Shoaib A, Arabi MIE (2006) Genetic diversity among Syrian cultivated and land races wheat revealed by AFLP markers. Genet Resour Crop Evolut 53:901–906
Sönmez F, Ülker M, Yılmaz N, Ege H, Bürün B, ve Apak R (1999) Tir Buğdayında Tane Verimi ve Bazı Verim Öğeleri Arası İlişkiler. Turkish J Agric For 23:45–52
Tan A (2002) In-situ comservation of landraces from transitional zone of Turkey. Final report for project no: TOGTAG-2347. Turkish Association for Scientific and Technical Research, Izmir, Turkey
Teshome AA, Brown HD, Hodgkin T (2001) Diversity in landraces of cereal and food legume crops. Plant Breeding Rev 21:221–261
van Zeist W, de Roller GJ (1995) Plant remains from Asikli Hö yük, a pre-pottery site in central Anatolia. Veget Hist Archeobot 4:179–185
Vavilov NI (1926) Studies on the origin of cultivated plants. (Russian) Bulletin Appl Bot Plant Breeding 14:1–245
Vinu V, Singh N, Vasudev S, Yadava DK, Kumar S, Naresh S, Bhat SR, Prabhu KV (2013) Assessment of genetic diversity in Brassica juncea (Brassicaceae) genotypes using phenotypic differences and SSR markers. Revista de Biologia Tropical 61:1919–1934
WIEWS (2012) World Information and Early Warning System on Ppant Genetic Resources for Food and Agriculture. http://apps3.fao.org/wiews/wiews.jsp. Acessed October, 2013
Xiaobai Li, Wengui Yan, Hesham Agrama, Biaolin Hu, Limeng Jia, Melissa Jia, Aaron Jackson, Karen Moldenhauer, Anna McClung, Dianxing Wu, (2010) Genotypic and phenotypic characterization of genetic differentiation and diversity in the USDA rice mini-core collection. Genetica 138 (11–12):1221–1230
Yıldırım A,Sönmezoglu ÖA, Gökmen S, Kandemir N, Aydm N (2011) Determination of genetic diversity among Turkish durum wheat landraces by microsatellites. African Jour of Biotech. Vol 10(19):3915–3920
Zencirci N, A Karagoz (2005) Effect of developmental stages length on yield and some quality traits of Turkish durum wheat lanraces: Influence of developmental stages lenght on yield and quality of durum wheat. Crop Gen. Res. Evol.52:765–774
Zeven AC (1999) The traditional inexplicable replacement of seed and seed ware of landraces and cultivars: a review. Euphytica 110:181–191

Zhukovsky P (1951) Türkiye'nin Zirai Bünyesi. Cevirenler: C. Kıpçak, H. Nouruzhan, S. Turkistanli, Sayı: 29, iSTANBUL
Zhukovsky PM (1927) Expedition to West Asia. Proc Appl Bot Breeding 17:135–158
Zohary D, Hopf M (2000) Domestication of plants in the old world. ' nd. edition. Clarendon Press, Oxford,UK

Chapter 3
Conservation of Plant Genetic Resources

Kürşad Özbek and Nusret Zencirci

3.1 General View on the Wheat Landraces

Turkey is one of the few countries that are self-sufficient in terms of food availability. It is an important agricultural country where agriculture has a wide social and cultural coverage. It has also a special place in terms of plant genetic resources available in the world. Thanks to its geographical position, climatic characteristics, topography, and soil structure, it covers many plant species and a higher endemism rate. Agriculture can be practiced in every region of the country including high mountainous villages. Turkey is the primary diversity center for many important species of agriculture such as wheat (*Triticum* spp.), barley (*Hordeum vulgare*), oats (*Avena sativa*), peas (*Pisum sativum*), and lentils (*Lens culinaris*) (World Bank 1993; Nesbitt 1995; Diamond 1997). The most prominent species among them is wheat.

Turkey, which is the homeland of wheat, is also one of the world's leading wheat producers. Wheat is the first plant species that human has grown for its important role in the global food security. It also has importance in Turkish economics and in Turkish cuisine with various dishes and consumption types. Turkey is well suited to wheat agriculture for its environmental and cultural features; consequently, wheat has been grown in Anatolia for thousands of years. For that reason, Turkey is home to many wheat landraces. Although registered wheat varieties are often cultivated in Turkey, landraces are still grown in rural areas, especially in mountainous villages or small farms for people's own consumption. Studies have shown that the average

K. Özbek
Department of Biodiversity and Genetic Resources, Field Crops Central Research Institute, Ankara, Turkey

N. Zencirci (✉)
Department of Biology, Faculty of Science and Art, Bolu Abant Izzet Baysal University, Gölköy, Bolu, Turkey

N. Zencirci et al. (eds.), *Wheat Landraces*,
https://doi.org/10.1007/978-3-030-77388-5_3

elevation of currently grown wheat landrace populations is 1,133 m. Ninety-three percent of the landrace production is for household uses (Kan et al. 2015).

Landraces are defined as; in contrast to breeding varieties, which possess a very large genetic base in population structure and "are dynamic population(s) of cultivated plant(s) with historical origin, distinct identity and without formal crop improvement, often genetically diverse, locally adapted, and, moreover, associated with traditional farming systems" (Camacho Villa et al. 2005). Genes resistant to biotic and abiotic stress factors, which are especially haunted by landraces, are of great importance for plant breeders in increasing the global wheat production. In addition, gene pools made of different landraces grown in different ecological conditions for wheat breeding within the scope of quality, yield, and other agricultural parameters are also used as the main source. According to 2017 CGIAR Genebank Platform Annual Report, CGIAR Genebank distributed a total of 109,339 germplasm samples (88,749 accessions) to users in 2017. Of these, 47,963 samples (44%) were provided to CGIAR Research Programs (CRPs) and 61,376 (56%) were distributed to recipients outside the CGIAR in 95 countries; the majority of resources distributed is traditional landraces and its wild relatives (53%). This is an important indicator as to what extent these landraces attract the attention of researchers all over the world.

In Turkey, the first wheat breeding activities started in 1925 and registration of varieties have been resorted by means of using landraces. Today, many landraces are still used as genitors for breeding studies carried out in Turkey. Although landraces grown in Turkey still play an important role in many breeding activities carried out (Fırat 1997), landraces have been replaced by foreign-origin wheat materials in wheat breeding studies in recent years (Karagöz and Zencirci 2005).

Many studies have been conducted on wheat landraces in Turkey. The first and the most important one among these studies was carried out countrywide by Mirza Gökgöl (Gökgöl 1939). In the study named "Turkish Wheats," he identified about 18,000 types of wheat, and among them, he identified 256 new varieties (Zencirci et al. 2018). In his study, Gökgöl stated that landraces in Turkey constitute an endless treasure for breeders. Even though small-scale studies were conducted in the following years, the most extensive study conducted nationwide is "Wheat Landraces in Farmers' Fields in Turkey National Survey, Collection and Conservation, 2009–2014" which was carried out in collaboration with CIMMYT, ICARDA, and Turkey. In this study, conducted in 65 provinces, it was stated that especially landraces were grown in high plateau areas that are that distant from city centers under difficult conditions and primitive agricultural methods. In this study, a total of 1,587 wheat landrace samples were collected from farmers' fields, and 162 local wheat landrace names were identified. The wheat landrace producers were divided into two groups as "Producers of only wheat landraces (1,320 farmers)" and "Producers of both wheat landraces and modern varieties (468 farmers)" (Kan et al. 2015).

Crop genetic resources are of vital importance for the present and future generations. However, the cultivation of landraces, which cannot compete with the registered wheat varieties with higher yield capacity, has been unfortunately decreasing day by day. Landraces are under threat for many reasons such as advancing science and technology, globalization, population growth, and pollution. At the same time, landraces have become an important part of social life in the same region since they are grown there for many years and also maintain very important traditional information. In other words, the decrease in the use of landraces brings about the risk of extinction for the traditional knowledge in which these landraces maintain. Every day, the use of local wheat varieties in Turkey has been decreasing. Landraces constitute less than 1% of this amount (Mazid et al. 2009) although a higher number of wheat cultivars are cultivated in Turkey.

A variety of conservation studies are carried out on biodiversity in order to prevent future erosion of genetic resources and to secure future food safety. The two main methods implemented for the protection of genetic resources are ex situ and in situ conservation studies.

3.1.1 Ex Situ Conservation

Conservation of genetic resources outside their natural environment under special conditions is called ex situ conservation. Ex situ conservation is implemented through organizations such as seed banks, land gene banks, in vitro banks, cryobanks (kept in ultracold conditions), DNA banks, and botanic gardens. However, the process of evolution in the ex situ conservation is intercepted since the interaction between species and environment is not sustained. On the other hand, natural or other potential risks that cannot be avoided during in situ protection require varieties to be protected outside these areas. Therefore, ex situ and in situ conservation activities should be applied as complementary programs.

Organizations that have special storage conditions at internationally accepted standards, where the plant material of genetic nature is collected from nature and stored for the purpose of conservation and maintenance of genetic diversity, are called as "seed gene banks." Conservation of genetic resources and biodiversity and their utilization and transformation into economic value are of great importance. Seed gene banks, therefore, are holding an important place in the world agenda in recent years. To state briefly, these institutions are the centers where the seeds are stored under appropriate conditions, when required, to be put at the disposal of the researchers. Seed gene banks should be research institutions where scientific researches are carried out as well as serving for conservation purposes. The more descriptive and explanatory information is produced about the samples taken under conservation, the more useful the stored sample will be available. In particular, morphological and molecular studies conducted in gene banks contribute to the accumulation of knowledge about existing genetic resources and prevent duplications (storage of the same material in the form of more than one sample).

Genetic resources have been used in research and development of new products in many fields after entering into many sectors, thanks to the advancement of technology each passing day. Therefore, the economic dimension of biodiversity and genetic resources has become prominent and has turned into a strategic issue in international policy development whose importance has gradually increased in recent years. Particularly in the international arena, states have made claim to sovereignty rights over their genetic resources, and this issue has been regulated by various legal legislations. The party that will provide the genetic resource must obtain the necessary prior permission from the state in order to gain access to a country's genetic resources in accordance with applicable national and international legislation. The same rules apply to access and use of traditional information that are in connection with genetic resources.

The magnitude of risk caused by erosion on plant genetic resources has been understood, and in particular, the political and scientific studies have been focusing on this globally in the last 50 years (Yong-Bi 2017). Worldwide, more than 1,750 seed gene banks and more than 2,500 botanic gardens are working on ex situ conservation of genetic resources. The majority of seed gene banks are working in accordance with the recommendations of IBPGR (The International Board for Plant Genetic Resources). Approximately 7.4 million samples, 45% of which consists of grain landraces (24% of wheat), are all stored in seed gene banks (Miguel et al. 2013; FAO 2010).

Today's collections have been formed because the rapid erosion in genetic resources has been perceived as a serious problem and seed gene banks have gained importance in solving this problem (Baur 1914; Harlan and Martini 1938; Frankel and Bennett 1970; Harlan 1972; Brush 1999; Scarascia-Mugnozza and Perrino 2002; Hammer and Teklu 2008; Fu and Somers 2009). For this reason, the mission descriptions of the gene banks have been improved, and subsequent changes have been made in the areas such as material collection and documentation. In this respect, modern gene banks have diversified research missions that include introduction of research on plant genetic resources, collection, conservation, characterization, documentation, valuation, and use of germplasm (Damania 2008a; b).

Within the framework of the collection program organized by The International Board for Plant Genetic Resources between 1975 and 1995, many different types of samples, i.e., more than 200,000, from 136 countries were collected and taken into conservation in many gene banks around the world. Of the collected samples, 27% were crop wild relatives and 61% were landraces (Thormann et al. 2015).

The number of samples taken under conservation in seed gene banks is increasing day by day. In a global-scale research conducted by Bettencourt and Konopka (1990) on seed genetic sources in the past years, 102 collections present in 47 countries were examined, and a total of 529,577 wheat samples were identified including 83,377 (15.7%) landraces. In the subsequent years, a study conducted over 191 seed banks on landraces' genetic resources worldwide identified 732,262 wheat accessions in 223 institutes (Knüpffer 2009). Of these, 167,133 were defined at the genes level, while 565,129 were defined at the species level. According to FAO data for 2010, a total of 856,168 wheat accessions belonging to 229 institutes were identified, in which 24% consisted of landraces (FAO 2010). As it is seen, the number of seed gene banks and the amount of material they store are increasing every day. Specialized or duplicated gene banks of different species in the centers have been established for different purposes: to multiply the number of materials and to characterize their data rapidly.

Seed gene banks differ in their properties, such as their purposes, on species they are specialized and their storage capacities. The largest public genes that were established by agricultural research institutes are the rice collection at the International Rice Research Institute (IRRI, Philippines) under the Consultative Group on International Agricultural Research (CGIAR) and wheat and maize collection at the International Wheat and Maize Improvement Center (CIMMYT, Mexico) (Sara Peres 2016).

The CIMMYT seed gene bank has the world's richest wheat collection with 150,000 samples from more than 100 countries around the world (https://www.cimmyt.org/germplasm-bank/). There are also approximately 596,000 accessions, 42,000 of which are wheat, stored in seed gene banks by the end of 2018 under US National Plant Germplasm System, which is one of the largest collections in the world (https://npgsweb.ars-grin.gov/gringlobal/query/summary.aspx). Between 1916 and 1933 composed of genetic resources collected from many places in the world N. I. Vavilov Institute of Plant Genetic Resources (VIR) in St Petersburg, Russia, seed gene bank, whose foundations consist of genetic resources collected by Vavilov from many places in the world between 1916 and 1933, contains 325,000 specimens regarding these samples (https://www.croptrust.org/blog/vavilov-collection-connection/). According to the Crop Trust 2017 annual report, 11 gene banks affiliated to CGIAR have the largest and the most widely used grain collection in the world. This collection plays a key role in the ex situ conservation globally. These gene banks have 774,000 accessions, which are available to everyone (Crop Trust 2017).

The Millennium Seed Bank is a facility that has the largest and the most diverse collection of wild plant species in the world. The vast majority of this collection was collected by a global network (the associated global network) linked to Millennium Seed Bank Partnership (MSBP) which operates in more than 80 countries and is the world's largest ex situ plant protection program (https://www.kew.org/science/collections/seed-collection). The Millennium Seed Bank collection located in Kew contains 39,100 species and 2.25 billion seeds from 189 countries (Gollin 2018).

Apart from all these mentioned seed banks, the Seed Vault was inaugurated in order to create a backup of world grain collections in 2008. The Seed Vault has the capacity to store 4.5 million varieties of grain. Given that each variety contains 500 seeds, 2.5 billion seeds can be stored here. This place has approximately 968,000 samples, about 175,000 of which are wheat, representing almost all countries in the world by the end of 2018. As a security backup, Svalbard Global Seed Vault has 60 seed gene bank materials, including 11 gene banks affiliated to CGIAR as well. This collection contains 23,000 samples from Turkey, in which about 8,000 are wheat seeds (https://www.croptrust.org/our-work/svalbard-global-seed-vault/).

As the seed collections and the relevant data in the world increase, it becomes challenging to reach the right material for researchers. Various databases have been established in order to guide the researchers to the right material and to prevent possible unwanted duplications. Among these databases, GENESYS is an online portal with more than 11 million characterization and evaluation data for more than 28 million accessions, in which more than 440 organizations are represented. In this way, users can access the grain genetic resources collected around the world on a single website. Through this system, hundreds of gene banks are able to share their accession data. At the same time, users can get information about samples of seed gene banks and can place orders. Approximately 283,000 units of the said material are composed of wheat samples, and there are about 13,000 materials collected in Turkey (https://www.genesys-pgr.org).

To consider the subject in terms of wheat, there are more than 80 wheat collections in the world, consisting of over 800,000 accessions that are conserved on the basis of ex situ method by the end of 2018. CIMMYT, Mexico (>100,000 accessions), and USDA-NSGC, Aberdeen, Idaho (nearly 40,000 accessions), have quite large wheat collections. Collections from the Vavilov Research Institute (VIR); Russian Federation; ICARDA; NBPGR, India; and Instituto del Germplasm, Bari, Italy, each contains about 30,000 accessions of wheat. In addition to this, collections of many small, specialized genetic stocks on wild wheat relatives are of importance for future breeding studies. As it is assumed that there are many duplications within these collections, advanced level of research is needed (https://www.genebanks.org/resources/crops/wheat/).

3.1.2 Ex Situ Conservation Studies of Genetic Resources in Turkey

Turkey is one of the leading countries in terms of plant genetic resources conservation studies and has gained great experience in the field of ex situ conservation since the 1960s. Ex situ conservation is implemented for both generative and vegetative collections, and these collections are stored in seed and land gene banks, respectively. In Turkey, the legal framework for the establishment of a strategy aimed at the continuation of biological diversity/genetic diversity is provided by the Constitution of the Republic of Turkey and the laws, regulations, and international conventions on biodiversity and the environment.

The research on the conservation of plant genetic resources was launched in the Ministry of Agriculture and Forestry research institutes that started in the 1960s. As a result, the first seed gene bank was established in 1972 called "National Seed Gene Bank" within the structure of the Directorate of Aegean Agricultural Research Institute, İzmir. Later, a gene bank was established in Ankara to back up the material within the Directorate of Field Crops Central Research Institute in 1988. As the number of materials in this gene bank increased, Turkish Seed Gene Bank was inaugurated in Ankara and the existing material was transferred to this gene bank in 2010.

3.1.3 *Seed Gene Banks in Turkey*

There are currently two seed gene banks in Turkey, as abovementioned, that are working affiliated to General Directorate of Agricultural Research and Policies of Ministry of Agriculture and Forestry. These are Turkish Seed Gene Bank in Ankara and National Seed Gene Bank in İzmir. These institutions are mainly responsible for the collection, conservation, and distribution of plant genetic resources, in particular landraces with their wild relatives.

Safety backup of the materials (duplication) is mutually performed between these institutions. Renovation and reproduction activities are planned in accordance with monitoring of samples in terms of vitality and quantity. The renovation activities of the material are carried out by the institutes with the studies that are subject to the project. All processes of collecting, conserving, producing, and distributing genetic resources in these seed gene banks are carried out in accordance with FAO's Gene Bank Standards (FAO 2013). Samples that are collected or received from the studies under a project are produced until the required amount is obtained for the conservation. At the same time, the material whose germination rate remains below the standards is produced and renovated. Documentation information for ex situ and in situ activities of seed and land collections is exchanged between institutions and carried out in harmony.

3.1.3.1 National Seed Gene Bank

The seed collection facilities of the National Seed Gene Bank, which are operating under the Aegean Agricultural Research Institute, are designed and equipped to meet the requirements of the long- and medium-term storage of both basic (base) and active collections, respectively. Three cold chambers consisting of 191 m^3 area are operated at -18 °C for long-term storage, and seven cold chambers consisting of 361 m^3 area are operated at 0 °C for medium-term storage. Two chambers consisting of 127 m^3 area operated at 4 °C provided facilities for temporary storage. In total, there are 12 cold chambers consisting of 679 m^3 area. Collections are always kept safe under the same conditions. In cold chambers, humidity control is not performed. Seeds are dried so that a moisture content of 5–6% is obtained and kept in moisture-tight tins, aluminum, or glass boxes for basic and active collections. Aluminum-coated foil is used for temporary and short-term storage. All the conditions in the National Seed Gene Bank comply with internationally recommended standards. The National Seed Gene Bank's cold chambers are equipped with new, modern cooling equipment and automatic generators in order to maintain these standards. The insulation of the cold chambers has been renewed.

The storage capacity of the National Seed Gene Bank is not close to the limit, and the spare chambers are set to operate at -18 °C and 0 °C. The material accepted to the gene bank is prepared for storage immediately after drying process. Room air-drying facility is used. Viability tests are carried out in all materials, and materials with high viability at the international standards are stored. The viability of the stored material is monitored every 5 years for active collections and once every 10 years for the basic collection. For some species, where no information is available on viability testing methods and dormancy mechanism, germination test methods are investigated (Tan 2010). The National Seed Gene Bank has around 50,000 materials as of 2018. A total of 35,000 of these, including 9,000 grains approximately, consist of landraces.

3.1.3.2 Seed Gene Bank of Field Crops Central Research Institute

In 1988, a seed gene bank consisting of small deep freezers was established within Field Crops Central Research Institute in Ankara in order to establish the safety backups of the material available at the National Seed Gene Bank. Up to 10,000 accessions coming from the studies under a project and collection programs had been maintained in this gene bank, which operated until 2009. The majority of these samples were landraces. The samples were stored as basic collection and safety backup. As a result of the studies conducted, a new seed gene bank was planned, and construction works started in 2009 when the existing infrastructure was deemed as insufficient. The material of aforementioned seed bank was transferred to newly opened Turkish Seed Gene Bank and the former seed bank was closed.

3.1.3.3 Turkish Seed Gene Bank

Turkish Seed Gene Bank, which is the biggest seed gene bank in Turkey, was inaugurated in 2010. The mission of Turkish Seed Gene Bank is to identify, collect, and conserve the landraces/village populations that make up plant genetic resources cultivated in Turkey, wild relatives of cultivated plants, other wild species and transition forms available in nature, bred/improved varieties and breeding lines with some important characters, and rare, endemic, and endangered plant species and to characterize these materials morphologically and molecularly. In addition, another mission of the Turkish Gene Bank is to raise awareness of the public for the conservation and sustainable use of biodiversity and to take part in educational activities, to cooperate and contribute to the academic studies, and to follow and to implement the studies conducted on this subject in our country and in the world.

Turkish Seed Gene Bank is among the world's leading seed gene banks with a total volume of 1,040 m^3 of cold chamber storage. Within its organization, Turkish Seed Gene Bank has Documentation, Seed Preparation Unit, Drying and Packaging Unit, seven Cold Storage Chamber, Seed Physiology Laboratory, Molecular Biology Laboratory, DNA and Tissue Gene Bank, Herbarium, and Imaging Room. Samples are stored as active and basic collections in the institution.

The number of materials kept under conservation is approximately 50,000 as of 2018. Approximately 16,000 of these samples are comprised of local wheat varieties. The main objective of the studies conducted in this institution is to contribute to the improvement of varieties in plant breeding by making morphological and molecular definitions of genetic resources. Genetic material also provides very important references and resources for research studies conducted in cooperation with universities, research institutes, and other relevant organizations.

The current projects of Seed Gene Bank include research studies conducted on collection programs, landraces, and endemic and endangered species. At the same time, the institution is making efforts to carry out more powerful and effective works by producing ideas, suggestions, and policies on issues related to conservation and sustainable use of biodiversity. Turkish Seed Gene Bank operated at a level that will lead to many gene banks around the world in terms of the bank's infrastructure and the number of materials stored here.

3.1.3.3.1 Units of Turkish Seed Gene Bank

Seed Test Chamber: In the Seed Test Chamber of Turkish Seed Gene Bank, physical examination and preliminary admission are performed on the material coming to the gene bank. Determination of moisture and physical examination of the material that enter the drying chamber are also performed in order to identify the moisture level of the samples before and after drying process. Samples coming out of the drying chamber are weighed in this room and prepared for packaging.

Seed Physiology Laboratory The material that comes to Turkish Seed Gene Bank is firstly tested in terms of viability in this laboratory. At the same time, in this laboratory, tests are carried out to measure the viability of the seeds stored periodically. Viability changes in seeds are determined in the long-, medium-, and short-term storage. After determining the viability of the seeds, these samples are taken into the renewal/production program if necessary.

Biotechnology Laboratory Studies that are aimed at determining the characterization of the samples at the molecular level are carried out in this laboratory. For instance, whether any examined variety has a gene of resistance to drought, cold, or any disease or pest is determined here by means of biotechnological analyses. The kinship degree of the samples is also tested in this laboratory.

Drying Chamber This is a room with controlled conditions where moisture levels are reduced to the levels of 5–6% before the seeds are put into packaging and for long-term storage without losing their viability. The humidity-heat changes in this chamber which has a volume of 80 m^3 are digitally controlled and recorded in a digital database.

Herbarium Preparation Chamber This is the chamber where the samples to be stored in the herbarium are prepared and then diagnosed. These samples are kept in the freezer for 2–3 days at -20 °C in the freezer and sterilized for the cleaning of the insects and their eggs which may damage the samples before placing the samples in the herbarium.

Herbarium (Plant Museum) Turkish Seed Gene Bank has a contemporary herbarium with a capacity of 60,000 samples. The plant samples collected from the field are pressed and dried and then stored here as herbarium material for reference. Researchers visit the herbarium collection and examine the plant variety they want to collect before carrying out a field study. As a result of the studies, approximately 40,000 herbarium samples, including samples from National Seed Gene Bank, have been taken into conservation. This collection serves all the researchers on the web page of "www.herbarium.tagem.gov.tr," which is created within the body of General Directorate of Agricultural Research and Policies (TAGEM). The existing digital database could be accessible by the users, and they can reach to high-resolution photos and relevant information of the samples in the herbarium collection.

3.1.3.3.2 Seed Conservation Rooms

Long-Term Conservation Chambers The dried seeds of the basic collection are stored in airtight packages at -18 °C in the two long-term conservation chambers within the gene bank. The samples in this chamber are subjected to viability tests every 10 years. Samples of the basic collection are not distributed to the researchers. Samples of this collection are only conserved for the provision of material to the active collection.

Medium-term conservation chambers: These are the chambers where the samples are stored at +5 °C. The material in question is the distribution material and is subject to necessary tests on period basis. The material of this collection is provided to the research studies under a project upon the requests of the researchers. When the number of materials is reduced in the active collection, the material in the basic collection is reproduced and made available to the researchers again.

3.1.4 Osman Tosun Gene Bank

In addition to the two seed gene banks operating under the Ministry of Agriculture and Forestry, there is also a seed gene bank named after Prof. Dr. Osman Tosun, Osman Tosun Gene Bank in the Department of Field Crops, Faculty of Agriculture in Ankara University.

This seed bank was opened in 1982 by Prof. Dr. Osman Tosun and his colleagues for the preservation of the materials that were obtained both inside and outside the country between the years of 1938 and 1975. In addition to the samples collected during domestic field visits conducted with national and international researchers and samples taken during seed control and certification studies, the samples collected from all over Turkey between the years of 1960 and 1972 according to the "new cultivars project" are under conservation. Seed samples and plant herbariums, which are very rich in terms of the landraces of cool climate grains, edible legumes and feed legumes, which have been created as a result of long-lasting works and stored under controlled conditions, are stored in this gene bank. Majority of the conserved samples were collected during the countrywide visits of Turkish and foreign scientists on different dates. In addition, there are also samples obtained from foreign gene banks. 459 bread wheat samples, 2064 durum wheat samples, 6106 barley samples, 754 oat samples, 159 rye samples, 261 triticale samples, 2215 chickpea samples, 451 faba bean samples, and 290 lentil samples are available in this gene bank (Adak and Akbaba 1986).

Although some universities, nongovernmental organizations, and private sector already have some local wheat varieties that have been taken into conservation both as ex situ and in situ, it is not possible to specify definite figures since there is no inventory study conducted on these varieties.

3.1.5 Botanic Gardens

Botanic gardens also occupy an important place in the ex situ conservation of genetic resources. In the botanic gardens, wild species are represented with more than one third of the plant species. These collections of wild species are kept as a viable material in the form of a combination and as seeds under conservation. Seed banking can provide efficient protection for the genetic diversity of wild plants. Information from Botanic Gardens Conservation International's (BGCI) data (GardenSearch, PlantSearch, ThreatSearch, and GlobalTreeSearch) has been analyzed in order to report global, regional, and national seed banking trends. There are at least 350 seed banking botanic gardens in 74 countries according to BGCI's database. In these collections, 56,987 taxa have been banked of which more than 9,000 have been threatened with extinction. 6881 tree species have been under ex situ conservation in these seed banks. More than half of these species (3562) are endemic species of 166 countries (O'Donnell and Sharrock 2017).

The studies of botanic gardens, which mostly contain wild species in their collections, are evaluated under the subject of ex situ conservation. National Botanic Garden of the Republic of Turkey which is in the process of establishment, Ege University Botanic Garden, Istanbul University and Ankara University Botanic Gardens, and Ataturk Arboretum of Istanbul University are some of the examples.

In recent years, private botanic gardens and arboretums such as Nezahat Gökyiğit Botanic Garden and Karaca Arboretum have been established by private enterprises. In relation to these botanic gardens, the infrastructure, on which production, conservation, and awareness studies will be carried out about the landraces, has been prepared in the National Botanic Garden of the Republic of Turkey, which will become operational in 2019, and in addition, training activities aiming to increase public awareness about landraces, especially for local wheat varieties, are also conceptualized under projects.

3.2 In Situ Conservation

3.2.1 In Situ/On-Farm Conservation

One of the conservation methods of genetic resources is in situ conservation. In situ conservation is the activity of protecting genetic resources on farm (by the farmers) or in natural habitats (Brush 1991; Maxted et al. 2000). The main objective of the on-site conservation of genetic resources is to follow the natural process taking place on genetic resources. Ex situ and in situ conservation are in the form of complementary elements. Collective studies conducted for conservation in gene banks are only referred to the genetic diversity of the collection area at that moment. In this way, instant genetic diversity can be conserved in seed gene banks for many years, but the processes such as genetic evolution arising out of interactions with biotic and abiotic stress factors from the area of genetic resources and natural selection or selection applied by the farmer stop. For these reasons, studies on the conservation of genetic diversity in their natural habitats should be carried out together in addition to ex situ conservation.

Although the wild relatives of the cultivated plants can be conserved in situ in the nature, human beings are required so that landraces could be protected in nature where they are grown. First, genetic resources should be kept in their natural habitats where farmers perform the agricultural activities. In situ conservation, i.e., conservation studies in the natural habitats, should be focused on landraces and their wild relatives that cannot compete with registered varieties and are under risk due to many reasons. These studies should be supported by civil society movements and individual, national, or international projects and programs. The conservation of these genetic resources in their natural habitats means not only the conservation of the species but also the conservation of genetic diversity in the region. Because genetic resources do not only contain genes that are important for scientists, they are also necessary for the species so that they can continue their relationship with other living organisms in nature. The areas protected by in situ conservation are a valuable reserve for the wild relatives of crop plants, medicinal plants, and herbaceous plants as well as other components of the ecosystem (Maxted et al. 2000; Tan and Tan 2002; Tan and Tan 2004). Since the landraces have been cultivated in the same ecology for many years, they also bring about the traditional agricultural knowledge that comes from many local technologies and traditional practices. This important traditional knowledge of genetic resources will be preserved through On-Farm (In Situ) Conservation Works.

Landraces are more resistant to biotic and abiotic stresses and are able to adapt to environmental conditions better than modern varieties. Ensuring the continuity of landraces will not be possible only by applying ex situ conservation, but it will be necessary to appreciate the landraces used under in situ conservation and breeding programs (Morgounov et al. 2016). The success of conserving the genetic resources on-site depends on the continuity of evolution and diversity among populations and within populations as well as the utilization of this evolution and diversity. In order to determine the effects of the in situ conservation, long-term and qualified studies are needed. In addition, the reactions of landraces against changing climatic conditions and pathogens such as abiotic and biotic stress conditions contain important evidences that support in situ conservation (Bellon et al. 2015a).

Farmers conserve the landraces with their traditional agricultural methods for many different reasons. First of all, the fields that have been divided and downsized by various reasons are not suitable for the application of modern agricultural systems. In higher areas where more challenging climate and soil conditions predominate, the heterogeneity of steep slopes and soil, especially in mountain farming practices, allows the landraces to compete with the cultivated crop plants in, at least, some part of the agricultural system. Also, the fact that such areas are distant from large-scale commercial markets reduces the competitive advantage of the bred crop plants. In addition to all of these, the landraces, which have an important place in the tables of Anatolia for many years, are produced by the farmers due to preferences for traditional and cultural consumption practices. These farmers often produce their own seeds. Other than that, sometimes, although rare, commerce and gift-giving are other ways of providing seeds. Farmers choose their seeds from the product according to their own criteria, and these seeds are stored in primitive conditions (in cloth or plastic bags or bottles or jars, sacks, etc.). Seed is passed on to the next years without experiencing any problems in the short-term preservation, thanks to this practice.

The potential value of genetic diversity in the evolutionary process is very important for the use of landraces that are grown in different regions, under different conditions, or in changing conditions. Therefore, it is preferable to grow these species in special places. The main question is how to make use of this value. It is considered that it would be very useful to prepare a value-defining guide in relation to variations that are deemed important in the subject of landraces for farmers who are going to make production in changing or different conditions, breeders, or other users. This requires harmonious and systematic efforts and collaboration between farmers, scientists, other social actors (e.g., workers, activists), and institutions (e.g., NGOs, local governments, schools). Methodologies and mechanisms should be decided on how and by whom a useful variation will be identified, and such studies should be carried out within the framework of a system (Caldu-Primo et al. 2017).

In the recent years, many projects have been carried out on the on-farm conservation of different grains in the world. In a comprehensive review conducted on this topic (Jarvis et al. 2011), it has been stated that 59 different types of initiatives have been supported worldwide about on-farm conservation; however, there is little experimental evidence that these different initiatives have made a difference beyond achieving actual conservation. The efforts about on-farm conservation unfortunately tend to be temporary, small scale, divided, and uncertain in terms of their impacts (Bellon et al. 2015a).There is, however, some recent systematic evidence that interventions implemented to support on-farm conservation can lead to higher levels of phenotypic diversity and livelihood benefits than would have been possible without them for Andean crops (Bellon et al. 2015b) and for phenotypic diversity only in the case of fruit trees in Central Asia (Gotor et al. 2017). To the best of our knowledge, there is still a lack of evidence that studies about on-farm projects have further implications for genetic diversity and its evolutionary process in grain genetic resources, and further studies are required in this area (Bellon et al. 2015a).

When local people are able to manage genetic resources in their own interests, they can become excellent resource managers. Landraces are the result of a good adaptation of the farmers' own preferences and long-standing production of these varieties in the region. Therefore, the participation of farmers in the conservation of biodiversity is highly crucial to determine how the more complex traditional systems can be adapted to contemporary needs while preserving the biodiversity of the ecosystem and their environments. It is not possible for biological/agricultural biological conservation to be successful, unless rural areas take a share of the benefits provided and do play a greater role in managing their own resources and understanding/measuring the impact of these resources on local biodiversity (Tan 2010).

Recently, the popularity of landraces has increased considerably. Landraces, which are superior to registered varieties in many aspects such as taste and aroma, are now preferred by people living the modern city life. In this context, people are producing landraces for their own consumption and exchanging their seeds through different activities especially in small gardens with the initiatives of NGOs and individual hobbyists. Although it causes some wrong applications, the production carried out by this means especially increases the awareness of the landraces and, although rather partially, ensures the continuity of the production.

3.2.2 Recommendations for In Situ/On-Farm Conservation of Landraces

Conservation of landraces under farmer conditions contains quite different characteristics than other conservation methods. First of all, it is necessary to determine the landraces grown by farmers and to understand the ecological and socioeconomic status of the region for a successful on-farm study. It is also necessary to identify the main factors affecting farmer's decisions, i.e., the storage needs for these varieties in order to continue the production of landraces. On the other hand, it is required to determine the effectiveness and direction of farmers' decisions affecting the variation in local diversity populations over time, to increase farmers' market opportunities and to look for whether there are ways to help the use of landraces or village varieties in order to create new varieties according to farmers' requests with the help of using landraces (Tan 2002).

In order to conserve the landraces in the farmer conditions, the conservation areas should be increased, and their legal status should be supported. Efforts for the conservation of the landraces that are about to be lost due to the spread of modern wheat varieties by the farmers should be encouraged as well as conservation of these varieties in gene banks. It is observed that the genetic resources have enriched constantly because landraces have been cultivated for a long time, the farmers have been sharing these varieties with one another, and adaptation process and the diversity have increased. In 2015, International Treaty on Plant Genetic Resources (http://www.planttreaty.org) gave an award to CIMMYT for on-farm conservation project of wheats in the fields of farmers across Turkey, Afghanistan, and Iran. As a result of the research studies, it has been observed that in situ conservation efforts in Turkey should be focused on three main regions. These are the richest provinces in terms of landraces according to the identified number of morphotypes and Shannon diversity index (Adana, Adıyaman, Aksaray, Bitlis, Diyarbakır, Hatay, Manisa, Tokat), the provinces hosting rare species such as einkorn and emmer wheat (Bolu, Karabük, Kars, Kastamonu, Kütahya, Samsun ve Sinop), and the provinces with the highest share of farmers producing both local and modern varieties (primarily the provinces of the Western Black Sea and Central Black Sea regions) (Morgounov et al. 2016).

Also, the study mentions that the utilization and conservation efforts for local wheat varieties in Turkey need to be directed to two goals; the first is to conserve the existing landraces and to increase cultivation areas and product variety. In addition to important on-site conservation practices and policies, it may be an appropriate option to achieve the genetic diversity in landraces and return them to farmers. Modern breeding techniques allow for the combination of selection and desired features quickly while maintaining the overall integrity of landraces. Competitiveness of landraces with modern varieties can be improved, and this can lead to more and more permanent uses preferred by farmers. The second goal is that great diversity of the collected wheats should have more explanation, classification, evaluation, and usage applications in the development and research programs. So far, >1000 landraces have been selected and superior genotypes have been identified. These long-standing landraces are highly valued genetic resources in order to overcome the challenges faced by modern agriculture (Morgounov et al. 2016).

Special markets aimed at landraces should be created to support the producers of landraces. This is essential for farmers to quit producing modern varieties. These can be local markets or sales of such niche products can be provided in the store chains. In particular, if the products to be sold in these chain stores are produced by well-organized NGOs such as cooperatives and are supported by popular topics such as ecotourism, gastronomy tourism, or organic production, a continuous and standard product entry is provided to the markets, and also an important step is taken in rural development by providing continuous income to the farmers. Today, the biggest reason why landraces cannot be included in the store chains is that the store chains' standard quality and quantity of product needs cannot be met. This need can only be achieved through well-organized farmers' associations. Again, local products originating from a particular region with specific characteristics can gain special protection status, such as geographical indications in this way. Popularity of landraces should be increased, especially by using national media or social media, which is becoming more and more effective every day. In this way, awareness-raising activities should be supported by reaching out to different segments of society. Nongovernmental organizations, universities, and other educational institutions carrying out these studies should be supported and encouraged to organize courses on awareness-raising.

Continuity is the most important and the most difficult element to maintain in situ conservation. Appropriate management programs should be prepared and implemented, by taking into account a great number of possibilities for such areas. Traditional farming methods for the production of landraces should be continued within the framework of management programs. Research and monitoring studies on these areas should be carried out on a regular basis; ministries and relevant research and implementation institutions such as universities and institutes should provide specific training to local people and NGOs on this subject; and exchange of information and cooperation at international and national levels should be improved.

On-farm conservation has many advantages. The adaptation process of the varieties that are under conservation to the environmental conditions in the region where they are grown has been continuing. In this way, biodiversity is conserved at different levels such as ecosystem, interspecific and intraspecific, while farmers are actively engaged in conservation activities. Thus, a sustainable indirect contribution is made to the agroecosystem because less fertilizers and drugs will be used with the application of traditional farming methods. Sustainable economic support would be provided for farmers engaged in agriculture with scarce resources. In addition, it is always easier for genetic resources to be protected, monitored, and accessible through farmers (Tan 2002).

Prior to making policies to protect traditional varieties and to ensure sustainability, necessary research should be conducted to identify which enterprises are involved in the process and where landraces are produced. The socioeconomic structure of farmer enterprises is an important factor in the conservation and production of landraces that are involved in on-site conservation under farmer conditions. If these varieties are determined to be grown under subsistence farming practices in rugged terrains away from cities, the policies should be prepared in consideration of special policies for specific areas rather than more general policies.

Ensuring the continuity of landraces in traditional agricultural systems will be possible with the collaboration of large-scale awareness activities with much participation, local NGOs, and the formal and informal sectors, including farmers. First of all, it should be well explained to the farmers that they are undertaking a significant task by means of describing the values of the landraces to them. It is not possible for biological/agricultural biological conservation to be successful, unless rural areas take a share of the benefits provided and do play a role in managing their own resources with regard to the on-site conservation of the landraces. This is the only possible way local people can become the perfect resource managers when they are able to manage their own resources in the direction of their own interests.

Conservation of traditional agricultural systems and their associated natural ecosystems is recommended as the most appropriate strategy for on-site conservation of cultivated plants and wild plant varieties. Conservation efforts should be linked to rural development programs that take into account the ethnobotanical knowledge of the rural people and emphasize both self-sufficiency in nutrition and local resource conservation. As long as local people, being a member of their own culture, can see the benefits of ensuring the conservation with the necessary applications for the continuation of plant heritage, knowing that conservation of this heritage is essential to their nutrition, such efforts are likely to be successful. The reasons behind the abandonment of neglected and underutilized varieties are in general due to the fact that these plants are grown in poor conditions and often in problematic areas and the competitiveness with other products in traditional agriculture (Tan 2010).

3.2.3 In Situ Studies in Turkey

The importance of in situ areas for maintaining the natural environment in Turkey has been considered from the early years of the Republic Period. The establishment of the first national park in the 1950s when environmental problems have yet to be started in Turkey is a good indication of the importance attached to the conservation of the nature. At the beginning of the 1970s, environmental protection policies began to be institutionalized in Turkey. In the early 1980s and in 1990s, environmental conservation became legally binding and was included in national programs. Turkey has signed international conventions aiming at the conservation of biological diversity, and this could be regarded as a reflection of its policies for the protection of nature. The Development Plans and Annual Programs, which were developed subsequently, addressed issues related to biodiversity in the environmental and agricultural sectors, and identified the necessary measures together with the policies for the conservation of biological diversity and sustainable development and economic value (Tan and Tan 2002). The tenth Development Plan, covering the years of 2014–2018, identifies the activities for conservation, improvement, and increasing the economic value of Turkey's biological diversity as a matter of priority. In situ conservation programs such as National Parks, Nature Conservation Areas, Nature Parks, Wildlife Development Areas, Special Environmental Protection Areas, Natural Sites, Natural Asset, and Gene Conservation and Management Areas have been implemented for many years in Turkey. Areas that have been granted status under the name of "Protected Area" are the fields which have been taken under conservation upon announcement by National Parks Law Numbered 2873 and Land Hunting Law Numbered 4915. There are 43 national parks, 229 nature parks, 111 nature monuments, 30 nature conservation areas, and 81 wildlife development areas by the end of 2018 within the scope of this framework which is determined as "a geographic area defined and managed by legislation with the aim of long-term conservation and continuity of ecosystem services and cultural values" by the laws. In this context, approximately 3.4 million hectares of land is under protection as of the end of 2017 (http://www.milliparklar.gov.tr/korunan-alanlar).

At the national level, various projects are carried out about in situ conservation in cooperation with the Ministry of Agriculture and Forestry, the Ministry of Environment and Urbanization, NGOs, and universities. One of the first studies, which aims to conserve the genetic diversity in Turkey, is the In Situ Conservation of the Genetic Diversity Project which was implemented within the framework of GEF Program of World Bank in 1993. Within the scope of this project, in situ conservation studies were conducted for the genetic resources of wild chestnut and plum genetic resources in Kaz Mountains and Central Taurus Mountains (Bolkar and Aladağlar) and for the genetic resources of wild wheat and legumes in Ceylanpınar region. Under this project, seven Gene Conservation and Management Areas (GEKYA) were determined for five wild wheat relatives within the boundaries of Ceylanpinar Agricultural Enterprise, and National Plan for In Situ Conservation of Plant Genetic Diversity was prepared for the first time in Turkey (Karagöz 1998; Karagöz et al. 2010). GEKYA management and research studies should be continued. GEKYAs are one of the most important outputs of the In Situ Conservation of the Genetic Diversity Project, and the focus should be on conservation and management efforts in seven GEKYAs, where the wild ancestors of cereals in Ceylanpinar are heavily populated. GEKYAs should be expanded to include Karacadağ region and other regions where wild relatives of wheat are densely populated. GEKYA programs should be proposed for other target species with the cooperation of the surrounding villages and the participation of local people (Ertekin 2002).

Another important study in which the conservation studies are addressed is "In Situ Conservation Protection and Management of Endangered Plant Species" aiming the in situ (on-site) conservation of endangered endemic plant species that have propagated in region of lakes and other wetlands in the west and Sultan reeds and Karataş Delta and other wetlands between these regions in the east mainly starting from the borders of Lake Tuz in the provinces of Konya, Aksaray, Ankara (Sereflikoçhisar), Isparta, and Burdur. Public awareness constitutes one of the main components of these projects (Tan et al. 2003; Tan 2010).

In our world, the conservation of agricultural diversity as in situ and on farm is also actively promoted in addition to various nature conservation activities. The supports provided by international and national donors and public institutions such as the Ministry of Agriculture and Forestry and the Ministry of Environment and Urbanization in Turkey back up in situ conservation of biodiversity and the development of on-farm conservation efforts. In this way, it is possible to evaluate and conserve the landraces by taking into consideration the threating factors. Reserves of different statuses have been established for the conservation of biological diversity. In addition to public institutions, some nongovernmental organizations have also been working for this purpose. Approaches have been developed in order to promote in situ/on-farm conservation and preservation and management of the ecosystem with the implementation of "National Plan for In Situ Conservation" and "National Biodiversity Strategies and Action Plans" (UBSEP).

In 1995, International Plant Genetic Resources Institute (IPGRI, Bioversity International) developed a global project to strengthen the scientific basis for in situ conservation of agricultural biodiversity together with the National Programs of nine countries. The nine countries involved in this project are Burkina Faso, Ethiopia, Nepal, Vietnam, Peru, Mexico, Morocco, Turkey, and Hungary (Jarvis et al. 1998; Jarvis et al. 1998). Under this program, the project titled "In Situ Conservation of Genetic Diversity in Turkey" was commenced by Aegean Agricultural Research Institute with national budget support in cooperation with local organizations of the study area, universities, and farmers' unions. The main objectives of the project were (1) to support the development of an information framework for decision-making processes affecting farmers' in situ conservation of agricultural biological diversity, (2) to strengthen the linkages between formal and informal sectors and farmers in order to plan a new implementation for the conservation of agricultural biological diversity, and (3) to expand the use of agricultural biological diversity and the participation of farmers' communities and other groups in conservation activities (Tan 2002a; b; Tan 2009). This project comes to the fore with its characteristic of transferring the research studies carried out by national research institutes/universities with the applications of pilot tests/demonstrations under farmer conditions for feedback purposes.

The results of the survey conducted under the National Program have shown that farmers have benefited from morphological and gastronomic characteristics, habits of their own lives, and functional selection criteria to differentiate between cultivated plants and landraces. The first two of these have been the most commonly used criteria. Gastronomic criteria are more important for people performing agriculture for commercial purposes in terms of their own consumption, while morphological criteria are more important than the gastronomic criteria in terms of the farmers who earn their livelihood from agriculture. Female farmers refer to more criteria in the development of landraces than male farmers. Similarly, numerous criteria play a role in the selection stage for landraces grown in home gardens. The agricultural sector in remote and distant areas prefers the traditional cultivated plants (landraces) to modern varieties because they are able to adapt to the environment without the need for external inputs as well as their taste and nutritional values (Tan and Taşkın 2009).

GEF/SGP has been operating in Turkey since 1993. The program has supported more than 100 projects implemented by more than 40 local and national NGOs to this date. This program has created a rich resource consisting of good practices, lessons learned, and experiences from the implementation of sustainable development in the local community and in real socioeconomic conditions. In addition to this, an "environmental community" involving nongovernmental organizations receiving support from this program was created in the virtual environment. It provided support to projects that are striving to protect the global importance of biodiversity in all types of ecosystems – arid and semiarid, coastal and marine, freshwater, forest, and mountain – and to eliminate the threats against these. In situ conservation studies of biodiversity that are important in terms of agriculture (agrobiodiversity) are also among the working areas of SGP. GEF/SGP Projects promote public awareness about the environment. Some of the GEF/SGP Projects that are directly related to conservation in farmer conditions/home gardens are as follows: "Networking and Participation for Conservation and Sustainability of Traditional Varieties," "Conservation of Local Races in Kars through Sustainable Village," "Conservation of Agricultural Biological Diversity in Kirazli Village," and "Local Fruit Varieties of Muğla: Cultural Heritage, Database and Conservation Project." All in situ conservation efforts are aimed at complementing ex situ protection. Therefore, it is mandatory to integrate these programs with National Plant Genetic Resources Program and to take under conservation in National Gene Bank by collecting the varieties in the study areas (Tan 2010).

Research programs are applied in order to improve the in situ and ex situ conservation methods of genetic resources. A great effort has been exerted on this issue in order to determine the status of important landraces in the northern regions of Turkey and to ensure the conservation of these varieties. Although there are many examples of successful conservation efforts for ex situ conservation such as seed gene banks across Turkey, in particular in situ conservation studies for landraces have been unfortunately insufficient in the country. In situ conservation studies for local wheat varieties have started very recently in Turkey, which has a longer history and adequate infrastructure for ex situ conservation.

Although local wheat production is carried out in smaller areas in Turkey, the distribution and diversity of landraces throughout the country are quite high. This diversity, which is important for global heritage, has been well documented, collected, and protected. Compared to the 1920s, the number of morphotypes decreased, especially in some regions, by 70%. Also, wheat fields in the past are known to be very different from 90 to 100 years ago when we make a comparison between the wheat fields of today and the past (Morgounov et al. 2016).

In a very comprehensive study conducted by CIMMYT (International Maize and Wheat Improvement Center) for on-farm conservation, a total of 65 provinces, 172 districts, and 523 villages were surveyed between 2009 and 2014 in Turkey. At the end of the study, 162 local wheat fields were identified. This research has demonstrated that genetic erosion is a continuous process and that many wheat fields have been lost in Turkey. It has been observed that many reported wheat landraces have not been cultivated when the obtained findings were compared to the collections of Mirza Gökgöl, Harlan, and Metzger. According to the results of this research, the 11 most common wheat varieties that are grown are as follows: Zerun, Ak (White) Wheat, Kırmızı (Red) Wheat, Sarı (Yellow) Wheat, Karakılçık Wheat, Kirik Wheat, Siyez Wheat, Koca (Big) Wheat, Topbaş Wheat, Şahman Wheat, and Üveyik Wheat. The continuity of conservation of landraces on on-farm conditions has shown that the sociocultural preferences of the society, especially the household, depend on parameters such as the utilization of the landraces and the economic uncertainties in general (Morgounov et al. 2016).

3.3 Conclusion

Genetic erosion in the world has been fast in recent years. Therefore, there are so many studies going on in that direction. From these studies, both protection of and providing genetic resources to breeding programs have been expected. Large budgets, longer times, and more labor in the protection of genetic resources have been needed whatever the method has been applied to. The collections into which large budgets and labor were put are always still under risk. Besides possible human fault, natural disasters and wars always create risk on these materials. A backup of any collection is always made available and their safety is always assured by national land international laws.

References

Damania A (2008a) History, achievements, and current status of genetic resources conservation. Agron J 100:9–21. https://doi.org/10.2134/agrojnl2005.0239c

Adak MS, Akbaba G (1986) Osman Tosun Gen Bankası. Bilim ve Teknik Dergisi Ekim Ayı Sayısı:28–29

Baur E (1914) Die Bedeutung der primitiven Kulturrassen und der wilden Verwandten unserer Kulturpflanzen f¨ur die Pflanzenz¨uchtung. (In German.). Jahrbuch der DLG 29:104–110

Bellon R, Gotor E, Caracciolo F (2015a) Conserving landraces and improving livelihoods: How to assess the success of on-farm conservation projects. Int J Agric Sustain 13:167–182

Bellon MR, Gotor E, Caracciolo F (2015b) Assessing the effectiveness of Project supporting on-farm conservation of native crops: evidence from the High Andes of South America. World Dev 70:162–176

Bettencourt E, Konopka J (1990) Directory of crop germplasm collections. 3. Cereals: Avena, Hordeum, millets, Oryza, Secale, Sorghum, *Triticum*, Zea and pseudocereals. International Board for Plant Genetic Resources, Rome

Brush S (1991) A farmer-based approach to conserving crop germplasm. Econ Bot 45:153–165

Brush SB (1999) Genetic erosion of crop populations in centers of diversity: A revision. In: Serwinski J, Faberová I (eds) Proceedings of the technical meeting on the methodology of the FAO world information and early warning system on plant genetic resources. Prague, Czech Republic, 21–23 June 1999, Rome, pp 34–44

Caldu-Primo JL, Mastretta-Yanes A, Wegier A et al (2017) Finding a needle in a haystack: distinguishing Mexican maize landrace suing a small number of SNPs. Front Genet 8:45. https://doi.org/10.3389/fgene.2017.00045

Camacho Villa TC, Maxted N, Scholten MA et al (2005) Defining and identifying crop landraces. Plant Genet Resour Characterization and Utilization 3(3):373–384

Croptrust Global Genebank Platform Annual Report (2017) (https://report.croptrust.org/2017/what-we-do/global-genebank-platform/)

Damania AB (2008b) History, achievements, and current status of genetic resources conservation. Agron J 100:9–21

Diamond J (1997) Guns, germs, and steel. the fates of human societies. W. W.Norton and Co., New York, USA

Ertekin S (2002) Karacadağ Bitki Çeşitliliği ve Tehditler. http://www.surkal.org.tr/karaca.htm

FAO (2010) The second report on the state of the world's plant genetic resources for food and agriculture. Food and Agriculture Organization of the United Nations, Rome

FAO (2013) Genebank standards for genetic resources for food and agriculture. FAO, Rome, Italy

Fırat E, Tan A (1997) In situ conservation of genetic diversity in turkey. In: Maxted N et al. Plant genetic conservation: the in situ approach. Chapman & Hall, London, UK.
Frankel OH, Bennett E (1970) Genetic resources in plants: Their exploration and conservation. Blackwell, Oxford, UK
Fu YB, Somers DJ (2009) Genome-wide reduction of genetic diversity in wheat breeding. Crop Sci 49:161–168. https://doi.org/10.2135/cropsci2008.03.0125
Gökgöl M (1939) Turkish wheat. Istanbul, Turkey
Gollin D (2018) Economic implications of emerging science conserving genetic resources for agriculture: Oxford University XX Annual BIOECON Conference King's College, Cambridge 13 September
Gotor E, Bellon MR, Turdieva M et al (2017) Livelihood implications of in situ-on farm conservation strategies of fruit species in Uzbekistan. Agrofor Syst. https://doi.org/10.1007/s10457-017-0069-6
Hammer K, Teklu Y (2008) Plant genetic resources: Selected issues from genetic erosion to genetic engineering. J Agric Rural Dev Trop Subtrop 109:15–50
Harlan HV, Martini ML (1938) The effects of natural selection in a mixture of barley varieties. J Agric Re 57:189–199
Harlan JR (1972) Genetics of disaster. J Environ Qual 1:212–215. https://doi.org/10.2134/jeq1972.00472425000100030002x
Jarvis DI, Hodgkin T, Sthapit BR et al (2011) An heuristic frame work for identifying multiple ways of supporting the conservation and use of traditional crop varieties within the agricultural production system. Crit Rev Plant Sci 30:125–176
Jarvis ID, Hodgkin T, Eyzaguirre P et al (1998) Farmer selection and genetic diversity: the need for a basic data set. In: Jarvis DI, Hodghin T (eds) Strengthening the scientific basis of in situ conservation of agricultural biodiversity on-farm. Options for data collecting and analysis. Proceedings of a workshop to develop tools and procedures for in situ conservation on-farm, 25-29, August 1997. IPGRI, Rome, Italy
Kan M, Kucukcongar M, Keser M et al (2015): Wheat Landraces in Farmers' Fields in Turkey: National Survey, Collection and Conservation, 2009-2015 FAO Ankara ISBN: 978-92-5-109048-0
Karagöz A (1998) In-situ conservation of plant genetic resources in Ceylanpınar State farm. In: Zencirci N, Kaya Z, AnixterY AWT (eds) The proceeding of the international symposium on in-situ conservation of plant genetic diversity. CRIFC, Ankara. IBSN 975-407-026-1,87-91
Karagöz A, Zencirci N (2005) Variation in wheat (*Triticum* spp.) landraces from different altitudes of three regions of Turkey. Genet Resour Crop Evol 52:6775–6785
Karagöz A, Zencirci N, Tan A et al (2010) Bitki Genetik Kaynaklarının Korunması ve Kullanımı. Türkiye Ziraat Mühendisligi VII, Teknik Kongresi, Ankara
Knüpffer H (2009) Triticeae genetic resources in ex situ gene bank collections. In: Feuillet C, Muehlbauer GJ (eds) Genetics and genomics of the Triticeae, plant genetics and genomics: crops and models 7. Springer, New York. https://doi.org/10.1007/978-0-387-77489-3-2
Maxted N, Tan A, Amri A et al (2000) In situ conservation. In: Maxted N, Bennet S (eds) Plant genetic resources of legumes in the mediterranean, Kluver, Dordrecht. ISBN 0-7923- 6707-3. 292–308 p. 1–386
Mazid A, Amegbeto KN, Keser M et al (2009) Adoption and impacts of improved winter and spring varieties in Turkey. ICARDA, Aleppo, Syria
Miguel A, Carvalho P, Bebeli PJ, Bettencourt E et al (2013) Cereal landraces genetic resources in worldwide Gene Banks. Agron Sustain Dev 33:177–203. https://doi.org/10.1007/s13593-012-0090-0
Morgounov A, Keser M, Kan M et al (2016) Wheat landraces currently grown in Turkey: distribution, diversity, and use. Published in Crop Sci 56:1–13. https://doi.org/10.2135/cropsci2016.03.0192
Nesbitt M (1995) Plants and people in ancient Anatolia. Biblic Archaeol 58:68–81

O'Donnell K, Sharrock S (2017) The contribution of botanic gardens to ex situ conservation through seed banking. Plant Diversity 39(6):373–378

Peres S (2016) Saving the gene pool for the future: seed banks as archives. Stud Hist Phil Biol Biomed Sci 55(96e):104

Scarascia-Mugnozza GT, Perrino P (2002) The history of ex situ conservation and use of plant genetic resources. In: Engels JMM, Ramantha RV, Brown AHD, Jackson MT (eds) Managing plant genetic diversity. Proceedings of an international conference, Kuala Lumpur, Malaysia, 12–16 June 2000. CABI Publ, Wallingford, UK, pp 1–22. https://doi.org/10.1079/9780851995229.0001

Tan A (2002a) In situ (on farm) conservation of landraces from transitional zone in Turkey, Project No: TOGTAG-2347. TUBİTAK, İzmir, Turkey

Tan A (2002b) Türkiye (Geçit Bölgesi) Genetik Çeşitliliğinin In Sıtu (Çiftçi Şartlarında) Muhafaza Olanaklarının Araştırılması (In-situ On-farm Conservation of Landraces grown in North-Western Transitional Zone of Turkey). Sonuc Raporu. (Final Report). TUBITAK-TOGTAG-2347. TUBITAK. Ankara

Tan A (2009) Türkiye Geçit Bölgesi Genetik Çeşitliliğinin İn situ (Çitçi Şartlarında) Muhafazası olanakları. Anadolu, J of AARI 19(1):1–12

Tan A (2010) Türkiye Gıda ve Tarım Bitki Genetik Kaynaklarının Durumu Gıda ve Tarım için Bitki Kaynaklarının Muhafazası ve Sürdürülebilir Kullanımına İlişkin TÜRKİYE İkinci Ülke Raporu Dr. Ayfer TAN Ege tarımsal Araştırma Enstitüsü Menemen, İzmir 2010

Tan A, Taşkın T (2009) Ege bölgesinde Sebze Olarak Kullanılan Yabani Bitki Türleri ve Kullanım Amaçları. Ege Tarimsal Araştırma Enstitüsü Müdürlügü, Yayın no: 136. 174 s. ISBN: 978–975–407–278–5

Tan A, Duman H, İnal N et al (2003) Ecosystem conservation and management for threatened species projects: description of target species and their habitats. TACN (Turkish Association for the Conservation of Nature)

Tan A, Tan AS (2002) In situ conservation of wild species related to crop plants: the case of Turkey. In: Engels JMM, Brown AHD, Jackson MT (eds) Managing plant genetic resources. CAB International, UK, pp 199–204

Tan A, Tan AS (2004) In situ Conservation for Genetic Diversity: Reserve Designation and Design. In: Goodman RM (ed) The encyclopedia of plant and crop science. Marcel Dekker, Inc., New York. www.dekker.com

Thormann I, Fiorino E, Halewoo M et al (2015) Cite as Plant genetic resources collections and associated information as a baseline resource for genetic diversity studies: an assessment of the IBPGR-supported collections. Genet Resour Crop Evolut December 2015 62(8):1279–1293

World Bank (1993) Republic of Turkey: in-situ conservation of genetic diversity. Global Environment Coordination Division 11295-TU, Washington DC-USA

Yong-Bi F (2017) The vulnerability of plant genetic resources conserved ex situ. Crop Sci Rev Interpret 57(5):2314–2328

Zencirci N, Yılmaz H, Garaybayova N et al (2018) Mirza (Hacızade) Gökgöl (1897–1981): the great explorer of wheat genetic resources in Turkey. Genet Resour Crop Evol 65(3):693–711

Chapter 4
Characterisation of Genetic Diversity in Wheat Landraces

Özlem Özbek

4.1 Introduction

Wheat is the most revolutionist crop in the world, because it resulted in conversion of our hunter-gatherer ancestors to sedantic societies, who became traditional farmers around 10,000 years ago. The other revolution in the history of wheat was the domestication process, by which the brittle rachis of wild emmer (*Triticum turgidum* ssp. *dicoccoides* Körnick.) and *Triticum monococcum* ssp. *aegilopoides* evolved to non-brittle rachis of emmer wheat [*Triticum turgidum* ssp. *dicoccon* Schrank Thell. ($2n = 4X = 28$, *AABB*)] and einkorn wheat [*Triticum monococcum* ssp. *monococcum* ($2n = 2x = 14$, A^mA^m)], respectively. *Triticum monococcum* and *Triticum dicoccon* were the most popular crops until the early Bronze Age. Then, they started to be replaced by the high-yielding and free threshing wheat varieties (*Triticum aestivum* $2n = 6X = 42$, *AABBDD* and *Triticum durum* $2n = 4X = 28$, *AABB*). After domestication process, domesticated wheat varieties started to be cultivated by traditional farmers, and the seeds have been sown for thousands of generations since then. Although traditional farmers did not apply formal breeding programmes, selection was still under progress due to the natural selection in the environment and farmers' personal interest on the wheat varieties they grew. Farmers used these primitive relatives of wheat for their domestic uses such as feeding their livestock and home use for making bulgur, erişte (homemade macaroni) and so on. When they grow the wheat landraces, they made selections on the wheat varieties, they grown about their resistance to biotic and abiotic stress factors, and amount of the yield and yield stability in low input agricultural system (Zeven 1999). Until the last century, traditional agriculture has been some kind of heritage from generation to generation; therefore, the seeds have been sown for thousands of generations about 10,000 years ago. Those were called the wheat landraces, which

Ö. Özbek (✉)
Department of Biology, Faculty of Art and Science, Hitit University, Çorum, Turkey

N. Zencirci et al. (eds.), *Wheat Landraces*,
https://doi.org/10.1007/978-3-030-77388-5_4

were described as "variable plant populations adapted to local agroclimatic conditions which are named, selected and maintained by the traditional farmers to meet their social, economic, cultural and ecological needs" (Teshome et al. 1997) (see some other definitions in Zeven 1998). The landraces have enormous amount of genetic diversity covering unique genotypes enhancing their adaptability to different environmental conditions particularly to extreme conditions in remote mountainous places. The landraces grown on farms by traditional farmers have the opportunity to continue their evolution, which provides a dynamic and diverse genetic structure, and they can easily adapt to the changing environmental conditions in their habitats. The seed exchange system which is carried out by traditional farmers increases the genetic diversity and extend the germplasm of the landraces. The seeds can be exchanged between farmers, friends and neighbours in the same village or purchased from a commercial seed market (Zeven 1999). In the last century, the wheat breeding studies started and the new wheat varieties, which have high yield and adapted to specific environmental conditions, were developed and replaced the landraces. The wheat landraces germplasm has lost 75% of their genetic diversity (Jaradat 2013). Therefore, the extent of variability and the characterisation and partition of genetic diversity within the local germplasm collections are important criteria to determine the status of wheat landraces particularly for the future interests of their use and for the improvement and efficient genetic diversity maintenance and the utilisation of plant species (Desheva 2014). The wheat landraces were investigated for the characterisation of genetic diversity in terms of morphological, phenological and agronomic traits, proteins, enzymes and molecular aspects.

4.2 Variation in Morphological, Phenological and Agronomic Traits

Climate change causes the change in the environment in which all organisms live. The plant species could manage these new conditions through phenotypic plasticity and adaptation by natural selection or moving to find conditions to which they are used to (Nicotra et al. 2010). When a genotype is expressed differently, it produces different phenotypes in different environments called phenotypic plasticity (Bradshaw 1965 cited in Turcotte and Levine 2016). Thus, it is important to determine the different phenotypes, which are examined as variations in the morphological and agronomical traits which are due to the phenotypic plasticity or genetic diversity. The characterisation of morphological and agronomical traits is an important issue for estimating the genetic diversity in the natural, local and cultivated plant populations for the breeding programmes and agricultural demands.

The variation in morphological characters and agronomic traits was investigated in diploid, tetraploid and hexaploid wheat species and their landraces in the several previous studies. The agronomical traits studied are biological yield (BY), grain yield per plant (GYP), harvest index (HI) and yellow pigment. The morphological traits studied are the number of spikelet per spike (SS), the number of kernels per spike (NKS), the number of kernels per spikelet or fertility, plant height (PHt), tiller number (TN), number of tillers per plant, fertile tiller, 100-kernel weight (g) (HKW), 1000-kernel weight (TKW), peduncle length (PL) and spike length (SL), spike width (SW), awn length (AL), beard length (BL), upper node length (UNL), width of truncation (WT) and barb length (BbL, root length (RL), root fresh weights (RFW) and dry weights (RDW), shoot fresh (SFW) and dry weights (SDW), seed colour (SC) and seed shape (SS). The phenological traits studied are days to heading (DTH), days to maturity (DTM), grain filling period (GFP) and glume pubescence (GP).

*4.2.1 Einkorn Wheat (***T. monococcum** *ssp.* **monococcum***) Landraces*

Einkorn ($2n$ = 14, *AA*) includes both wild and cultivated forms originated in Southeast Anatolia near Karacadağ (Heun et al. 1997). Wild einkorn *Triticum boeticum* is divided into single-grain *T. agilopoides*, and two-grain *T. thoudar* and *T. urartu*, while the cultivated einkorn is named as *Triticum monococcum* ssp. *monococcum*. *T. monococcum* has annual habit, 30–70 cm culm from the base to the upper end, indeterminate, bilaterally compressed spike with two rowed, 8–10 cm in length and awned. The spike disarticulates into the individual spikelets with rachis segments at maturity. The brittle rachis differentiates the wild einkorn from the cultivated einkorn (Stallknecht et al. 1996; Kimber and Feldman 1987).

There has been no much research interest in morphological or agronomical traits of einkorn wheat according to the previous studies. Variability in agronomic traits, sprouting, frost resistance, heading, maturity time, plant height, number of spikelets per spike, grain weight and TKW in einkorn (*T. monococcum*) landraces was evaluated. Einkorn landraces generally displayed late maturity and small seed prevalent character; thus, einkorn had limiting cultivation in many places all over the world, but in Metalliferous Mountains, it is still a common cultivar (Butnaru et al. 2003). The most relative variable agronomical character was the grain yield, followed by the spike length and the thousand kernel weight in a set of 15 einkorn wheat landraces, and cluster analysis indicated that the landraces were grouped into six groups (Uzundzalieva et al. 2016).

4.2.2 *Emmer Wheat (***Triticum turgidum** *ssp.* **dicoccon** *Schrank Tell.) Landraces*

The geographic distribution of genetic diversity and the population structure of tetraploid wheat landraces in the Mediterranean basin have also received relatively little interest. This is complicated by the lack of consensus concerning the taxonomy of tetraploid wheats and by unresolved questions regarding the domestication and spread of naked wheats. These knowledge gaps hinder the crop diversity conservation efforts and the plant breeding programmes (Oliveira et al. 2012). Emmer wheat cultivation has been drastically reduced during the last century as a consequence of replacement with the high-yielding wheat varieties. However, more recently, the increase in interest for the healthy foods, along with its agronomic and nutritive values, has caused an increase in the cultivation area, which is now more than 2000 ha (Pagnotta et al. 2005). Thirty-nine Italian ecotypes and cultivars of *Triticum turgidum* L. spp. *dicoccum* Schrank ex Schübler (emmer wheat) displayed a huge amount of diversity not only between varieties but also within the varieties according to agro-morphological and molecular analysis (Pagnotta et al. 2005). Former studies, on Italian (and foreign) emmer accessions, have been carried out to assess variation in the agronomic and quality traits by the morphological field evaluations (Damania et al. 1992; Piergiovanni et al. 1996; Galterio et al. 1998; D'Antuono and Minelli 1998), in order to select the material among the old landraces. The analysis of morphological characters (plant height, spike length, spike weight, distance between spikelets, number of spikelets per spike, number of grains per spike, number of grains per spikelets, grain length, thousand grain weight and weight of grain per spike) of a collection of Spanish emmer wheat lines indicated that there were seven different botanical varieties, which were less than previous records of ten, thus displaying a wide diversity (Alvarez et al. 2007). These kinds of studies are important for the future registration of the material, germplasm conservation and use of this valuable source of emmer germplasm for the future breeding programmes (Pagnotta et al. 2005).

4.2.3 *Durum Wheat (***Triticum turgidum** *var.* **durum** *Desf.) Landraces*

The morphological characters have correlations with some agronomical traits. Some morphological characteristics such as the number of spikelet per spike (SS), the number of kernels per spike (NKS), the number of kernels per spikelet or fertility, tiller number (TN), the number of tillers per plant, 100-grain weight (HGW) and 1000-kernel weight (TKW) positively correlated with agronomical traits such as biological yield (BY), grain yield per plant (GYP) and harvest index (HI) (Al-Ajlouni and Jaradat 1997; Jaradat 1991). Variation in these morphological traits will directly affect the agronomical traits, which are important for the agricultural production.

Variation in the morphological and agricultural traits in durum wheat landraces was analysed by comparing them with the commercial cultivated durum wheat checks. The reason for this was that the traits were superior in landraces to the commercial durum wheat varieties, which were determined to have been exploiting for the improvement of commercial varieties' germplasm as a genetic resource for future durum wheat breeding programmes. On the other hand, the traits, which the commercial durum wheat varieties were superior to landraces, might be considered to improve the landrace germplasm by crossing studies (Getachew et al. 1993).

The variation in the morphological, phenological and agronomical traits was reported in the durum wheat landrace germplasm from the different regions of the world in the previous studies.

The variation in glume colour, glume pubescence, spike density, spike length, seed colour, seed virtuousness, seed size, seed shape, beak length, spikelet per spike and seeds per spike in durum wheat landraces from four regions and five altitudinal gradients in northern and north-central regions of Ethiopia was used to estimate the genetic diversity. Most of the traits were found to be polymorphic, while monomorphism was common in many of the populations for the dense spike, long beak and glabrous glume. The highest mean diversity was observed for seed colour, seed shape and glume pubescence, whereas the spike density displayed the lowest diversity index (Bechere et al. 1996).

Seven tetraploid wheat populations from Shoa and Gojem Administrative Regions of Ethiopia were examined for variation of eight morphological characters (glume colour, glume pubescence, awn condition, awn length, awn colour, beak length, spike density and seed colour). Monomorphism was common for the awn length and glume pubescence in many of the populations, and the awn condition was found to be a fixed character in the entire collection, whereas the rest of the characters exhibited polymorphism in varying degrees. The lowest level of diversity was determined for glume pubescence (excluding awn condition). The analysis of variance of diversity for individual characters indicated that most of the variation was due to differences among the districts rather than among the populations within the districts such as the glume pubescence, which showed significant differences only among the districts. It might be considered for the future studies and that would be better if the study samples are collected from more different areas than having more samples from similar areas (Tesfaye et al. 1991).

Durum wheat (*Triticum turgidum* var. *durum* Desf.) landraces from the Syrian Arab Republic were characterised in terms of morphological characters (pubescence of leaves, spike density, spikelet attitude, glume colour, lemma colour, awn colour and glume hairiness) and agronomic traits (grain yield, straw yield, crop growth duration (days), spikes/m^2, seeds per spike, spike weight and 1000-seed weight). According to farmers' decisions, average grain yield was the lowest in western mountainous regions, while it was the highest with overestimated yield level in southern parts of the country, which displayed the tendency of landraces to produce more straw rather than grain dry matter under high rainfall conditions. Glume, lemma and awn colour were found to be highly variable, both within and among the landrace groups. The plants both with glabrous glumes and with pubescent glumes were observed in three landrace groups (Bayadi, Shihani and Surieh), while the other landrace groups were identified having only one form of glume hairiness. The yellow colour was the predominant colour for kernel in the groups, whereas the variation in spike density was clearly the first character to differentiate landraces. Distribution patterns of the various landrace groups indicated that only a few landrace groups were widely distributed and most others were regionally concentrated. The results indicated that Syrian durum wheat landraces had genetic diversification due to the heterogeneous nature of landraces and different landraces adapted to per region or village. However, durum landraces were found in mixture with *T. aestivum* at large proportions in the mountainous regions in the west of the country, where farmers prefer a species mixture (Elings and Nachit 1991).

Ethiopia is one of the countries where the wheat landraces are still growing. In a study, 34 tetraploid wheat (*Triticum turgidum* L.) landrace populations from four regions in Ethiopia were investigated to determine the diversity and distribution of these traits on the basis of administrative regions and altitudinal gradients for some morphological characters (glume colour, glume pubescence, beak awn, seed colour and spike density). All characters displayed polymorphism, except spike density, in all regions and most altitude groups. The highest variable character was seed colour and the lowest variable character was spike density. On the other hand, the diversity was increasing as with the increasing better climatic conditions and in optimal altitude ranges. All the result showed that the diversity has not changed considerably within the past 25 years or so, when compared to previous estimates (Belay et al. 1997).

The landrace genotypes of durum wheat from Jordan have been evaluated for 18 morphological- and yield-related traits (plant height, peduncle length, plant height ratio, spike length, awn length, awn/spike length ratio, the number of spikelets per spike, the number of seeds per spikelet, spike node, the number of seeds per main head, seed weight/main head, the number of tillers per plant, number of seeds per tiller, seed weight per plant, kernel weight (main head), kernel weight (tillers), kernel weight (average) and protein content). The study results indicated that the magnitude of phenotypic divergence in these landrace genotypes is large, especially when they are compared with a world collection of durum wheat. According to experimental evidences, the variation in altitude and long-term average rainfall of the collection sites gives rise to the phenotypic differentiation. The landrace genotypes are clustered into five groups on the basis of altitude and long-term average rainfall of collection site. Three canonical factors accounted for 92% of total variance in these clusters. The phenotypic diversity found in these landrace genotypes could help to identify genetically different genotypes for durum wheat improvement (Jaradat 1991). Mac Key (1966) noted the continuity of morphological traits among *T turgidum* convar., which is parallel in the morphological variation among the *T. durum* landrace groups to such an extent that clear distinction between groups could not always be made. The grouping should therefore merely be considered as a systematic description of visible variation.

4.2.4 *Bread Wheat (*Triticum aestivum *L.) Landraces*

Omani wheat landraces were analysed based on 15 qualitative and 17 quantitative characters and they displayed variations. Quantitative characters had higher diversity index ($H' = 0.66$) than qualitative characters ($H' = 0.52$) in tetraploid wheat, and 0.63 and 0.62, respectively, in hexaploid wheat. The morphological data showed that Omani wheat landraces had considerably high diversity and even simple morphological characters could be used for an effective characterisation of diversity in Omani wheat (Al Khanjari et al. 2008).

Fifty-three pure lines of bread wheat (*Triticum aestivum* L.) derived from seven landraces collected from southeastern Iran were analysed to determine genetic variation and heritability for 13 developmental and quantitative characters. The landrace genotypes displayed lower values for the number of grains per spike, 1000-grain weight, grain yield and harvest index, while they were late in days to heading and taller than the cultivars. On the other hand, some landrace genotypes showed similar grain yield with the modern cultivars, and for number of grains per spike, number of spikes per plant, 1000-grain weight and harvest index, they showed moderate to high genetic variation. Wheat landraces have genetic variation, which is an important source for agronomic characters. Therefore, the landraces with higher genetic variation for agronomical characters might be used for improvement of landrace germplasm by inter-crossing (Moghaddam et al. 1997).

In a previous study, it was reported that wheat (*Triticum* spp.) landrace populations were mixture of different tetraploid and hexaploid wheat species in Ethiopia. The tetraploid ($2n = 4x = 28$) wheat species were identified in mixtures of varying proportions of *Triticum durum* Desf., *Triticum turgidum* L., *Triticum aethiopicum* Jakubz., *Triticum polonicum* L. and *Triticum dicoccon* Schrank, and the hexaploid ($2n = 6x = 42$) wheat species was mixed with *Triticum aestivum* L., and *Triticum durum* was determined as the most predominant species, while the hexaploid *Triticum aestivum* was determined in nine populations from Wollo (Eticha et al. 2006).

Buerkert et al. (2006) revealed interesting results about the genetic composition of farmer's wheat (*Triticum* spp.) landraces from Afghanistan. They selected randomly 21 cereal fields on both sides of the Panjshir River in the upper Panjshir valley of northern Afghanistan. They surveyed morphological differences on morphological characters, and after that, they collected information about the field size and grain yield and a formal interview with the landowner on the cropping sequence and the inputs used. The results of morphological evaluation displayed that the collection included 19 taxonomically different varieties of bread wheat (*Triticum aestivum* L.); in addition, barley and triticale (*Triticosecale* Wittm.) were found in the mixtures. Farmers were not aware of morphological differences within these mixtures; however, they recognised their populations according to grain colour, cooking properties and resistance to mildew and frost. The most interesting thing about the result was the most widespread occurrence of *T. aestivum* var. *subferruginінflatum*, which so far have only been reported together with var.

subgraecinflatum from Mongolia. The landrace populations were clustered into the different groups according to the cluster analyses based on the isozymes and agronomic data, which were the prime target of artificial selection and under the effects of different evolutionary forces.

The *ex situ* collection of common wheat (*Triticum aestivum* L.) landraces stored for more than 10 years in IPGR-Sadovo were analysed for the variation in the morphological and agronomic characteristics. The wheat landraces were characterised with the most relative variable character which was the spike length (C.V. % = 15.09%), followed with 1000-grain mass (C.V. % = 8.04%) and the number of spikelets per spike (C.V. % = 7.66%) (Desheva 2014).

4.3 Variation in Seed Storage Proteins

The seed storage or endosperm proteins offered great opportunity to reveal genetic variation within and between the populations of a species or interspecies. By means of seed storage, proteins, gliadins and glutenins which are also known as prolamins provide nutrients (amino acids) for embryo during germination of seeds (Ciaffi et al. 1993). The germplasm of wheat landraces at all ploidy levels was investigated in terms of seed storage proteins (gliadin and glutenin) and also for conferring their relationship with technological properties (Ciaffi et al. 1991; Ciaffi et al. 1992).

4.3.1 *Einkorn Wheat (*Triticum monococcum *L. ssp.* monococcum*) Landraces*

Einkorn wheat (*Triticum monococcum* L. ssp. *monococcum*) has only A^mA^m genome with $2n = 14$ chromosomes. The expression of HMW-glutenin subunits and gliadin proteins differs in diploid, tetraploid and hexaploid wheat species; particularly, the subunits encoded by *AA* genome decreased during the evolution of wheat. The screening of seed storage protein composition of Spanish-cultivated einkorn wheat (*Triticum monococcum* L. ssp. *monococcum*) indicated that three and up to six allelic variants were detected for the *Glu-A1*^{m} and *Glu-A3*^{m} loci, respectively, while 7 and 14 alleles were detected for the *Gli-A1*^{m} and *Gli-A2*^{m} loci, respectively, among the accessions analysed. The Spanish einkorn wheat collection displayed 48 different genotypes based on the origin, and seed storage protein compositions have been determined (Alvarez et al. 2006).

Evaluation of seed endosperm proteins (gliadins and glutenins) in Turkish-cultivated einkorn wheat (*Triticum monococcum* ssp. *monococcum*) landrace populations displayed great genetic diversity ($H_e = 0.65$). The gliadin proteins showed higher genetic diversity and allele combinations than HMW-glutenin proteins (Keskin et al. 2015). The increasing trend in the world as well as in Turkey is the great interest for healthy foods particularly organic farming products or food products produced by the local farmers. Turkish-cultivated einkorn wheat farming is the most popular crop landraces particularly in Bolu and Kastamonu provinces in Turkey recently. Its local name is "Siyez" and used to make bread, bulgur and pastry products such as cookies, savoury roll covered with sesame seed, etc.

A large germplasm collection, including *Triticum monococcum*, *T. boeoticum*, *T. boeoticum* ssp. *thauodar* and *T. urartu*, were investigated for their protein composition at the *Gli-1* and *Gli-2* loci (Ciaffi et al. 1997). The results indicated that *T. monococcum* and *T. boeoticum* were very similar in their gliadin patterns, and they were different distinctly from *T. urartu*, which was resembled to the *A* genome of polyploid wheats more than did *T. boeoticum* or *T. monococcum* in the gliadin pattern. The study confirms that *T. monococcum* and *T. boeoticum* are different subspecies of the same species, but *T. monococcum*, while it supports the hypothesis of *T. urartu*, is the donor of the *A* genome in cultivated wheats. Diploid wheat germplasm showed high level of variation for gliadin proteins, which might be further analysed as to whether the loci coding them have any linkage with the genes encoding for the desirable traits to determine parental candidates and to transfer alien genes into cultivated polyploid wheats (Ciaffi et al. 1997).

Einkorn and emmer wheat were the most popular cereal crops until the early Bronze Age, and they were replaced by high-yielding wheat varieties. The formal breeding studies speeded up the replacement of landraces with the modern bread wheat and durum wheat varieties, which are the products of breeding programmes. This replacement has caused a 75% loss of genetic diversity in the last century (Jaradat 2014).

4.3.2 *Emmer Wheat (***Triticum turgidum** *L. ssp.* **dicoccon** *(Schrank) Thell.) Landraces*

The cultivated emmer wheat was domesticated around the Karacadağ Mountain in the southeastern part of Turkey (Özkan et al. 2002; Özkan et al. 2005). Emmer wheat germplasm is important for revealing the domestication of wheat and for the improvement of modern wheat varieties due to its rich germplasm harbouring the high level of genetic diversity and different gene combinations. Polymorphism for gliadin proteins of Turkish-cultivated emmer wheat (*Triticum turgidum* L. ssp. *dicoccon* (Schrank) Thell.) populations was evaluated, and the results indicated that emmer wheat populations had great genetic diversities ($H_e = 0.92$) and showed 27 different patterns, which were the combinations of different gliadin proteins (Fig. 4.1). Most of the Turkish-cultivated emmer wheat populations had the α-45 and w-35 gliadins closely associated with dough quality; thus, emmer wheat germplasm is bearing the desirable traits related to quality. Pearson's correlation analysis displayed that the latitude had strong influence ($r_P = 0.510$; $p = 0.026$ at $<0.05\%$) on the genetic diversity estimates (Özbek et al. 2011).

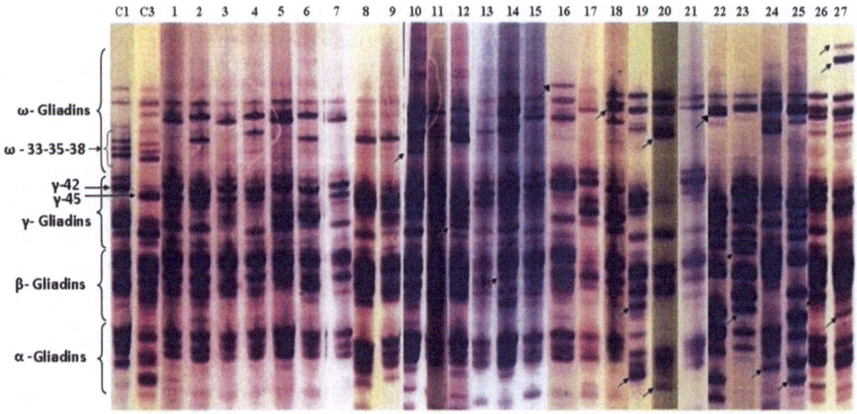

Fig. 4.1 Gliadin band patterns observed in Turkish emmer wheat populations. (Source: Özbek et al. 2011)

Glutenin proteins are part of seed storage proteins and are composed of two different types of protein groups according to their molecular weights as high molecular weight (HMW) and low molecular weight (LMW) glutenin subunits (Payne and Lawrence 1983). HMW-glutenin subunits have polymeric protein structures because of cross-linkages by disulphide bonds and rich in glycine residue in content (Payne and Corfield 1979).

Investigation of genetic diversity in Turkish emmer wheat landrace populations collected from different provinces in terms of HMW-glutenin proteins indicated that the mean values of expected heterozygosity (gene diversity) and average heterozygosity among the populations were estimated as 0.31 and 0.12, respectively. On the other hand, actual genetic differentiation (*D*) reveals that the partition of genetic diversity was between 24% and 76% within populations. These results infer that emmer wheat landraces are well adapted to the different environmental conditions and have dynamic evolutionary history still in progress. The results of this study showed the significant influence of ecogeographical variables on HMW-glutenin diversity (Özbek et al. 2012). The band patterns of HMW-glutenin subunit of a Turkish emmer wheat population from Kastamonu province are given in Fig. 4.2.

*4.3.3 Durum Wheat (*Triticum turgidum *var.* durum *Desf.) Landraces*

A collection of *Triticum durum* wheat comprising 25 cultivars from different regions in Iran and 10 cultivars from different European countries were investigated in terms of HMW-glutenin subunits. In the collection, HMW-glutenin subunit 1 encoded by *Glu-A1* was not detected, while the prevalence of the null allele (52%)

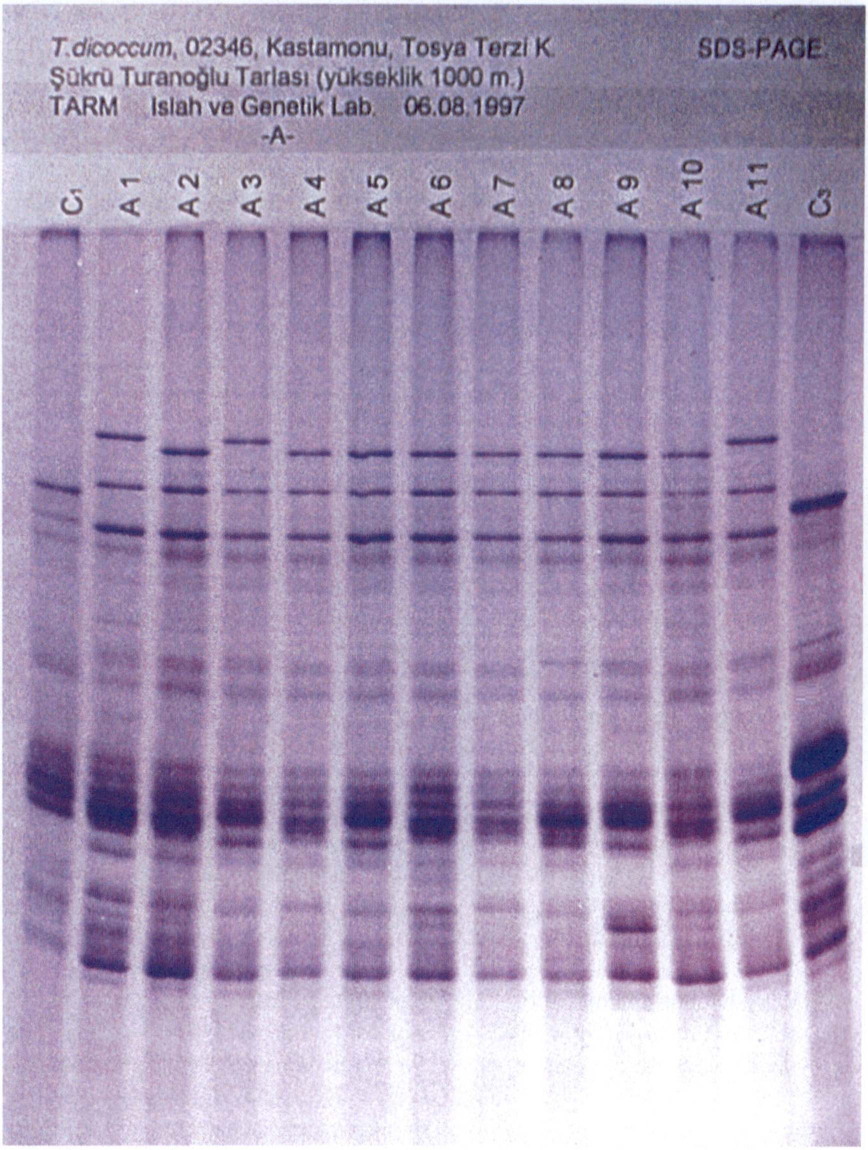

Fig. 4.2 High molecular weight glutenin subunits band patterns observed in Turkish emmer wheat population from Sinop province. (Photo by Özbek Ö., 2006)

and 2* subunit (48%) at the *Glu-A1* was observed. For *Glu-B1* locus, about 80% of the alleles composed of 14 + 15 (32%), 13 + 16 (28%) and 7 + 20 (24%) in the Iranian durum wheat compared to European durum wheat (Motalebi et al. 2007).

Iranian durum wheat (*Triticum turgidum*) landraces presented higher frequency for HMW-glutenin subunit in the null allele as reported in Iranian bread wheat

(*Triticum aestivum* L.) landraces (Kamali et al. 2011). This might be related to the existence of bread wheat, which took place in the area including Iran. Therefore, null allele is the common and the most frequent allele in both durum and bread wheat landrace germplasm. The allele number observed for *Glu-A1* and *Glu-B1* loci was 3 and 11 alleles, respectively. Iranian durum wheat landrace germplasm might have different combinations of HMW-glutenin subunits, and this might be considerable for breeding programmes particularly for improving pasta-making quality (Naghavi et al. 2009).

Anatolian durum wheat "Kunduru" landrace germplasm displayed high polymorphism in terms of gliadin and glutenin proteins. In the expense of high-yielding wheat varieties, wheat landraces are neglected, and many of the landraces are under extinction risk from the local farmers' fields. Therefore, the studies on the landraces are revealing the importance of maintaining and conserving these valuable genetic resources (Alsaleh et al. 2016).

4.3.4 Bread Wheat (Triticum aestivum L.) Landraces

The electrophoresis profiles of HMW-glutenin subunits in some of the Iranian bread wheat (*Triticum aestivum* L.) landraces displayed null allele which was the predominant compared to subunits 1 and 2* at the *Glu-A1* locus, while *Glu-B1* locus displayed higher allelic variation as 7, 7 + 8, 6 + 8, 14 + 15, 7 + 9, 17 + 18, 13 + 16 and 20. The highest and the lowest frequencies were observed for the subunits 7 + 8 (56%) and 13 + 16 (2%) at *Glu-B1* locus and 2 + 12 subunits (74%) and the rare 2*** + 12′ (2%) subunits at *Glu-D1* locus, respectively. Based on Payne scoring method, three landraces were identified as superior. It sounds that Iranian wheat landraces have high variation for the quality traits and HMW-glutenin subunits and present potential sources for the desirable quality traits to be used in bread wheat breeding programmes to improve bread-baking quality (Kamali et al. 2011).

Terasawa et al. (2009) investigated variations in the morphological characters and HMW-glutenin subunit composition of an Afghan wheat collection maintained in Kyoto University together with 65 accessions of Iranian and Pakistani wheat. The higher frequencies were observed for the alleles encoding for HMW-glutenin subunit of *Glu-A1c* (encoding subunit null), *Glu-B1b* (7 + 8) and *Glu-D1a* (2 + 12). The Afghan wheat landraces represented typical morphological characters of landraces, and for the genetic diversity in terms of HMW-glutenin subunits and AFLP molecular markers, they had equal to or lower than neighbouring countries and lower in Afghan wheat landraces, respectively. The results of the study displayed the existence of a decrease in the genetic diversity in Afghanistan wheat landraces.

4.4 Variation in Isoenzymes

Analysis of diversity in terms of isozymes and agronomical traits in ten tetraploid wheat landrace populations from different localities in the central highlands of Ethiopia indicated that the variation observed within populations was higher than between populations. The differentiation between populations was not high among the populations which were attributed to the landrace populations who were sharing a common ancestral population and/or adaptation to similar climatic conditions. The pattern of genetic divergence was determined as independent from geographic distance. The landrace populations were clustered into the different groups according to the cluster analyses based on the isozymes and agronomic data, which were the prime targets of artificial selection and under the effects of different evolutionary forces.The clustering based on the agronomic traits resulted in the populations grouping together due to the similar agronomic performance. Therefore, taking more samples within a locality or population would be a better approach to capture the range of variation in the landrace populations of the central highlands of Ethiopia (Tsegaye et al. 1996).

4.4.1 *Emmer Wheat (***Triticum turgidum** *L. ssp.* **dicoccon** *(Schrank) Thell.) Landraces*

Isoenzymes are important tools for the characterisation of genetic diversity in the natural and cultivated cereal crops. Turkish emmer wheat landraces were investigated by means of three isozyme (endopeptidase-1 (*Ep-1*), aminopeptidase-1 (*Amp-1*) and aminopeptidase-2 (*Amp-2*)) systems, and considerably high level of genetic diversity ($H_e = 0.23$) was estimated and eco-geographical variables had significant influence on genetic diversity of isoenzymes in Turkish emmer wheat populations according to statistical analysis (Özbek et al. 2013).

Isoelectric focusing (IEF) polyacrylamide gel electrophoresis is a method used for the analysis of isozyme diversity, which provides important information about the evolutionary history of plant species. The tetraploid wheat landraces (*Triticum dicoccon* Shrank, *T. turgidum* L., *T. durum* Desf., *T. pyramidale* Percival and *T. aethiopicum* Jakubz.) grown in Ethiopia were investigated to determine the genetic diversity in terms of α-amylase isozymes and to get inferences about the evolutionary histories of Ethiopian wheat landraces. Two zymogram types, band 18 (*α-Amy-B1*) and band 1 (*α-Amy-B3*), of the malt types were identified in *T. dicoccon*, while in the rest of the landraces, four zymogram phenotypes were identified. The overall results displayed that the genetic diversity was low in cultivated tetraploid wheats from secondary centres for α-amylase isozymes, which might be due to the founder effect or selection. It is supposed that among the tetraploid wheat species, *T. dicoccon* was the first wheat arriving to the Ethiopian highlands ca. 5000 years ago. It is contradictory whether the feral type Ethiopian tetraploid wheat landraces are direct descendants of *T. dicoccon*, or were introduced independently.

The feral types and *T. dicoccon* were sharing common α-amylase zymogram pattern band 1 that is showing the gene flow between them (Belay and Furuta 2001).

Landraces have desirable traits, which can be exploitable for improvement of wheat varieties or developing new wheat cultivars. In previous studies, it was reported that emmer wheat germplasm has some desirable traits such as resistance to the leaf diseases and common bunt (Corazza et al. 1986), resistance to yellow rust (Damania and Srivastrava 1990), powdery mildew (Jakubziner 1969) and Fusarium head blight or scab (Oliver et al. 2008).

*4.4.2 Durum Wheat (*Triticum turgidum *var.* durum *Desf.) Landraces*

The genetic diversity has been analysed between origins, and within origins, of a durum wheat world collection according to 13 isozymes. The comparison of the isozyme frequencies in wild emmer and durum wheat could provide the knowledge to understand the effect of domestication process. According to the origins, Iran, Mexico, Ethiopia, Egypt and Afghanistan had the highest genetic diversity for durum wheat. The geographical or political lines were the effective factors for the grouping of the landraces along within-variability of the origins. According to gene frequencies, Egypt might be considered a microcentre of diversity for durum wheat within the Mediterranean centre, although it is certainly related to Ethiopia, while Mexico has become a new microcentre of diversity, quite likely man-made, and is distant from other centres of durum wheat diversity (Asins and Carbonell 1989).

*4.4.3 Bread Wheat (*Triticum aestivum *L.) Landraces*

Analysis of five isozymes to determine the genetic diversity and genetic structure of 324 Chinese wheat landraces indicated that the landraces from the western part had higher diversity than from the eastern part, and populations were clustered into three major groups particularly neighbouring populations clustered together; the first group included most of the populations from western China and two populations from Xinjiang and Gansu and Ningxia, the second group included the northern populations from Mongolia to Japan and the third group consisted of the southeastern populations from Shaanxi, China and Japan. The results suggested that the Silk Road had important role through the transmission of wheat on both northern and southeastern populations. The genetic differentiation between eastern, northern and southern populations was determined as well, and it was reported in Korea and Japan (Ghimire et al. 2005).

Isozymes have functional roles in plant metabolism and they play differential activities in different parts and stages of plants. Esterase or peroxidase isozymes were investigated in both roots and shoots of seedling among landraces of wheat (*Triticum aestivum* L.) from Sichran. In root tissues, esterase had two isozymes, while peroxidase had single isozyme pattern and displayed no variation. Both esterase and peroxidase showed variation displaying 6 and 15 isozymes for shoots, respectively, and peroxidase had higher variation than esterase in shoots (Zuli et al. 1999).

The isozymic variation of *peroxidases*, *malic dehydrogenase*, *alcohol dehydrogenase*, *acid* and *alkaline phosphatase* were assessed in dry kernels and of *α-amylases* in germinating kernel endosperms, and of *phosphor-glucose mutase* and *isomerase*, *esterases*, *leucine aminopeptidase*, *glutamic oxaloacetic transaminase*, *malic dehydrogenase* and *peroxidases* in 15-day-old seedling leaves, were analysed. The overall results indicated that variation in isozymes analysed was not successful to identify the cultivars and classification based on the isozymes that did not infer the ancestor-descendant relationships among related cultivars (Salinas et al. 1982), because isozymes don't have high genetic variation.

4.5 Genetic Diversity

4.5.1 Einkorn Wheat (L. ssp. monococcum*) Landraces*

The diploid wheat *Triticum monococcum* L. (einkorn) was replaced by the high-yielding tetraploid and hexaploid wheat varieties and largely forgotten by the modern breeders. Einkorn germplasm was not subjected to breeding programmes; therefore, it was devoid of breeding bottlenecks, and it has conserved the genetic variation that existed during its domestication period (Kilian et al. 2007).

The molecular analysis based on the nuclear and chloroplast microsatellites of 50 einkorn wheat (*Triticum monococcum* L.) accessions from Europe, North Africa and Near East indicated that there were two main gene pools, one was from Morocco and the Iberian Peninsula and the other was from Europe and Near East in einkorn. Gene diversity ranged between $H = 0.411$ in Iberia Peninsula and $H = 0.594$ in other einkorn accessions (Oliveira et al. 2011).

Heun et al. (1997) investigated the origin of domestication site of *T. monococcum* using molecular markers (AFLP). The molecular data suggested that a wild group of *Triticum monococcum* ssp. *boeoticum* lines from the Karacadağ mountains (southeast Turkey) was the likely progenitor of cultivated einkorn varieties along with the evidence from archaeological excavations of early agricultural settlements near the Karacadağ mountains, where domestication of einkorn wheat began.

The wild einkorn underwent a process of natural genetic differentiation, most likely an incipient speciation, and prior to domestication. It was determined that three genetically, and to some extent morphologically, distinct wild einkorn races existed, and they were designed as a, b and c. Race b was used by humans for domestication (Kilian et al. 2007). The observations of higher genetic diversity in domesticated einkorn are inferring that domestication process had no effect on the reduction of genetic diversity in einkorn wheat. A specific wild einkorn race that arose without human intervention was subjected to multiple independent domestication events (Kilian et al. 2007).

The genetic diversity analyses in einkorn wheat (*Triticum urartu*, *T. boeoticum* and T. *monococcum*) and *Aegilops* ssp. (*Ae. speltoides* and *Ae. squarrosa*) in terms of DNA markers (AFLP and SSLP) clearly indicated that *T. urartu* was greatly differentiated from the other two A genome species. The observations of less intraspecific DNA variations of the nuclear genomes within the einkorn wheat ssp. were consistent with Kilian et al. (2007) and were smaller than those within the two *Aegilops* species that displayed the largest nuclear genome variation, while its chloroplast genome variation was the least (Mizumoto et al. 2002).

The analysis of the germplasm of Iranian einkorn group (*T. monococcum*, *T. boeoticum* subsp. *boeoticum*, *T. boeoticum* subsp. *thaoudar* and *T. urartu*) using DNA markers (IRAP) produced great polymorphism, 84% of which was attributed to total variation within population and the remaining 16% was among the species according to AMOVA (Farouji et al. 2015). The close genetic similarity among the species revealed the high affinity, gene flow and genetic relationships between species belonging to einkorn. The genetic distance value was high between *T. monococcum* and *T. urartu* and low between *T. boeoticum* subsp. *boeoticum* and *T. boeoticum* subsp. *thaoudar*. A centre of high diversity in the west and the northwest of Iran clearly exposed patterns of two distinct geographic regions (Farouji et al. 2015). Microsatellite markers (SSRs) are a very powerful new tool to support the determination of critical races in diploid wild wheat species (Hammer et al. 2000).

4.5.2 *Emmer Wheat (***Triticum turgidum** *ssp.* **dicoccon** *Schrank Tell.) Landraces*

Turkey is the country where emmer wheat originated and travelled through the Balkans, Italy, Spain and the North African countries. The Gene Bank material of emmer wheat (*Triticum turgidum* L. ssp. *dicoccon* (Schrank) Thell.) germplasm conserved in Aegean Agricultural Research Institute in Izmir, Turkey, was analysed to determine genetic diversity in terms of DNA markers (SSRs), which produced 100% polymorphic 497 alleles and displayed great genetic variation ($H_e = 0.9$). The genetic differentiation was between 15% and 85% within populations. Landraces displayed higher genetic diversity estimates for the *A* genome than the *B* genome, while the SSR loci at telomeric and sub-telomeric regions displayed lower genetic diversity than other regions on the chromosomes. x-gwm-312, a microsatellite marker reported having linkage with the salinity tolerance in wheat, displayed the highest polymorphism ($H_e = 0.97$) among the SSR markers used for analysis. Thus, Turkish emmer wheat germplasm conserved in the Gene Bank might have potential salinity tolerance that could be exploitable in formal wheat breeding programmes (Özbek and Demir 2019).

The molecular analysis of emmer wheat (*Triticum dicoccon* Schrank Thell.) accessions from India including 28 from a local collection and 20 Indian accessions obtained from CIMMYT, Mexico, using DNA markers (SSR) indicated that emmer wheat accessions had a high level of similarity, and Indian emmer wheats were not very diverse. The breeders could exploit the diversity from other ecogeographic groups or even from other wheat species to increase the diversity within the Indian emmer wheat ecogeographic group (Salunkhe et al. 2013).

Ethiopian tetraploid wheat landraces consisting of three species *Triticum durum* Desf., *T. dicoccon* Schrank and *T. turgidum* L. displayed a high level of polymorphism and a large number of alleles unique for each species based on microsatellite marker analysis. A higher genetic diversity was observed in *T. durum* compared to emmer (*T. dicoccon*) and popular (*T. turgidum*) wheats. This might be related with Ethiopia as one of the places, where durum wheat landrace cultivation is carried out in wide areas. The A genome was more polymorphic than the B genome in all the three species. Genetic distances were lower between *T. durum* and *T. turgidum* than between *T. durum* and *T. dicoccon* or between *T. turgidum* and *T. dicoccon* (Teklu et al. 2006).

4.5.3 *Durum Wheat (*Triticum turgidum *var.* durum *Desf.) Landraces*

The molecular analysis of Turkish durum wheat landraces by means of RAPD markers indicated that the landraces had high level of genetic diversity estimates for observed heterozygosity and gene diversity. Some morphological traits (plant height, spike length, grain number per spike, biological yield, resistance to lodging, etc.), pathological traits (stripe and leaf rusts) and technological traits (1000-kernel weight, hectolitre weight, protein ratio, SDS sedimentation, etc.) were also investigated along with RAPD markers and showed great variation. Altogether, these results display that Turkish durum wheat landraces have great genetic diversity not only for expected genetic diversity but also for observed genetic diversity along with variation in other characteristics, that is, exploitable in breeding programmes of development of commercial cultivars, which have lower genetic diversity than landraces in the present study, with higher yield, resistance to rusts and desirable quality traits (Akar and Ozgen 2007).

Greek landraces and cultivars of durum wheat (*Triticum turgidum* L. var. *durum* (Desf.)), commercial bread wheat (*Triticum aestivum* L.) cultivars and a genotype of *Triticum monococcum* L. were assessed in terms of DNA markers (RAPD) and produced great polymorphism, 125 polymorphic fragment (83.3%). The overall results indicated that durum wheat landraces were sharing some fragments with bread, while *T. monococcum* was standing apart from all other genotypes (Mantzavinou et al. 2005).

4.5.4 *Bread Wheat (***Triticum aestivum** *L.) Landraces*

Characterisation of genetic diversity based on molecular data in different plant species is the recent advancement to gain knowledge about genetic structure of natural or cultivated plant species nowadays. This is an important issue for explaining the phylogenetic relationships among the plant species particularly closely related species, to understand their evolutionary dynamics and to predict their future tendencies, to investigate origin of species, to explore the amount of genetic diversity in wild and primitive relatives of modern crop plant varieties and to exploit their germplasm, which have substantially high genetic diversity and harbouring different gene combinations for the biotic and abiotic stress factors, for improvement of high yielding commercial cultivars adaptable to specific environmental conditions. It was proved that some molecular markers are associated with some desired genes, which are concerned in quality, yield or resistance to the biotic and abiotic stress factors. Recent advances in molecular marker technology enabled the scientists to access the data stored in the gold mine of DNA in the organisms. The molecular markers have been used in different modern wheat species and varieties as well as in wild and primitive wheat species for several decades.

The hexaploid wheat (*Triticum aestivum* L.) landrace collection covering most of the cultivation areas in northern Oman had high genetic diversity. Omani wheat landraces from different districts showed that they have different allele combinations, which was revealed by the correlation values between genetic diversity and allele number. Omani wheat landraces had well adaptation in different environmental conditions and maintains the high level genetic diversity (Al Khanjari et al. 2007)..

The characterisation of genetic diversity in a collection of wheat landraces, whichwere collected from different regions of Turkey based on the data obtained from the analysis of microsatellite markers and some morphological traits, indicated that Turkish wheat landraces have been grouped into two distinct groups according to genotypes and phenotypes based on microsatellite markers and morphological traits, respectively. In both cases, Turkish wheat landraces displayed great variation (Sönmezoğlu et al. 2012).

Afghanistan has important agroecological zones as a secondary origin of wheat for investigation of the genetic diversity and novel alleles/allele combinations. A wheat landrace collection (400) of Dr. Hitoshi Kihara et al. was screened by using diversity array technology and single-nucleotide polymorphism markers, as well as diagnostic molecular markers at important loci controlling vernalisation (*Vrn*), photoperiod response (*Ppd*), grain colour (*R*), leaf rust (*Lr*), yellow rust (*Yr*), stem rust (*Sr*) and Fusarium head blight (*Fhb*). The results indicated that Afghanistan wheat landrace collection had 53% winter types, 43% either spring types or facultative types and 4% either unknown or had *Vrn-A1c*, which is a rare spring allele;

nevertheless, confirmation is needed with additional genotyping and phenotyping, and 97% of the lines represented a photosensitive allele for photoperiod response. For the characterisation of grain colour, 39% of landraces displayed white grain, and 17 unique landraces were determined as resistant to rust and *Fhb* (Manickavelu et al. 2014).

Omani durum wheat (*Triticum durum* Desf.) and bread wheat (*Triticum aestivum* L.) landraces displayed that the genetic diversity was conserved within populations rather than between. The durum wheat landraces displayed the higher genetic diversity than bread wheat landraces that might be due to their different domestication history. Omani bread wheat landraces were resembling Turkish and Mexican bread wheat landraces as reported in the previous studies. There was no close relation between Omani bread wheat landraces and today's landraces from Africa, Asia, Western Europe, Turkey and Central or South America. According to the cluster analysis, Omani bread wheat landraces clustered together with two landraces from Pakistan that might be a possible, previously unknown relationship (Zhang et al. 2006).

Using a SNP-based diversity map, Cavanagh et al. (2013) characterised the impact of crop improvement on the genomic and geographic patterns of the genetic diversity. Their results suggested that loci targeted by selection for wheat improvement have changed over time, potentially reflecting the breeding efforts aimed at developing higher yielding varieties that are adapted to the new or changing local conditions.

Next-generation sequencing (NGS) technologies will still open new undiscovered possibility for analysis of the genetic and management of genetic diversity and more precise, rapid and successful utilisation for wheat improvement.

References

Akar T, Ozgen M (2007) Genetic diversity in Turkish durum wheat landraces. In: Buck HT, Nisi JE, Salomón N (eds) Wheat production in stressed environments. Developments in plant breeding, vol 12. Springer, Dordrecht

Al Khanjari S, Filatenko AA, Hammer K, Buerkert A (2008) Morphological spike diversity of Omani wheat. Genet Resour Crop Evol 55(8):1185–1195

Al Khanjari S, Hammer K, Buerkert A, Röder MS (2007) Molecular diversity of Omani wheat revealed by microsatellites: II. Hexaploid landraces. Genet Resour Crop Evol 54(7):1407–1417

Alsaleh A, Baloch FS, Nachit M, Ozkan H (2016) Phenotypic and genotypic intra-diversity among Anatolian durum wheat "Kunduru" landraces. Biochem Syst Ecol 65:9–16

Alvarez JB, Caballero L, Ureña P, Vacas M, Martín LM (2007) Characterisation and variation of morphological traits and storage proteins in Spanish emmer wheat germplasm (*Triticum dicoccon*). Genet Resour Crop Evol 54(2):241–248

Alvarez JB, Moral A, Martín LM (2006) Polymorphism and genetic diversity for the seed storage proteins in Spanish cultivated einkorn wheat (*Triticum monococcum* L. ssp. *monococcum*). Genet Resour Crop Evol 53:1061–1067

Al-Ajlouni MM, Jaradat AA (1997) Diversity in durum wheat landraces collected from Jordan I. Quantitative Traits. Cereal Res Commun 25(2):169–175

Asins MJ, Carbonell EA (1989) Distribution of genetic variability in a durum wheat world collection. Theor Appl Genet 77:287–294

Bechere E, Belay G, Mitiku D, Merker A (1996) Phenotypic diversity of tetraploid wheat landraces from northern and north-central regions of Ethiopia. Hereditas 124:165–172

Belay G, Bechere E, Mitiku D, Merker A, Tsegaye S (1997) Patterns of morphological diversity in tetraploid wheat (*Triticum turgidum* L.) landraces from Ethiopia. J Acta Agric Scandinavica, Section B– Soil Plant Science 47(4):221–228

Belay G, Furuta Y (2001) Zymogram patterns of α-amylase isozymes in Ethiopian tetraploid wheat landraces: insight into their evolutionary history and evidence for gene flow. Genet Resour Crop Evol 48:507–512

Bradshaw AD (1965) Evolutionary significance of phenotypic plasticity in plants. Adv Genet 13:115–155

Buerkert A, Oryakhail M, Filatenko AA, Hammer K (2006) Cultivation and taxonomic classification of wheat landraces in the Upper Panjsher Valley of Afghanistan after 23 years of war. Genet Resour Crop Evol 53(1):91–97

Butnaru G, Sarac I, Blidar A, Holly L, Mar I (2003) Morpho-agronomic variability of *Triticum monococcum* L. landraces in the Timisoara area. ISIRR, Section IV, Hunedoara, Romania, p 167

Cavanagh C, Chao S, Wang S et al (2013) Genome-wide comparative diversity uncovers multiple targets of selection for improvement in a worldwide sample of hexaploid wheat landrace and cultivars. Proc Natl Acad Sci U S A 110(20):8057–8062. www.pnas.org/cgi/doi/10.1073/pnas.1217133110

Ciaffi M, Benedettelli S, Giorgi B, Porceddu E, Lafiandra D (1991) Seed storage proteins of *Triticum turgidum* ssp. *dicoccoides* and their effect on the technological quality in durum wheat. Plant Breed 107:309–319

Ciaffi M, Dominici L, Lafiandra D, Porceddu E (1992) Seed storage proteins of wild wheat progenitors and their relationships with technological properties. Hereditas 116:315–322

Ciaffi M, Lafiandra D, Porceddu E, Benedettelli S (1993) Storage-protein in wild emmer (*Triticum turgidum* ssp. *dicoccoides*) from Jordan and Turkey. II. Patterns of allele distribution. Theor Appl Genet 86:518–525

Ciaffi M, Dominici L, Lafiandra D (1997) Gliadin polymorphism in wild and cultivated einkorn wheats. Theor Appl Genet 94:68–74

Corazza L, Pasquini M, Perrino P (1986) Resistance to rusts and powdery mildew in some strains of *Triticum monococcum* L. and *Triticum dicoccum* Schubler cultivated in Italy. Genetica Agraria 40:243–254

Damania AB, Hakim S, Moualla MY (1992) Evaluation of variation in *Triticum dicoccum* for wheat improvement in stress environment. Hereditas 116:163–166

Damania AB, Srivastrava JP (1990) Genetic resources for optimal input technology ICARDA's perspectives. In: El-Bassam N, Dambrot M, Loughman BC (eds) Genetic aspects of plant mineral nutrition. Kluwer, Dordrecht, The Netherlands, pp 425–430

Desheva G (2014) Morphological and agronomical characterization of common wheat landraces (*Triticum aestivum* L.) from the National Wheat Collection of Bulgaria. Emir J Food Agric 26(2):164–169

D'Antuono LF, Minelli M (1998) Yield and yield components analysis of emmer wheat (*Triticum dicoccum* Schubler) landraces from Italy. In: Jaradat AA (ed) Proceedings of the 3rd international Triticeae symposium, May 1997, Aleppo Syria. Science Publisher Inc, USA, pp 393–404

Elings A, Nachit MM (1991) Durum wheat landraces from Syria. I Agro-ecological and morphological characterization. Euphytica 53(3):211–224

Eticha F, Belay G, Bekele E (2006) Species diversity in wheat landrace populations from two regions of Ethiopia. Genet Resour Crop Evol 53(2):387–393

Farouji E, Khodayari A, Saeidi H et al (2015) Genetic diversity of diploid Triticum species in Iran assessed using inter-retroelement amplified polymorphisms (IRAP) markers. Biologia 70:52–60

Galterio G, Codianni P, Novembre G, Saponaro C, Di Fonzo N, Pogna NE (1998) Storage protein composition and technological characteristics of F6 lines from the cross *Triticum turgidum* spp. *durum* × *Triticum turgidum* spp. *dicoccum*. In: Proceedings of the 9th international wheat genetics symposium, vol 4, Saskatoon, Canada, pp 148–150

Ghimire SK, Akashi Y, Maitani C, Nakanishi M, Kato K (2005) Genetic diversity and geographical differentiation in Asian common wheat (*Triticum aestivum* L.), revealed by the analysis of peroxidase and esterase isozymes. Breed Sci 55(2):175–185

Hammer K, Filatenko AA, Korzun V (2000) Microsatellite markers – a new tool for distinguishing diploid wheat species. Genet Resour Crop Evol 47:497–505

Heun M, Schäfer-Pregl R, Klawan D, Castagna R, Accerbi M, Borghi B, Salamini F (1997) Site of einkorn wheat domestication identified by DNA finger printing. Science 278:1312–1314

Jakubziner MM (1969) Immunity of different wheat species. Agric Biol 4:837–847. (in Russian)

Jaradat AA (2013) Wheat landraces: a mini review. Emir J Food Agric 25(1):20–29

Jaradat AA (1991) Phenotypic divergence for morphological and yield-related traits among landrace genotypes of durum wheat from Jordan. Euphytica 52(3):155–164

Jaradat AA (2014) The vanishing wheat landraces of the Fertile Crescent. Emir J Food Agric 26(2):203–217

Kamali MRJ, Bozorghipour R, Moghaddam ME, Kamali MRJ, Kazemi S, Amini A, Bozorgipour R, Najafian G, Baghaei N (2011) Assessment of high molecular weight glutenin sub-units and baking quality related traits in some of the Iranian bread wheat (*Triticum aestivum* L.) landraces. Crop Breed J 1(1):29–40

Keskin SŞ, Özbek Ö, Eser V, Taşkin Göçmen B (2015) Polymorphism in seed endosperm proteins (gliadins and Glutenins) of Turkish cultivated einkorn wheat [*Triticum monococcum* ssp. *monococcum*] landraces. Cereal Res Commun 43(1):108–122

Kilian B, Özkan H, Walther A, Kohl J, Dagan T, Salamini F, Martin W (2007) Molecular diversity at 18 loci in 321 wild and 92 domesticate lines reveal no reduction of nucleotide diversity during *Triticum monococcum* (einkorn) domestication: implications for the origin of agriculture. Mol Biol Evol 24:2657–2668

Kimber G, Feldman M (1987) Wild wheats: an introduction. Special Report 353. College of Agriculture, Columbia, Missouri, pp. 1–142

Mantzavinou A, Bebeli PJ, Kaltsikes PJ (2005) Estimating genetic diversity in Greek durum wheat landraces with RAPD markers. Aust J Agric Res 56:1355–1364

Manickavelu A, Niwa S, Ayumi K, Komatsu K, Naruoka Y, Ban T (2014) Molecular evaluation of Afghan wheat landraces. Plant Genet Res 12(1):31–35

Mac Key J (1966) Species relationship in *Triticum*, Proc. of 2nd Int. Wheat Genetics Symp. Hereditas 2:237–276

Mizumoto K, Hirosawa S, Nakamura C, Takumi S (2002) Nuclear and chloroplast genome genetic diversity in the wild einkorn wheat, *Triticum urartu*, revealed by AFLP and SSLP analyses. Hereditas 137:208–214

Moghaddam M, Ehdaie B, Waines JG (1997) Genetic variation and interrelationships of agronomic characters in landraces of bread wheat from southeastern Iran. Euphytica 95(3):361–369

Motalebi M, Keshavarzi M, Naghavi MR (2007) Glutenin subunit composition in durum (*Triticum durum*) landraces and cultivars. Asian J Plant Sci 6:399–402

Naghavi MR, Monfared SR, Ahkami AH, Ombidbakhsh MA (2009) Genetic variation of durum wheat landraces and cultivars using morphological and protein markers. World Acad Sci Eng Technol 25:73–75

Nicotra AB, Atkin OK, Bonser SP et al (2010) Plant phenotypic plasticity in a changing climate. Trends Plant Sci 15:684–692

Oliveira HR, Jones H, Leigh F et al (2011) Phylogeography of einkorn landraces in the Mediterranean basin and Central Europe: population structure and cultivation history. Archaeol Anthropol Sci 3:327

Oliveira HR, Campana MG, Jones H, Hunt HV, Leigh F, Redhouse DI, Lister DL, Jones MK (2012) Tetraploid wheat landraces in the Mediterranean Basin: taxonomy, evolution and genetic diversity. PLoS One 7(5):e37063

Oliver RE, Cai X, Friesen TL, Halley S, Stack RW, Xu SS (2008) Evaluation of Fusarium head blight resistance in tetraploid wheat (*Triticum turgidum* L.). Crop Sci 48:213–222

Özbek Ö, Taskin BG, Şan SK, Eser V, Arslan O (2011) Gliadin polymorphism in Turkish cultivated emmer wheat (*Triticum turgidum* L. ssp. *dicoccon* (Schrank) Thell.) landraces. Plant Syst Evol 296(1–2):121–135

Özbek Ö, Taşkın BG, Şan SK, Eser V, Arslan O (2012) High-molecular-weight glutenin subunit variation in Turkish emmer wheat [*Triticum turgidum* L. ssp. *dicoccon* (Schrank) Thell.] landraces. Plant Syst Evol 298:1795–1804

Özbek Ö, Taskin BG, Şan SK, Eser V, Arslan O (2013) Genetic characterization of Turkish cultivated emmer wheat [*Triticum turgidum* L. ssp. *dicoccon* (Schrank) Thell.] landraces based on isoenzyme analysis. Cereal Res Commun 41(2):304–315

Özbek Ö, Demir S (2019) Characterization of genetic diversity in cultivated emmer wheat [*Triticum turgidum* L. ssp. *dicoccon* (Schrank) Thell.] landrace populations from Turkey by SSR. Russ J Genet 55:969–977

Özkan H, Brandolini A, Schafer-Pregl R, Salamini F (2002) AFLP analysis of a collection of tetraploid wheats indicates the origin of emmer and hard wheat domestication in Southeast Turkey. Mol Biol Evol 19:1797–1801

Özkan H, Brandolini A, Pozzi C et al (2005) A reconsideration of the domestication geography of tetraploid wheats. Theor Appl Genet 110:1052–1060

Pagnotta MA, Mondini L, Atallah MF (2005) Morphological and molecular characterization of Italian emmer wheat accessions. Euphytica 146(1–2):29–37

Payne PI, Lawrence GJ (1983) Catalogue of alleles for the complex gene loci, *Glu-A1*, *Glu-B1*, and *Glu-D1*, which code for high molecular weight subunits of glutenin in hexaploid wheat. Cereal Res Commun 11:29–35

Payne PI, Corfield KG (1979) Subunit composition of wheat glutenin proteins, isolated by gel filtration in a dissociating medium. Planta 145:83–88

Piergiovanni AR, Laghetti G, Perrino P (1996) Characteristics of meal from hulled wheats (*Triticum dicoccum* Schübler and *Triticum spelta* L.): an evaluation of selected accession. Cereal Chem 73(6):732–735

Salinas J, Vega MP, Benito C (1982) Identification of hexaploid wheat cultivars based on isozyme patterns. J Sci Food Agric 33(3):221–226

Salunkhe A, Tamhankar S, Tetali S, Zaharieva M, Bonnett D, Trethowan R, Misra S (2013) Molecular genetic diversity analysis in emmer wheat (*Triticum dicoccon* Schrank) from India. Genet Resour Crop Evol 60:165–174

Sönmezoğlu ÖA, Bozmaz B, Yildirim A, Kandemir N, Aydin N (2012) Genetic characterization of Turkish bread wheat landraces based on microsatellite markers and morphological characters. Turk J Biol 36:589–597

Stallknecht GF, Gilbertson KM, Ranney JE (1996) Alternative wheat cereals as food grains: einkorn, emmer, spelt, kamut, and triticale. In: Janick J (ed) Progress in new crops. ASHS Press, Alexandria, VA, pp 156–170

Teklu Y, Hammer K, Huang X et al (2006) Analysis of microsatellite diversity in Ethiopian tetraploid wheat landraces. Genet Resour Crop Evol 53:1115–1126

Terasawa Y, Kawahara T, Sasakuma T, Sasanuma T (2009) Evaluation of the genetic diversity of an Afghan wheat collection based on morphological variation, HMW glutenin subunit polymorphisms, and AFLP. Breed Sci 59(4):361–371

Tesfaye T, Getachew B, Worede M (1991) Morphological diversity in tetraploid wheat landrace populations from the central highlands of Ethiopia. Hereditas 114:171–176

Teshome A, Baum BR, Fahrig L, Torrance JK, Arnason TJ, Lambert JD (1997) Sorghum (*Sorghum bicolor* (L.) Moench) landrace variation and classification in North Shewa and South Welo, Ethiopia. Euphytica 97:255–263

Tsegaye S, Tesemma T, Belay G (1996) Relationships among tetraploid wheat (*Triticum turgidum* L.) landrace populations revealed by isozyme markers and agronomic traits. Theor Appl Genet 93:600–605

Turcotte MM, Levine JM (2016) Phenotypic plasticity and species coexistence. Trends Ecol Evol 31(10):803–813

Uzundzalieva K, Desheva G, Valchinova E, Kyosev B (2016) Comparative evaluation of einkorn accessions (*Triticum monococcum* L.) of some main agricultural characters. Agron J 17(1):69–80

Zeven AC (1998) Landraces: a review of definitions and classifications. Euphytica 104:127–139

Zeven AC (1999) The traditional inexplicable replacement of seed and seed ware of landraces and cultivars: a review. Euphytica 110:181–191

Zhang P, Dreisigacker S, Buerkert A, Alkhanjari S, Melchinger AE, Warburton ML (2006) Genetic diversity and relationships of wheat landraces from Oman investigated with SSR markers. Genet Resour Crop Evol 53(7):1351–1360

Zuli Y et al (1999) The study on esterase and peroxidase isozymes in landraces of wheat from to Sichuan. Journal of Sichuan Agricultural University. http://en.cnki.com.cn/Journal_en/D--D000-SCND-1999-04.htm

Getachew B, Tesemma T., Becker HC, Merker A (1993) Variation and interrelationships of agronomic traits in Ethiopian tetraploid wheat landraces. Euphytica 71(3): 181–188

Chapter 5
Macro-Microelements in Wheat Landraces and Their Use in Breeding

Ayten Salantur and Cuma Karaoğlu

5.1 Macro-Microelements in Wheat Landraces

Cereals provide 60% of the daily calorie requirement, especially for people living in developing countries (Awika 2011). Wheat is an important plant covering the largest cultivation area among cereals in the world. Nutritional quality of wheat (*Triticum* sp.), as one of the first cultivated plants in the world, has a significant impact on human health worldwide. Wheat is an important source of macro-minerals such as K, P, and Mg and micro-minerals such as copper, iron, magnesium, manganese, phosphorus, selenium, zinc etc. However, the mineral content of widely cultivated modern wheat varieties is reported to be significantly reduced (Jaradat 2011).

The concentration of minerals in wheat flour is genetically determined by cultivar and environment-soil, climate, and management practices. Wheat ancient species such as einkorn (*Triticum monococcum* ssp. *monococcum*), emmer (*Triticum dicoccon*) and landraces have been found to have higher nutritional values (Megyeri et al. 2014; Rachon et al. 2015). In previous studies, K, Mg, and P values have been reported to differ according to wheat varieties (Jakobsone et al. 2015; Kan 2015; Lyons et al. 2005).

Zinc, iron, copper, and magnesium concentrations remained stable in wheat cultivars from 1845 to the mid-1960s. Then, they have significantly decreased, which coincided with the introduction of semi-dwarf, high-yielding cultivars (Fan et al. 2008). This causes individuals fed on wheat-based diet to experience health problems called "hidden hunger." Especially in women and children, hidden hunger can cause blindness, premature death, and mental development problems (Ahmed et al. 2012). For example, more than 60% of the world's population has Fe deficiency, more than 30% has Zn deficiency, and approximately 15% has Se deficiency (Mayer

A. Salantur (✉) · C. Karaoğlu
Central Research Institute for Field Crops (CRIFC), Ankara, Turkey
e-mail: ayten.salantur@tarimorman.gov.tr

N. Zencirci et al. (eds.), *Wheat Landraces*,
https://doi.org/10.1007/978-3-030-77388-5_5

et al. 2008; Rayman 2008). Mg, Fe, and Zn are mainly found in the aleurone layer of common wheat grains, and these macro- and micro-minerals of whole wheat products are an important source of human daily needs (Piergiovanni et al. 1997). Increasing the nutritional properties of wheat is of great importance for human population worldwide. Increasing the amount of microelements such as zinc, iron, and protein content in planting areas without reducing wheat yield or keeping the yield at a reasonable level will decrease the diseases caused by these deficiencies.

For many years, large-scale commodity production of crops has focused on increasing yield, but increasing interest from health-conscious consumers has stirred interest in grains in order to improve nutrition and health. Depending on growing conditions and wheat variety, important macro-minerals like calcium, magnesium, and potassium and micro-minerals like iron, copper, and selenium are found in wheat kernels. These are distributed throughout the aleurone layer (55%), endosperm (20%), pericarp (10%), scutellum and testa (10%), and embryo (5%).

Summarized below are some of these essential minerals and their important roles in maintaining good health.

- Magnesium: Contributes to efficient metabolism as well as proper muscle and nerve functioning and is shown also to reduce diabetes and metabolic ailments
- Calcium: An essential component for the development of musculoskeletal, cardiovascular, and nervous systems and furthermore promotes overall physiological performance
- Phosphorus: Necessary for proper functioning of kidneys and heart muscle, contributes to bone and dental strength, and regulates protein reactions
- Potassium: Contributes to proper heart muscle contraction, neural impulse transmission, and fluid system balance
- Copper: Facilitates the functions of C-oxidase enzymes and promotes connective tissue development and iron metabolism
- Selenium: Inhibits some types of cancer cell formation and promotes essential antioxidant reactions
- Iron: Needed for hemoglobin synthesis and energy production
- Zinc: Regulates the function of many enzymes, glucose, and insulin and synthesizes proteins

Lyons et al. (2005) investigated whether the genotypic selenium variation in breeding is sufficient. The result indicated that there was no significant genotypic variation among commercial bread and durum wheat cultivars for selenium. Selenium is, on the other hand, an important micronutrient for animals and humans for its antioxidant, anticancer, and antiviral effect. It was found in the study that there was a little difference among commercial varieties for selenium content; moreover, they also discovered that diploid *Aegilops tauschii* wheat had more than 42% selenium than commercial varieties and 35% higher than rye. Ash-rich einkorn wheat was found to be richer in important minerals such as calcium, phosphorus, potassium, magnesium, manganese, and zinc except iron.

Some other studies have shown that genotype and environment interactions are important for Zn and Fe concentrations in wheat (OrtizMonasterio et al. 2007;

Morgounov et al. 2017). As identified in Jaradat (2011) study, Fe content in wheats collected from local wheat plants in Turkey is higher than in other areas, showing strong correlation between Fe content and the cereal. In addition, a high degree of inheritance was observed in both Fe and Zn concentrations (Velu et al. 2017), pointing out a strong genetic share in accumulation of these minerals in grain. It was as well specified that environmental factors were effective as much as genotype in these differences. Besides, multiple regression analysis showed that both increasing yield and harvest index were highly significant factors that explained the downward trend in grain mineral concentration (Fan et al. 2008).

5.2 Wheat Landraces in Breeding Studies

Cereals are grain seeds from plants of Gramineae family such as wheat, corn, rice, barley, oats, and rye, which have been the basis of human nutrition for thousands of years. Mineral deficiencies are common in many people fed by cereals on earth; therefore, improving mineral content in cereals represents an important strategy for increasing human mineral intake and health (Ficco et al. 2009). Widely grown modern wheat varieties have high yield capacity; therefore, they are cheap and important nutrient sources to meet the daily needs of less fortunate people, but these wheat varieties are poor especially in micronutrient sources such as Zn and Fe (Welch and Graham 2004).

In order to overcome this deficiency, local wheats which are mineral- and phytochemical-rich herbal sources (Arzani and Ashraf 2017) may have been utilized as genitors in wheat breeding programs (Hocaoglu and Akcura 2017). As the primary gene source of breeding programs, wheat landraces have been collected intensively since the 1900s, and many of them have been identified and started to be used in breeding programs (Morgounov et al. 2017).

With the increase in yield, mineral content of modern wheat varieties decreased proportionally including copper, iron, magnesium, manganese, phosphorus, selenium, and zinc. High levels of these nutrients can be found in soils and in old low-yield wheat landraces. In many breeding studies, new varieties were firstly selected from wheat landraces by selection and used directly in production or used as genitors in breeding programs. For example, in winter wheat breeding program in Turkey, many varieties of rootstock obtained by hybridization breeding have created wheat landrace varieties (Salantur et al. 2017).

Wheat landraces have also been used as genetic material in breeding programs in different countries. Norin-10 was developed in 1924 by giving Turkey red local variety to the short Daruma x Fultz hybrid made by Japanese breeders in 1917 (Reitz and Salmon 1968). Horarek, which was selected by Zhukovsky from local varieties in 1951, was superior to many varieties in Russia with its earliness, yield, and fusarium resistance (Qualset et al. 1996). Fifty-one lines selected from the Turkish local variety PI 178383 resistant to yellow rust were used as genitors in the development of new varieties in the USA and have made significant contributions to the US economy. Turkey red varieties were taken from Anatolia in 1874 in Kansas City. It has been the most common variety in Kansas for 70 years and formed the basis of the wheat industry. This wheat variety has a unique, rich, and complex flavor and excellent baking qualities (Anonymous 2019).

Utilizing wheat landraces to expand the existing gene pool in bread and durum wheat has been a new approach in recent years. Durum wheat, *Triticum durum* (AABB), is thought to be developed from wild emmer wheat, *Triticum dicoccoides (Körn. ex Asch. & Graebner) Schweinf.).* Locally grown emmer wheat (*Triticum dicoccon*) is one of the first examples. It is known that some of the important genetic features found in wild relatives in the process of obtaining cultured plants are lost and cannot be transferred to culture plants anymore (Tanksley and McCouch 1997).

Wheat landraces hulled or hulless are superior to cultivated wheats in many features. For instance, Kamut wheat, which is an old wheat type, contains a significant amount of selenium than cultured wheat (Piergiovanni et al. 2009). Hulled wheats are transitional forms between today's wheat and wild wheat relatives. Einkorn wheat (*Triticum monococcum* ssp. *monococcum*) is the first type of wheat cultivated on the foothills of Karacadağ, Diyarbakır (Heun et al. 1997). It is known to make important contributions to human nutrition and health (Hidalgo and Brandolini 2008; Pirgozliev et al. 2015).

In order to increase the mineral content of durum wheat, *Triticum dicoccoides* and *Triticum dicoccon* were used as female genitors in wheat breeding. Macro- and micro-mineral contents were determined in fixed F7 lines. In this study, it was observed that iron and zinc contents increased significantly than already produced durum wheat varieties. While iron content of durum wheat varieties ranged between 11.80 and 15.61 ppm, iron content ranged between 8.54 and 86.76 ppm. Kamut (Khorasan wheat; Figs. 5.1 and 5.2) or einkorn (Fig. 5.3) wheat also was used in this project as a female for breeding program (Anonymous 2016).

Einkorn wheat proved to have the highest levels of protein, fat, ash, phosphorus, potassium, magnesium, calcium, copper, zinc, iron, and manganese (Rachon et al.

Fig. 5.1 Kamut X bread wheat cross

Fig. 5.2 Kamut X bread wheat cross

Fig. 5.3 Einkorn X durum wheat cross

2015). It was also found that macro-micronutrient content of spelt (*Triticum spelta*; 2n = 42) was significantly higher than that of cultured wheat (Berecz et al. 2001; Ruibal-Mendieta et al. 2005).

Several studies have shown that some genotypes from landraces are directly related to breeding targets and can be used as a gene source in breeding, i.e., to increase resistance to biotic and abiotic stresses (Jaradat 2011), cold resistance (Küçüközdemir and Tosun 2016), drought resistance, coleoptile length (Öztürk et al. 2014), quality (Akar et al. 2009; Akçura 2011), and mineral content (Fan et al. 2008; Ficco et al. 2009; Hussain et al. 2010).

Wheat landraces of common wheat are important sources for increasing micronutrients in plant breeding programs. In a study, the macro- and micronutrient contents (Fe, Zn, B, K, Mn, Cu, Mg, Ca, Mo) of 37 pure lines and 9 bread wheat varieties obtained from wheat landraces collected from Western Anatolia (Eskişehir and Kütahya) and Thrace (Edirne) were evaluated. Higher correlation existed between iron and zinc contents with boron and molybdenum contents of genotypes. A pure line (L4) was the most prominent genotype for iron and zinc content, while it was superior for both boron and molybdenum contents. Copper content of cereals was negatively correlated with iron and zinc content. While wheat varieties have relatively higher Mo content, they can also be improved for Fe, Zn, B, K, and Ca contents. Fe, Zn, and Mn contents of many pure advanced lines improved based on landraces were usually higher than those of modern cultivars. Moreover, mean grain concentrations of Fe, Zn, and Mn in pure advanced lines improved based on landraces lines from wheat landraces were significantly higher than all cultivars, 9.25, 14.82, and 6.75%, respectively. Therefore, some pure lines could be recommended to be used as genetic material to enhance the genetic basis of bread wheat breeding programs worldwide (Akcura and Kokten 2017).

In another study, it was found that nine wheat landraces have the potential to be incorporated into the wheat gene pool. Of these, two wheat landraces, 782 and 528 ppm, have higher P and Fe composition, respectively, with good grain weight and ideal candidates for crop improvement (Kondou et al. 2016). While comparing the concentrations of 5 macro- and 15 microelements in the whole grain of spring lines of emmer, einkorn, spelt, and two common wheat cultivars, all grown under identical environmental conditions, *Triticum* species differed significantly for P, Mg, Zn, Fe, Mn, Na, Cu, Sr, Rb, and Mo. The grain of all hulled wheats, compared with common wheat, contained significantly more Zn (from 34% to 54%), Fe (from 31% to 33%), and Cu (from 3% to 28%). Significant positive correlations existed between the levels of Fe, Zn, and Mn, in particular in *T. monococcum* ssp. *monococcum* and *T. dicoccon*. A strong correlation between Zn, Fe, and Mn could have important implications for wheat quality breeding (Suchowilska et al. 2012).

Dietary Zn deficiency is widespread, especially in developing countries, and breeding (genetic biofortification) through the HarvestPlus program has recently started to deliver new wheat varieties to help alleviate this problem in South Asia (Khokhar et al. 2018). A study by Lyons et al. (2005) determined no significant genotypic variation in grain Se density among modern commercial bread or durum wheat or triticale or barley varieties. However, the diploid wheat, *Aegilops tauschii*, and rye have 42% and 35% higher, respectively, grain Se concentration than other cereals in separate field trials, and, in a hydroponic trial, rye was 40% higher in foliar Se content than two wheat landraces.

Wheat landraces are indispensable genetic resources for low-input agriculture and organic farming due to uncertainties caused by global warming, demand for good nutrition, and increased demand for organic products. Success in wheat breeding depends on the availability of genetic diversity for target traits in the present gene pool. Wheat landraces are important variation sources (Akçura et al. 2011; Hocaoglu and Akcura 2017). Wheat landraces have become indispensable as they are used directly in improvement as well as genetic material of wheat breeding programs and as such have made significant contributions to the genetic structure of today's wheat and will continue to do so.

In the future studies, wheat breeding will be shaped by attempts to increase the microelement contents such as iron and zinc which are important for human health besides being important quality features of the grain. Local varieties are, above all, part of the world's cultural heritage, an important guarantee of food safety and as such must be cultivated, protected, and inherited for future generations as genetic treasures.

References

Ahmed T, Mustafa M, Santhia I et al (2012) Nutrition of children and women in Bangladesh: trends and directions for the future. J Health Popul Nutr 30(1):1–11

Akar T, Mert Z, Yazar S et al (2009) Sustainable use of winter durum wheat landraces under Mediterranean conditions. Afr J Biotechnol 8(17):4108–4116

Akçura M (2011) The relationships of some traits in Turkish winter bread wheat landraces. Turk J Agric For 35:15–125

Akcura M, Kokten K (2017) Variations in grain mineral concentrations of Turkish Wheat landraces germplasm. Qual Assur Saf Crops Food 9(2):153–159

Akçura M, Partigoç F, Kaya Y (2011) Evaluating of drought stress tolerance based on selection indices in Turkish bread wheat landraces. J Anim Plant Sci 21(4):700–709

Anonymous (2016) TÜBİTAK 1003 project final report. No: 113O115. https://trdizin.gov.tr/publication/show/pdf/project/TVRjMU56VTE=

Anonymous (2019) https://www.slowfoodusa.org/ark-item/turkey-hard-red-winter-wheat

Arzani A, Ashraf M (2017) Cultivated ancient wheats (Triticum spp.): a potential source of health-beneficial food products. Comp Rev Food Sci Food Saf 16:477–488

Awika J M (2011) Major cereal grains production and use around the world. Advances in cereal science: implications to food processing and health promotion, 1089:1–13. ACS symposium series 1089. American Chemical Society

Berecz K, Simon-Sarkadi L, Ragasits I, Hoffmann S (2001) Comparison of protein quality and mineral element concentrations in grain of spelt (Triticum spelta L.) and common wheat (*Triticum aestivum* L.). Arch. Acker Pft. Boden 47:389–398

Fan MS, Zhao FJ, Fairweather TSJ, Poulton PR, Dunham SJ, McGrath SP (2008) Evidence of decreasing mineral density in wheat grain over the last 160 years. J Trace Elem Med Biol 22:315–324

Ficco DBM, Riefolo C, Nicastro G et al (2009) Phytate and mineral elements concentration. Field Crop Res 111(2009):235–242

Heun M, Schäfer PR, Klawan D et al (1997) Site of einkorn wheat domestication identified by DNA fingerprinting. Science 278(5341):1312–1314

Hidalgo A, Brandolini A (2008) Kinetics of carotenoids degradation during the storage of Einkorn (*Triticum monococcum L. ssp. monococcum*) and bread wheat (*Triticum aestivum* L. ssp. aestivum) flours. J Agric Food Chem 56(2008):11300–11305

Hocaoglu O, Akçura M (2017) Evaluating mineral contents of selected bread wheat landrace pure lines derived from West Anatolia and Marmara regions and cultivars by GGE biplot .the special issue of 2nd International Balkan Agriculture Congress May 16–18

Hussain A, Larsson H, Kuktaite R et al (2010) Mineral composition of organically grown wheat genotypes: contribution to daily minerals intake. Int J Environ Res. Public Health 2010 7(9):3442–3456

Jakobsone I, Kantane I, Zute S, et al (2015) Macro-elements and trace elements in cereal grains cultived references in Latvia. Proceeding of the Latvian Academy of Science, Section B, 69:152–157

Jaradat AA (2011) Wheat landraces: genetic resources for sustenance and sustainability. USDA-ARS, pp 1–20

Kan A (2015) Characterization of the fatty acid and mineral composition of selected cereal cultivars from Turkey. Rec Nat Prod J 9(1):124–134

Khokhar JS, Sareen S, Tyagi BS et al (2018) Variation in grain Zn concentration, and the grain ionome, in field-grown Indian wheat. PLoS One 13(1):e0192026

Kondou Y, Manickavelu A, Komatsu K et al (2016) Analysis of grain elements and identification of best genotypes for Fe and P in Afghan Wheat Landraces. Breed Sci 66:676–682

Küçüközdemir Ü, Tosun M (2016) Bazı Yerel Buğday Genotiplerinde Verim, Verim Unsurları ve Soğuğa Dayanıklılığın Belirlenmesi. Atatürk Üniversitesi Ziraat Fakültesi Dergisi 45(1):43–54

Lyons GH, Ortiz-Monasterio L, Stangoulis JCR et al (2005) Selenium concentration in wheat grain: sufficient genotypic variation for selection? Plant and Soil 269:369–380

Mayer JE, Pfeiffer WH, Beyer P (2008) Biofortified crops to alleviate micronutrient malnutrition. Current opinion in plant biology, genome studies and molecular genetics, edited by Juliette de Meaux and Maarten Koornneef/Plant Biotechnology, edited by Andy Greenland and Jan Leach, 11(2):166–70

Megyeri M, Miko P, Molnar I et al (2014) Functional compounds of einkorn and emmer genotypes antioxidants and trace elements. Strategies for organic and low-input integrated breeding and management diversity strategies for organic and low input agricultures and their food systems Solibam final congress Nantes, France, pp 96–97

Morgounov A, Keser M, Kan M et al (2017) Wheat landraces currently grown in Turkey: distribution, diversity, and use. Crop Sci 56:3112–3124

Ortiz-Monasterio JI, Palacios-Rojas N, Meng E et al (2007) Enhancing the mineral and vitamin content of wheat and maize through plant breeding. J Cereal Sci 46(3):293–307, ISSN 0733-5210. https://doi.org/10.1016/j.jcs.2007.06.005

Öztürk A, Taskesenligil B, Haliloğlu K et al (2014) Characterization for drought resistance at early stages of wheat genotypes based on survival, coleoptile length, and seedling vigor. Turkish J Agric For 38:824–837

Piergiovanni AR, Rizzi R, Pannacciulli E et al (1997) Mineral composition in hulled wheat grains: a comparison between emmer (*Triticum dicoccon* Schrank) and spelt (*T. Spelta* L) accessions. Int J Food Sci Nutr 48:381–386

Piergiovanni AR, Simeone R, Pasqualone A (2009) Composition of whole and refined meals of kamut under southern Italian conditions. Chem Eng Trans 17:891–896

Pirgozliev V, Rose SP, Pellny T et al (2015) Energy utilization and growth performance of chickens fed novel wheat inbred lines selected for different pentosan levels with and without xylanase supplementation. Poult Sci 94:232–239

Qualset CU, Zannata ACA, Keser M et al (1996) Agronomic performance of wheat landraces from Western Turkey. Basis for in-situ conservation practices by farmers. In 5. International Wheat Conference, June 10–14, 1996, Book of Abstracts, Ankara

Rachon L, Szumilo G, Brodowska M et al (2015) Nutritional value and mineral composition of grain of selected wheat species depending on the intensity of a production technology. J Elem 20(3):705–715

Rayman MP (2008) Food-chain selenium and human health: emphasis on intake. Br J Nutr 100(2):254–268

Reitz LP, Salmon SC (1968) Origin and use of Norin 10 wheat. Crop Sci 8:686–689

Ruibal-Mendieta NL, Delacroix DL, Mignolet E et al (2005) Spelt (*Triticum aestivum ssp. spelta*) as a source of breadmaking flours and bran naturally enriched in oleate and minerals but not phytic acid. J Agric Food Chem 53:2751–2759

Salantur A, Tekin M, Bağcı SA et al (2017) Türk buğday yerel çeşitleri ve bitki ıslahı. TÜRKTOB Dergisi 2017 Sayı: 24 Sayfa:18–20

Suchowilska E, Wiwart M, Kandler W et al (2012) A comparison of macro- and microelement concentrations in the whole grain of four Triticum species. Plant Soil Environ 58:141–147

Tanksley SD, Mc Couch SR (1997) Seed banks and molecular maps: unlocking genetic potential from the wild. Science 277(5329):1063–1066

Velu G, Tutus Y, Gomez-Becerra HF et al (2017) QTL mapping for grain zinc and Iron concentrations and zinc efficiency in a tetraploid and hexaploid wheat mapping populations. Plant and Soil 411(1–2):81–99

Welch RM, Graham R (2004) Breeding for micronutrients in staple food crops from a human nutrition perspective. J Exp Bot 55:353–364

Chapter 6
Nutritional and Technological Properties of Wheat Landraces

Asuman Kaplan Evlice

6.1 Introduction

Wheat has played a significant role as a main source of foodstuff since the early civilizations in the Fertile Crescent which includes some parts of Turkey. The domestication of wheat began about 12,000 years ago in Göbekli Tepe, southeastern Turkey, according to evidence from archaeological excavations (Schmidt 2007; Dietrich et al. 2012). During the migration, many traditional cultivars or landraces were chosen by farmers and nature to fit environmental and cultural niches (Hernández-Espinosa et al. 2019).

Until the beginning of the twentieth century, wheat cultivars were predominantly landraces, which were well adapted to their local environments. Since then, landraces have been used as a source of variability in the creation of modern wheat cultivars as breeding methods have developed. After World War II, intensive wheat breeding resulted in the entire replacement of landraces by new semi-dwarf and high-yielding wheat cultivars, resulting in a reduction in wheat genetic diversity (Bordes et al. 2008).

Although landraces were mostly displaced by the superior modern wheat cultivars, they have provided some opportunities for breeders, farmers, manufacturers, and consumers. Wheat landraces might behave as donors with significant features, such as drought and cold tolerance and grain quality. Wheat landraces generally represent considerably wider genetic diversity than modern wheat cultivars; therefore, they could lead to extending the genetic base of modern wheat cultivars (Azeez et al. 2018). Wheat landraces can precisely be adapted to their locality of origin and are frequently related with tolerance to biotic and abiotic stresses and higher grain yield under lower input management practices (Hernández-Espinosa et al. 2019).

A. Kaplan Evlice (✉)
Department of Quality and Technology, Central Research Institute for Field Crops, Ankara, Turkey

N. Zencirci et al. (eds.), *Wheat Landraces*,
https://doi.org/10.1007/978-3-030-77388-5_6

Wheat landraces and old cultivars could play a significant role in food safety not only as being an easily accessible gene source for breeders but also as being more resilient than modern wheat cultivars because they perform well in marginal areas (Migliorini et al. 2016). Besides, the landraces and old cultivars grown in marginal environments or under organic conditions can provide higher income to farmers as compared to modern wheat cultivars, as they are primarily used for the production of natural and healthy whole grain products (Kantor et al. 2001).

Healthy foods have been getting increasing popularity recently with a renewed interest in wheat landraces and old cultivars. Landraces cannot compete with modern cultivars for yield; however, they have generally higher desirable nutritional values (Marconi and Cubadda 2005). Landraces and old cultivars usually have higher protein contents (Dinelli et al. 2013; Giunta et al. 2019), minerals (Hussain et al. 2012a; Velu et al. 2019), and phenolic compounds (Dinelli et al. 2013) than modern wheat cultivars. In comparison with modern wheats, the landraces and old wheats tend to be richer in protein and linoleic acid, to be poorer in starch, and to have softer grains (Bordes et al. 2008). It is also reported that modern wheat cultivars possess the highest albumin and total starch contents, whereas landraces have the highest grain protein and gliadin contents, and old genotypes possess the highest glutenin and amylose contents (Boukid et al. 2018).

The local character of a landrace can derive from its grain quality features which are often appropriate for producing a particular local product according to the preferences of the consumer in a specified region (processing, baking, cooking, and tasting) (Hernández-Espinosa et al. 2019). The landraces are still being used in rustic areas worldwide to cook traditional foods. For instance, bulgur in different cities or regions of Turkey is produced with specific landraces including einkorn and emmer to obtain locally desired end-use product features.

Landraces have got increased popularity recently. Therefore, comprehensive studies comparing the nutritional values of landraces have recently been undertaken by scientists. However, limited studies have been carried out to screen the end-products' quality characteristics of wheat landraces. The aim of this chapter is to review and compare the results of the nutritional and technological characteristics of wheat landraces accessible in the literature.

6.2 Physical Properties

Grain size is one of the main quality traits subjected to selection. Grain size has a clear effect on many compositional and qualitative characteristics since large and heavy kernels have higher amounts of starchy endosperm and lower levels of aleurone layers and external pericarp (Brandolini et al. 2011). Beside kernel size, uniform kernel size is important for effective milling. Out of specification, small grain proportions decrease the market value of the wheat and the advantage of the higher grain yield. Therefore, wheat breeders should screen the distribution of kernel size during the breeding in order to prevent small kernel problems (Hare 2017).

In a comprehensive study comparing kernel weights of 37 Iranian and 42 Mexican durum wheat landraces, none of the landraces came close to the bread wheat check samples, with an average kernel weight of Iranian accessions slightly higher than that of Mexican (Hernández-Espinosa et al. 2019). Dotlačil et al. (2010) studied two different sample sets of combining obsolete cultivars and winter wheat landraces and found that Set I (n = 122) and Set II (n = 101) had lower thousand kernel weight values as mean compared to bread wheat samples. Nazco et al. (2012) compared 154 durum wheat landraces, originating from 20 Mediterranean countries, as three groups (Eastern, North Balkan, and Western), to 18 modern wheat cultivars. Landraces from the Eastern Mediterranean countries showed the highest variability (28.9–54.9 g) with a mean of 44.6 g, but lower kernel weight than Western (49.4 g) and North Balkan (50.7 g) Mediterranean countries as well as modern cultivars (46.7 g).

The variation in kernel weight between both tetraploid and hexaploid landraces was large; however, tetraploid wheat landraces had heavier kernel weight than hexaploids (Blum et al. 1987). Beside wheat species and cultivars, environmental conditions and agronomic applications can also affect the kernel size and weight. A relatively low heritability value (51.27%) was reported for thousand kernel weight (Heidari et al. 2016). In the same study, thousand kernel weight of wheat landraces varied from 26.8 to 55.0 g.

Grain hardness is one of the most significant quality characteristics for milling and baking quality of wheat. The attachment between the protein and the starch in the endosperm is stronger in hard wheat than in soft one. During the milling, hard wheat has a higher amount of starch damage, flour extraction proportion, and energy consumption (Bedõ et al. 2010). Based on kernel hardness, wheat is classified as soft, medium-soft, medium-hard, hard, and extra-hard. This classification creates a fundamental basis for differentiating the world trade of wheat grain (Pasha et al. 2010).

The hardness is inherited and controlled by *Ha* hardness genes (Pin a and Pin b) residing on the short arm of chromosome 5D but is also impacted by other small-effect loci (Pasha et al. 2010). Pin a and Pin b have various alleles in hexaploid wheat (Morris and Bhave 2008). The variability in Pin function considerably impacts the quality features of the milling and end-product in wheat (Pasha et al. 2010). Grain hardness is a parameter strongly linked to wheat species. The durum wheat has no D-genome and represents a harder grain texture (Morris and Bhave 2008). The meaning of durum is "hard" in Latin and the species is the hardest of all wheats (Hare 2017).

Einkorn kernels showed extra-soft texture (99–306 g) followed by spelt (205–214 g), bread (383–458 g), emmer (596–685 g), and durum (756–885 g) wheats, respectively (Brandolini et al. 2008). Migliorini et al. (2016) reported that grain hardness ranged from soft (34.0%) to hard (82.5%) among the landraces. Similarly, in a study conducted by Bordes et al. (2008), 372 hexaploid wheat accessions including 139 modern cultivars and 233 landraces and old cultivars were evaluated for grain hardness which showed a wide variation ranging from very soft (1.5%) to very hard (99.9%). No difference in grain hardness was reported between landraces and modern wheats in the same study. Cetiner et al. (2020) compared landraces and old and new bread wheat cultivars and stated that two of the three landraces were classified as hard. However, generally old cultivars had a softer grain structure compared to modern cultivars in the same study. The authors attributed this result to the harder wheat selection over the years to develop cultivars with better bread-making quality. The grain hardness (SKCS) scores of the 133 wheat landraces showed a wide range and varied between 28.0% and 99.3%, with an average of 67.9% (Black et al. 2000). Li et al. (2008) reported that frequencies of soft, mixed, and hard wheat samples were 3.9, 20.4, and 75.6%, respectively, in 431 landraces. In hard wheat landraces, distributions of Pin a-D1b, Pin b-D1b, and Pin b-D1p were determined as 38.0, 0.9, and 59.6%, respectively.

6.3 Protein Content and Quality

The functional characteristics of wheat for producing bread or pasta depend on protein content and quality (Cubadda and Marconi 1996). Therefore, improving grain protein content has been the main focus of wheat breeding (Mujeeb-Kazi et al. 2013). Grain protein contents of bread wheat landraces (13.8–16.7%, $n = 42$) were slightly higher compared to modern cultivars (13.9–15.2%, $n = 7$) (Akçura 2011). Raciti et al. (2003) reported that about 95% of accessions ($n = 116$) were characterized by grain protein content higher than (or equal to) the average protein content of control cultivars. It was also reported that the grain protein content of landraces ranged from 13% to 20%, and the variation was wider in tetraploid landraces than hexaploid landraces (Blum et al. 1987). Likewise, grain protein content displayed a declining trend over time of cultivar release, from ~18% in the old wheats to ~16.5% in the modern durum wheats (De Vita et al. 2007), comparable to results by Dinelli et al. (2013).

Landraces and old cultivars tend to have higher grain protein contents than modern cultivars at the same or even lower nitrogen (N) application levels and in soils with lower fertility (Nazco et al. 2012; Giunta et al. 2019). However, this does not justify the classification of them in common as protein-rich crops since their high protein content could be a result of lower grain yield (Čurná and Lacko-Bartošová 2017). An inverse relationship was reported between yield and grain protein content by some researchers (Akçura 2011; Shewry et al. 2013). Similarly, higher average protein content was obtained from 37 Iranian and 42 Mexican durum wheat landraces compared to bread wheat check with a higher thousand kernel weight (Hernández-Espinosa et al. 2019).

The characteristics of grain quality, including protein content, vary both within and across wheat species; however, they are highly affected by the environment (Arzani and Ashraf 2016). The variation on protein content strongly relies on the wheat cultivar, the growing circumstances, the fertility of the soil, and the fertilizer, especially nitrogen (Carson and Edwards 2009). The genotype effect was the strongest on protein content (Hidalgo and Brandolini 2017), and a high heritability value

(82.3%) was calculated for grain protein content by Heidari et al. (2016). However, Shewry et al. (2013) reported that nitrogen fertilization had a greater effect on protein content compared to genotype. As a general rule, grain protein content increases with increased nitrogen fertilization rate. Besides, late nitrogen application has a positive effect on quality since it leads to increased protein content (Ottman et al. 2000). Migliorini et al. (2016) found about 30% lower protein content in the second year compared to the first year in the grains of wheat landraces because of the effect of the environment. Sowing time also affected the grain protein content and higher protein content was obtained from delayed sowing time (Fois et al. 2011). Melash et al. (2019) reported that increasing seed rate, from 100 to 175 kg ha^{-1}, reduced the grain protein content, Zeleny sedimentation value, and wet gluten content by 8.7, 9.1, and 10.8%, respectively. It was also reported that the foliar application of $FeSO_4$ in the anthesis stage tended to increase the grain protein content, Zeleny sedimentation value, and wet gluten content compared to the foliar application of $ZnSO_4$.

Protein quality is an important quality characteristic of wheat breeding. There are several physical and chemical tests to evaluate the protein quality of wheat (Hare 2017). Among various predictive tests, SDS sedimentation has been extensively used in wheat breeding programs, being particularly useful when only small flour samples are available. The sedimentation test depends on the swelling and flocculating properties of glutenin protein in dilute lactic acid solution, and the results are correlated to gluten strength and baking quality of wheat. A significant correlation between the sedimentation volume and Alveograph energy (W) value confirmed that the sedimentation test, simple and fast, is a valuable tool for predicting the Alveograph energy (W) value (Vázquez et al. 2012). The positive correlation between SDS sedimentation and gluten index revealed that high SDS sedimentation value was related to strong gluten strength (Dick and Quick 1983).

The variation is high among wheat genotypes in terms of the sedimentation volume. The variation among landraces for SDS sedimentation values has been identified by several researchers, ranging between 43 and 58 ml ($n = 50$) (Heidari et al. 2016), 22 and 36 ml ($n = 20$) (Akar et al. 2009), 16 and 24 ml (Blum et al. 1987), and 52 and 62 ml ($n = 300$) (Sezer et al. 2019). The SDS sedimentation range was 23–83 ml when a group of genotypes ($n = 116$) was analysed. Of 116 genotypes, 55 genotypes had higher or equal SDS sedimentation values than the control mean value (40 ml) (Raciti et al. 2003). Hernández-Espinosa et al. (2019) reported that average sedimentation volume was slightly higher in Iranian landraces than the Mexican group, comparing 37 Iranian and 42 Mexican durum wheat landraces. Genotype effect was found the strongest on SDS sedimentation value reported by Hidalgo and Brandolini (2017), and quite high heritability values (> 93%) were determined for both SDS and Zeleny sedimentation parameters (Heidari et al. 2016).

Wheat gluten was isolated from flour in 1728 by Beccari and wheat proteins were identified based on their extractability in different solvents: globulins (salt-soluble), albumins (water-soluble), prolamins (gliadin; alcohol-soluble), and glutelins (glutenin; dilute acid-soluble) by Osborne (Wrigley 2010; Kiszonas and Morris 2018). In wheat flour, gluten consisting of gliadins and glutenins accounts for about 80% of the total protein (Hoseney et al. 1969; Shewry et al. 2009). Glutenins, polymeric proteins, are responsible for the strength and elasticity of the dough, while gliadins, monomeric proteins, contribute to dough viscosity and extensibility. Gluten plays an important role in the bread-baking quality by supplying a network formation during the mixing and hydration in the dough. The gluten network provides distinctive features to the dough, such as air holding through the creation of an impermeable membrane around gas cells, resulting in a foamlike baked end-product (Delcour and Hoseney 2010).

Wet gluten can be obtained from wheat flour or meal using the automated gluten washer equipment. It is known that wet gluten content correlates positively with dry gluten content (Desheva et al. 2014). The gluten index value on wheat indicates gluten elasticity and strength and does not firmly rely on protein content. The values higher than 80% represent strong gluten (Migliorini et al. 2016). In a study, including 50 wheat landraces, carried out by Heidari et al. (2016), the dry gluten content had a high heritability value (84.8%) and ranged from 9.9% to 19.7% with a mean of 15.8%. The dry gluten contents and gluten index values of old durum wheat cultivars ($n = 14$) grown in Italy varied from 10.7% to 14.3% and from 4.5% to 60.6%, respectively (Mefleh et al. 2019). Konvalina and Moudrý (2008) reported that wet gluten and gluten index values of six emmer landraces were between 34.0% and 50.9% and 10% and 48%, respectively. Landraces and old cultivars possess lower gluten index values than modern durum wheat cultivars (De Vita et al. 2017). Breeding resulted in a significant increase in the gluten index, reflected by the development of protein quality (Motzo et al. 2004).

The wheat quality is influenced by the protein content and composition in the endosperm, particularly the contents and proportions of two gluten fractions, gliadins and glutenins, and of their low molecular weight (LMW) and high molecular weight (HMW) subunits (Mefleh et al. 2019). HMW glutenins have a greater effect on dough elasticity and strength than LMW glutenins (Gupta et al. 1991; Gupta and MacRitchie 1994). The gluten subunits come together to form a strong gluten network during dough mixing; however, with continuous dough mixing, LMW glutenins first disassociate from the gluten network, followed by HMW glutenins (Bonilla et al. 2019). Fois et al. (2011) reported that modern durum wheat cultivars with HMW-GS 6 + 8 and 7 + 8 showed superior gluten strength than old (before 1950) and intermediate (1950–1973) durum wheat cultivars with HMW-GS 20. The HMW/LMW ratio ranged from 0.54 to 1.03, from 0.61 to 0.68, and 0.54 to 0.89 for landraces, old, and modern genotypes, respectively. The gliadin to glutenin (Gli/Glu) ratio also varied between 0.59 and 1.18, 0.44 and 0.57, and 0.66 and 0.78, for the aforementioned genotype groups, respectively (Boukid et al. 2018). The presence of γ-45 gliadin and the absence of γ-42 gliadin bands are associated with strong gluten. It was reported that Mediterranean durum wheat landraces ($n = 171$) possessed 68.9% of γ-45 gliadin, 11.1% of γ-42 gliadin, and 20.0% of both gliadin bands (Nachit et al. 1995).

The time and rate of nitrogen fertilization can influence the gluten fraction proportions. Albumins and globulins were hardly affected by nitrogen application; however, gliadins were more affected than glutenins (Pechanek et al. 1997; Wieser and Seilmeier 1998). Mefleh et al. (2019) reported that additional nitrogen fertilization improved the gliadin fraction by 12.4% and unaffected the glutenin fraction. The gliadin to glutenin (Gli/Glu) ratio ranged from 3.46 to 4.74 among old cultivars, which was similar to the result of the modern cultivar in the same study. Also, sowing time significantly influenced Gli/Glu ratio (Fois et al. 2011).

6.4 Starch and Lipid Contents

Starch provides up to 90% of the average calorie intake in the diet of developing countries and more than 50% in the developed countries (Wang et al. 2015). High-yielding wheat cultivars with high starch content were selected to feed the increasing population by breeders (Boukid et al. 2018). Starch forms 65–70% of the dry matter in wheat grain (Bedõ et al. 2010). The total starch content ranged from 54.0% to 66.8%, from 55.6% to 63.3%, and from 65.1% to 67.8% for landraces, old genotypes, and modern wheats, respectively (Boukid et al. 2018). Alfeo et al. (2018) reported an inverse relationship between starch and protein contents. The starch content of durum wheat landraces ranged from 59.7% to 67.2% in the same study.

Starch, the main component of endosperm, is made up of 25% amylose (a mixture of linear and lightly branched) and 75% amylopectin (monodisperse and highly branched) (Maningat et al. 2009). Although both starch fractions have the same basic structure, their length and degree of branching are different, which influences the physicochemical properties of starch (Sofi et al. 2013). The gelatinization and pasting properties of starch are affected by the amylose content of wheat (Zeng et al. 1997). The proportion of amylose to amylopectin and amylopectin structure affect the processing, organoleptic characteristics, and digestibility in starch-based foodstuffs (Bao et al. 2006). The proportion of these two starch polymers within the starch granules differs, relying on the cereal and its cultivar (McKevith 2004).

The contents of amylose and amylopectin and amylose/amylopectin ratio of 33 lines, belonging to an old cultivar named Bánkúti 1201, ranged from 14.4% to 24.2%, 75.8% to 85.6%, and 0.16 to 0.32, respectively, suggesting that old cultivars are heterogeneous for starch contents (Rakszegi et al. 2003). In a study, carried out by Black et al. (2000), the amylose content of wheat landraces ($n = 133$) varied between 23.4% and 30.2% with a mean of 27.9%, while those of two commercial wheat cultivars were 27.9% and 29.9%.

Increased amylose is linked with increased resistant starch which is essential in the prevention of diabetes and obesity (Hazard et al. 2014). Therefore, healthy nutrition trends for enriched fibre consumption with low glycaemic food have pushed the growth of high amylose starch as a source of resistant starch acting like dietary fibre (Bertolini 2009). The wider variation was observed in landraces (28.0–60.1%) than old genotypes (51.5–65.9%) and modern genotypes (30.5–58.3%) in terms of resistant starch by Boukid et al. (2018). Dinelli et al. (2013) reported the cultivar and growing season effects on the resistant starch content of wheat. The higher resistant starch contents were obtained from a landrace and an old durum wheat cultivar (6.1–8.2 g/kg) compared to modern durum wheat cultivars (3.0–6.0 g/kg). The cultivars grown in the first year possessed approximately twofold resistant starch content than the ones grown in the second year of the study. Similarly, the higher resistant starch contents were determined at a landrace (39.8 ± 3.8 mg/kg dmb) and an old (44.6 ± 1.2 mg/kg dmb) durum wheat genotypes than modern durum wheat cultivars (12.0–37.0 mg/kg dmb) and Kamut (17.7 ± 1.2 mg/kg dmb) (Marotti et al. 2012).

Lipids are rather a minor component in wheat; however, they play a key role in nutrition, grain storage, and processing like dough mixing and baking. The lipids associate with the gluten proteins to form complexes, which contributes to the stabilization of the gas-cell structure. Therefore, they have important effects on bread volume and final texture of the baked products (Uthayakumaran and Wrigley 2010). Interaction between lipids and starch can influence gelatinization, retrogradation, and pasting properties of wheat starch and the vulnerability of starch to enzyme attack (Copeland et al. 2009).

In wheat, most of the lipids are concentrated in the germ (28.5%) and aleurone (8.0%), with only small amounts in the endosperm (1.5%) (Delcour and Hoseney 2010). The distribution of the lipids within the wheat species varies narrowly. Wheat lipids make up 2.03–2.85%, 1.80–2.85%, 1.88–1.93, and 1.96–2.82% of the weight of the whole einkorn, emmer, bread, and durum grains, respectively (Giambanelli et al. 2013). However, high variability, ranging from 22.4 to 33.7 g/kg, among durum wheat genotypes including a landrace and an old durum wheat cultivar was found by Dinelli et al. (2013). The crude fat content of 30 einkorn landraces grown in Kastamonu province in Turkey ranged from 1.62% to 2.72% with an average of 2.19% (dry matter basis) (Emeksizoğlu 2016). Lipid content was affected by sowing date, higher for spring (1.92–2.85%) compared to fall (1.80–2.65%) sowing (Giambanelli et al. 2013).

6.5 Vitamins and Minerals

Wheat is considered to be a significant source of vitamin B, particularly B1 (thiamine), B2 (riboflavin), B3 (niacin), B6 (pyridoxine), and B9 (folate) (Shewry and Hey 2015). Consuming products made by whole grain contributes to 40% of the suggested daily allowance for B1, 10% for B2, 22% for B3, 33% for B6, and 13% for B9 (Uthayakumaran and Wrigley 2010).

Abdel-Aal et al. (1995) reported that thiamine contents among wheat species were not large, ranging from 0.50 mg/100 g to 0.60 mg/100 g. Riboflavin content was relatively high in einkorn (0.45 mg/100 g) and bread (0.55 mg/100 g) wheat, but was low in spelt (0.14–0.17 mg/100 g). However, spelt had higher content of niacin (2.0–5.7 mg/100 g) compared to einkorn (3.1 mg/100 g) and bread wheat (2.3 mg/100 g). The amount of pyridoxine varied from 0.35 mg/100 g to 0.49 mg/100 g among species. In another study carried out by Stehno et al. (2011), variability ranges were as follows: thiamine (0.29–0.44 mg/100 g), riboflavin (0.108–0.135 mg/100 g), niacin (8.4–10.6 g/100 g), and pyridoxine

(0.27–0.45 mg/100 g) contents for eight emmer genotypes compared to bread wheat, with 0.36 mg/100 g thiamine, 0.071 mg/100 g riboflavin, 6.8 mg/100 g niacin, and 0.37 mg/100 g pyridoxine (dry matter basis).

Cereal and cereal products are important sources of folate, a water-soluble form of vitamin B9 and also known as folic acid or folacin. Folate is important for the prevention of neural tube defects, anemia, and cardiovascular disease (Scott et al. 2000; De Wals et al. 2007). In the HEALTHGRAIN project, only the folate content of ancient and modern wheat species was determined. The folate concentrations were found higher in durum (0.74 μg/g dmb) and emmer (0.69 μg/g dmb) wheats compared to einkorn, spelt, and bread wheats (0.58, 0.58, and 0.56 μg/g dmb, respectively) (Piironen et al. 2008). Heritabilities of these vitamins in a G × E study including 26 lines were found quite low, with the highest values being for thiamine (31%) followed by folate (24%), riboflavin (16%), pyridoxine (12%), and niacin (7%) (Shewry et al. 2013).

Major micronutrients in wheat are vitamin E, some B vitamins, and several minerals. These minerals are distributed unequally in the seed and are mostly localized in germ and bran (Uthayakumaran and Wrigley 2010). The recommended daily intake of vitamin E, an important antioxidant, is 10 mg/day according to the European Union Council (EC). Comparing genotypic groups, the highest vitamin E activity was determined as 12.3 mg/kg from landraces, followed by old cultivars (10.8 mg/kg), modern cultivars (10.6 mg/kg), spelt wheat (7.7 mg/kg), and primitive wheat (6.1%). Wheat contributed 12.2–24.5% vitamin E of the daily intake based on EC. Vitamin E activity is known to diminish by heating (Hussain et al. 2012b).

Functional and nutritional properties of wheat are important components of grain quality; however, increasing mineral concentration in grain did not have a priority in genetic improvements. Therefore, modern wheat cultivars have generally lower mineral contents than older cultivars (Fan et al. 2008; Hussain et al. 2012a) because landraces and old cultivars have generally lower thousand kernel weight and grain yield. In a study comparing Zn, Ca, and Fe contents of bread wheat cultivars from obsolete to current, cultivars released between 1965 and 1976, compared with the current cultivars (2001–2008), contained significantly more Zn (18%) and Ca (14%) but similar Fe content (Hussain et al. 2012a). Some mineral element contents of 86 bread wheat landraces grown in Turkey varied from 35.53 to 53.08 mg/kg for Fe, from 22.66 to 38.57 mg/kg for Zn, from 30.92 to 48.58 mg/kg for Mn, from 8.63 to 15.77 mg/kg for B, from 4.12 to 6.69 mg/kg for Cu, from 0.85 to 1.78 mg/kg for Mo, from 2.25 to 5.41 g/kg for K, from 1.02 to 1.69 g/kg for Mg, and from 0.34 to 0.55 g/kg for Ca (dry matter basis). Landraces had especially higher Zn, Fe, and Mn contents than the bread wheat cultivars in the same study (Akcura and Kokten 2017). In a study conducted by Manickavelu et al. (2017), the landraces (n = 267) were more variable and contained higher average values (Fe, Zn, Mn, Mg, P, and K) except for Fe than check cultivars of Japan and Afghanistan. Similarly, Kondou et al. (2016) observed that the landraces showed greater variability than the check cultivars regarding K, P, Mg, and Fe.

Humans require more than 22 mineral elements to meet their metabolic needs. Some of them are needed in large amounts such as Na, Ca, K, and Mg, but others, like Zn, Fe, Cu, Mn, I, and Se are needed in trace amounts (Welch and Graham 2004). Cereals and cereal products provide on average >40% of the daily intake of Fe, >10% of K, 27% of Mg, 30% of Ca, 25% of Zn, and 33% of Cu (Swan 2004). Recommended daily intakes of Zn, Fe, and Ca are usually not achieved in the developing countries (Brown et al. 2001; Gibson 2006). Among the micronutrient insufficiencies, deficiencies of Fe and Zn are mainly important for affecting human health (Ozkan et al. 2007). As a sustainable solution, biofortifying grains with essential minerals that are insufficient in peoples' diets are recommended (Bouis and Welch 2010; Ficco et al. 2009). Biofortification depends on agronomic and genetic methods to increase the bioavailable amount of minerals in cereals (Hawkesford and Zhao 2007; Hussain et al. 2010). Breeding for biofortification, genetic engineering for more uptake from the soil, and fertilizer application are the main methodologies to increase the contents of mineral elements in grain (Cakmak 2008; Waters et al. 2009; Wang et al. 2008). For instance, there is large genotypic variation in contents of Fe and Zn among 54 einkorn wheat genotypes ranged from 0.21 to 2.16 μg seed^{-1} for zinc and from 0.54 to 3.09 μg seed^{-1} for iron, and these variations might be used for biofortification in wheat breeding (Ozkan et al. 2007). Wild emmer is also a significant genetic resource for increasing the concentration of Fe and Zn in modern wheats (Cakmak et al. 2004). Selenium fortification provided more Se accumulation in landraces and obsolete cultivars than in modern cultivars. Besides, Se content in the durum wheat was raised by up to 35-fold that of the untreated application (De Vita et al. 2017). It was also reported that wheat accumulated more Fe than Zn (Manickavelu et al. 2017).

The milling process affects mineral content because the outer layer of the grain is removed. Milling of durum wheat grain into semolina can cause a 40–80% loss of Fe, Zn, Cu, and Mg (Cubadda et al., 2009; Lyons et al., 2005). De Vita et al. (2017) reported that the Se concentration diminished during the milling (11%), while the processing of pasta did not display significant decreases.

There is a strong genetic effect on Zn and Fe accumulation in the grain although there is a significant genotype × environment (G × E) interaction effect on Zn and Fe contents. Further research also showed that there was not an inverse linkage between yield and Zn and Fe contents in the grain. Therefore, it should be possible to increase Zn and Fe contents in wheat grain by breeding (Welch and Graham 2004). High zinc wheat lines were obtained from the biofortification breeding program, using wild relatives and landraces (Velu et al. 2014). Di Silvestro et al. (2012) reported that old bread wheat cultivars had higher mineral contents than modern cultivars when grown under low input management practices. Beside breeding, the biosynthesis and accumulation of minerals are affected by genotype and environment (Migliorini et al. 2016) and farming practices (Rizzello et al. 2015).

It is also possible to increase the bioavailability of minerals in wheat grain by reducing phytic acid (White and Broadley 2009), because phytic acid is an antinutrient and binds positively charged mineral cations such as iron, zinc, and calcium to create insoluble complexes, which inhibits the absorption of the minerals into the body (Weaver and Kannan 2002). Most of the total phosphorus present in wheat grain (75%) is stored as phytic acid, particularly in the germ and aleurone layers of the wheat kernel (Lott and Spitzer, 1980). Phytase activity decreases the phytic acid breakdown in wheat. Hence, the mineral bioavailability is attached to mineral and phytase concentrations. These should be taken into consideration in wheat improvement for biofortification (Mujeeb-Kazi et al. 2013). The phosphorus content is controlled to a large extent by the environment (62%). Wider ranges in inorganic phosphorus content were reported among the modern durum wheat genotypes (0.47–0.76 mg/g), compared to landraces (0.48–0.69 mg/g) and advanced lines (0.46–0.66 mg/g) (Ficco et al. 2009). Çetiner et al. (2018) reported that bread wheat cultivar Tosunbey possessed lower phytic acid content (797 mg/100 g) than old cultivars and landraces, ranging from 1125 to 1606 mg/100 g.

6.6 Phytochemicals and Antioxidants

Wheat is an important source of health-promoting components, particularly phytochemicals and antioxidants as well as the main components of protein, carbohydrate, and lipid (Arzani 2019).

Ferulic acid is the main phenolic component of both the insoluble-bound and the soluble-conjugated fractions in different wheat species (Yilmaz et al. 2015). According to Li et al. (2008), the average total ferulic acid concentrations were similar for spelt, durum, and bread wheat samples (about 400 μg/g dmb), but higher in emmer wheat samples (476 μg/g dmb) and lower in einkorn wheat samples (298 μg/g dmb). This finding is corroborated by the results of Serpen et al. (2008): the ferulic acid content of einkorn wheat was about twofold lower than that of emmer wheat. Ferulic acid is a distinctive trait of old and modern wheat genotypes that the landraces possessed the lower ferulic acid content (0.64–0.85 g/kg) than modern wheat cultivars (1.21–1.36 g/kg) (Piergiovanni 2013).

Alkylresorcinols, one of the main groups of phenolic compounds, are mainly located in the external layers of the wheat grain with high levels (Landberg et al. 2008). A comprehensive study carried out by Ziegler et al. (2016) using whole grain flour of 15 genotypes each of five species grown at four environments showed that the contents of alkylresorcinol varied greatly among the genotypes within each species, and the overall average concentrations of the species were 761 ± 92.3, 743 ± 56.7, 654 ± 47.9, 697 ± 93.6, and 737 ± 90.9 μg/g dmb in bread, spelt, durum, emmer, and einkorn wheat samples, respectively. Ciccoritti et al. (2013) reported that alkylresorcinol mean values were 344 ± 8, 377 ± 17, 321 ± 18, and 286 ± 11 $\mu g\ g^{-1}$ dmb for einkorn, emmer, bread, and durum wheats, respectively. Similarly, the results of the HEALTHGRAIN project displayed significant variations in the total alkylresorcinol content in wheat species, and alkylresorcinol contents were higher in ancient wheats (emmer, einkorn, and spelt) compared to modern wheats (bread and durum) (Andersson et al. 2008).

High heritability values were found for tocochromanols (h^2 = 0.88–0.97), steryl ferulates (h^2 = 0.88–0.94), and alkylresorcinols (h^2 = 0.69–0.97) regarding five wheat species. These results demonstrated that lipophilic antioxidant contents in einkorn, emmer, spelt, bread, and durum wheats were under strong genetic control (Ziegler et al. 2016).

Tocols are a class of lipid-soluble liquids, viscous, synthesized only by photosynthetic plants, and classified as tocopherols and tocotrienols. They both occur as a polar chromanol ring and a hydrophobic 16-carbon side chain. In tocopherols, the side chain is a saturated isoprenoid group, while in tocotrienols, it has three double bonds. Tocopherols and tocotrienols both consist of four derivatives: α-, β-, γ-, and δ-tocols and collectively known as tocochromanols (Hidalgo et al. 2006; Lampi et al. 2008; Okarter et al. 2010; Lachman et al. 2013; Ziegler et al. 2016). But α- and β-tocols are the major derivatives present (Lampi et al. 2008). There are more tocotrienols than tocopherols, and β-tocotrienol is the predominant tocol followed by α-tocotrienol, α-tocopherol, and β-tocopherol in wheat (Hidalgo et al. 2006; Hidalgo and Brandolini 2017).

Though all tocopherols and tocotrienols have antioxidant activity, α-tocopherol is the most effective antioxidant for the breaking free radical driven-chain reactions (Packer 1995). Besides having antioxidant activity, only α-tocopherol has vitamin E activity (Schneider 2005). In addition to their antioxidant properties, the tocochromanols of cereals could have positive health effects such as lowering LDL cholesterol in the blood and the risks of cancer and cardiovascular diseases (Tiwari and Cummins 2009). Furthermore, tocotrienols have potential as neuroprotective dietary factors (Frank et al. 2012). In a study comparing the amount of total tocochromanol among the wheat groups, the landraces with 32.9 ± 3.37 mg/kg was followed by modern cultivars (32.5 ± 0.99 mg/kg), old cultivars (30.3 ± 4.41 mg/kg), spelt wheat (28.9 ± 3.47 mg/kg), and primitive wheat (28.0 ± 5.39 mg/kg) (Hussain et al. 2012b).

Wheat has antioxidant activity because of its lipophilic (carotenoids, tocopherols) and hydrophilic (phenolics, selenium) antioxidant contents (Konvalina et al. 2017). Lachman et al. (2012) reported that spring wheat genotypes possessed lower antioxidant activity (195.8–210.0 mg Trolox/kg dmb) than einkorn (149.8–255.8 mg Trolox/kg dmb) and emmer (215.4–257.6 mg Trolox/kg dmb) wheats. In a study conducted with 26 genotypes of einkorn, emmer, spelt, bread wheat landraces, and spring wheat in three growing seasons, the average antioxidant activity ranged from 225.45 to 400.83 mg Trolox/kg dmb, displaying a broad range among wheat species and genotypes (Konvalina et al. 2017). These results are about twofold higher than the findings of Lachman et al. (2012). This difference in antioxidant activity content is explained by the weather or stress conditions during the growing season and genotype effects by Konvalina et al. (2017). A linear relationship ($r = 0.74, p < 0.05$) existed between antioxidant activity and total polyphenols (Lachman et al. 2012).

Polyphenols are the most indicative antioxidant compounds in wheat kernel (Migliorini et al. 2016). Polyphenols consist of flavonoids and phenolic acids, and they might be found in the bound insoluble and the free soluble forms (Dinelli et al. 2009; Migliorini et al. 2016). In a study in which five bread wheat landraces grown at two different years and locations, significant differences were found between years and cultivars, showing that the second year and cultivar Gentil Rosso possessed higher amounts of the total, free, and bound polyphenols. These results demonstrated that polyphenols were affected by wheat cultivars and environmental conditions such as abiotic and biotic stresses (Migliorini et al. 2016). It was also reported that the environment was the main source of variation in the total soluble phenolic content although genotype, environment, and their interaction had significant effects on the parameter (Bellato et al. 2013). Dinelli et al. (2009) found that there were no significant differences between the mean values of old and modern wheat cultivars in terms of phenolic and flavonoid compounds. However, old cultivars had slightly higher free (181.8 ± 37.8 μmol GAE/100 g), bound (696.4 ± 53.5 μmol GAE/100 g), and total (878.2 ± 19.0 μmol GAE/100 g) phenolic compounds than modern cultivars. These values were 178.4 ± 51.9, 687.4 ± 91.0, and 865.9 ± 128.9 μmol GAE/100 g for the modern cultivars, respectively. In terms of flavonoid compounds, old cultivar had higher free flavonoid compounds (52.5 ± 22.7 μmol CE/100 g), while modern cultivars possessed higher bound (80.0 ± 15.9 μmol CE/100 g) and total (122.6 ± 25.4 μmol CE/100 g) flavonoid compounds on average. Though the range in mean values of bound, free, and total phenolic contents between modern and old cultivars did not vary significantly, differences between the cultivars were significant.

Phytosterols, consumed with the diet, may have a role in preventing colon cancer (Rao and Janezic 1992) and cardiovascular diseases (Piironen et al. 2000). In a screening study of 175 genotypes of different wheat types, which represent current, uncommon, and obsolete, the highest average total phytosterol content was obtained from einkorn (1054 μg/g dmb), followed by durum (987 μg/g dmb), spelt (928 μg/g dmb), spring (864 μg/g dmb), emmer (857 μg/g dmb), and winter (841 μg/g dmb) wheats. The difference between the lowest (670 μg/g dmb) and highest (1187 μg/g dmb) total phytosterol contents in all wheat genotypes was determined as 77% by Nurmi et al. (2008). The total sterol contents of einkorn and emmer wheats were found similar (554.3–828.5 and 500.8–816.4 mg kg^{-1} dmb, respectively), higher than bread wheat (440.8–661.8 mg kg^{-1} dmb) and lower than durum wheat (614.8–929.0 mg kg^{-1} dmb) by Giambanelli et al. (2013). The most abundant phytosterol in wheat is β-sitosterol (34.2–42.7% of phytosterols) followed by campesterol, sitostanol, and campestanol (Giambanelli et al. 2016). It was also confirmed that the most plentiful phytosterol in wheat types, which contain current, uncommon, and obsolete, was sitosterol (40–61% of total phytosterols), while the highest variation was shown in total stanols (7–31% of total phytosterols) (Nurmi et al. 2008).

Carotenoids, lipid-soluble antioxidants, are produced by most photosynthetic organisms and are accountable for the orange, red, and yellow colours in numerous fruits, flowers, and bird feathers. Lutein is the predominant component of carotenoids followed by zeaxanthin in wheat; however, other carotenoids like α-carotene and β-carotene only present in minor amounts (Hidalgo et al. 2006; Abdel-Aal et al. 2007; Digesù et al. 2009). The total yellow pigment content is a widely used test in durum wheat breeding programs. Significant positive relations were observed between total yellow pigment content and lutein ($r = 0.94$, $p < 0.01$) and total carotenoid ($r = 0.99$, $p < 0.01$) contents. This shows that the total yellow pigment content or colorimetric method is a reliable test to predict the lutein and total carotenoid contents in wheat (Abdel-Aal et al. 2007). It is also reported that the ratio of carotenoids was 33.2% of the yellow pigment concentration by Digesù et al. (2009).

Since yellow or amber pasta colour is usually desired by consumers, yellow pigment content is an important quality parameter in the evaluation of semolina colour, especially in determining the end-product quality of durum wheat (Digesù et al. 2009). For this reason, wheat breeders have focused on high yellow pigment content during the selection of new wheat cultivars for the last two decades. Therefore, modern durum wheat cultivars have generally higher yellow pigment content than older cultivars (Digesù et al. 2009). The total carotenoid contents of 102 tetraploid wheat accessions ranged from 1.18 to 4.42 μg/g dmb, with a mean of 2.46 μg/g dmb. Durum wheat cultivars (released after 1991) possessed a higher average value (3.11 μg/g dmb) than the older ones (released in the period 1971–1991) (2.56 μg/g dmb) and landraces (before 1971) (2.33 μg/g dmb) (Digesù et al. 2009). Dinelli et al. (2013) observed significant differences between durum wheat genotypes ($n = 8$), including a landrace and an old durum wheat cultivar, in terms of lutein and total carotenoid contents. Total carotenoid contents of genotypes varied between 3.28 and 6.09 μg/g, while lutein contents of those changed from 1.50 to 3.23 μg/g which was almost half of the total carotenoid amount. Nazco et al. (2012) compared the yellow colour indexes of 154 durum wheat landraces, originating from 20 different Mediterranean countries, as three groups (Eastern, North Balkan, and Western), to those of 18 modern durum wheat cultivars. The landraces from the Eastern Mediterranean countries showed the widest variability (11.5–17.6) with a mean of 15.3, but lower than modern cultivars (15.9). North Balkan (14.6) and Western (14.8) Mediterranean countries possessed similar yellow index values determined at whole grain flour. Large and significant variation was determined in carotene contents of tetraploid landraces, ranging from 1.5 to 5.5 ppm by Blum et al. (1987). Similarly, Akar et al. (2009) reported that there was a large variation among the tetraploid landraces ($n = 20$) in terms of semolina b colour value, varying between 22 and 29.

Carotenoid concentration and lutein content were under genetic control (Lachman et al. 2013; Ziegler et al. 2016). High heritability values were calculated for total carotenoid (0.94), lutein (0.93), and yellow pigment (>0.91) concentrations, intermediate values for α-carotene (0.79) and β-cryptoxanthin (0.72) concentrations, and relatively low values for β-carotene (0.57) and zeaxanthin (0.48) concentrations by Digesù et al. (2009). Yellow pigment content is present at different levels in wheat cultivars and species. Einkorn had the highest lutein content ranging from 6.37 to 8.46 μg/kg dmb in whole flour with an average value of 7.41 μg/kg dmb, higher than durum (5.41 μg/kg dmb), emmer (3.97 μg/kg dmb), bread (2.11 μg/kg dmb), and spelt (1.47 μg/kg dmb) wheat species (Abdel-Aal et al. 2007).

Starr et al. (2015) investigated the volatile compound profiles of 64 wheat cultivars and 17 landraces. A large diversity in volatile profiles happened among wheat samples that landraces had higher levels of alcohols, esters, and some furans, while modern cultivars possessed higher levels of pyrazines, terpenes, and straight-chained aldehydes.

6.7 Dietary Fibre and β-Glucan

Dietary fibre is described as the edible part of plants or similar carbohydrates which resist digestion and absorption in the small intestine while they are completely or partially fermented in the large intestine (Gebruers et al. 2008).

In the HEALTHGRAIN cereal diversity screening program, among wheat species, bread wheat genotypes with 11.5–18.3 g/100 g contained the highest level of dietary fibre compared to other wheat species, such as durum (10.7–15.5 g/100 g) and spelt (10.7–13.9 g/100 g) wheats, whereas wild wheats such as einkorn (9.3–12.8 g/100 g) and emmer (7.2–12.0 g/100 g) had the lowest values (Gebruers et al. 2008). Similarly, the highest total dietary fibre content was found in bread wheat (12.3% dmb), followed by spelt, einkorn, and emmer wheats with mean values of 10.3, 8.7, and 7.9% dmb, respectively (Løje et al. 2003). The total, soluble, and insoluble dietary fibre contents of durum wheat genotypes, including two landraces and one old wheat, ranged between 127.4 and 199.7, 18.1 and 37.1, and 102.3 and 180.8 g/kg dmb, respectively (Marotti et al. 2012). The result of total dietary fibre obtained from the previous study was about 42% higher on average than that reported by Gebruers et al. (2008), where the range in total dietary fibre content was 107.0–155.0 g/kg dmb. This variation was explained by the effects of both genotype and environment on dietary fibre content (Gebruers et al. 2010; Shewry et al. 2010). Beside genotype and environment effects, the cultivar × year interaction effect was also reported on the total dietary fibre contents of durum wheat genotypes including one landrace and old cultivar by Dinelli et al. (2013).

The most essential dietary fibre components are the non-starch polysaccharide arabinoxylans (AX), which are the most plentiful dietary fibre, mixed-linkage β-glucans, cellulose, and the non-polysaccharide compound lignin, which are all cell wall components (Gebruers et al. 2008; Bedõ et al. 2010). AX have many health benefits such as immunomodulatory activity, attenuate type II diabetes, cholesterol-lowering activity, faecal bulking effect, enhanced absorption of certain minerals, and prebiotics effect (Mendis and Simsek 2014). Beside health benefits, AX affect water-binding capacity, rheology, starch retrogradation, and gas retention in dough (Simsek et al. 2011).

In a comprehensive study carried out by Gebruers et al. (2008), beside total dietary fibre, bread wheat genotypes had the widest variation in total arabinoxylan content (TO-AX) varying between 1.35 and 2.75% dmb. TO-AX ranging from 1.70% to 2.35% dmb, from 1.60% to 2.15% dmb, from 1.45% to 2.35% dmb, and from 1.40% to 1.95% dmb were reported for durum, spelt, einkorn, and emmer wheat flours, respectively. According to Marotti et al. (2012), the TO-AX of durum wheat genotypes, including two landraces and an old wheat, varied between 26.9 and 35.6 g/kg dmb, with a mean value of 32.7 g/kg dmb. In a study including old and modern bread wheats and landraces, the modern wheats possessed the highest mean TO-AX value (8.03%) compared to old wheats (7.60%) and the landraces (6.41%). It was also stated that modern breeding had no negative effects on the contents of AX components when comparing the wheat groups (Cetiner et al. 2020).

Based on the literature, it is obvious that there is a significant variation in TO-AX of wheat genotypes. This phenomenon was explained by the effects of genotype and environment on TO-AX (Li et al. 2009; Gebruers et al. 2010; Simsek et al. 2011). Finnie et al. (2006) concluded that cultivar was the main source of variability for TO-AX although the effects of cultivar and environment were statistically significant on TO-AX.

The water-extractable arabinoxylan (WE-AX) accounts for 25–30% of TO-AX, while water-unextractable arabinoxylan (WU-AX) accounts for the rest of TO-AX (Meuser and Suckow 1986). The variation for WE-AX in wheat species was wide. The largest variation in WE-AX in wheat flour was observed for bread wheat (from 0.30% to 1.40% dmb), while narrow variation (from 0.50% to 0.65% dmb) was found in einkorn. Durum and spelt wheats had a similar range from 0.25% to 0.55% dmb and from 0.30% to 0.45% dmb, respectively. The lowest value (0.15–0.55% dmb) was obtained from emmer wheat in the same study (Gebruers et al. 2008). In a study, comparing the landraces and old and modern wheats in terms of WE-AX and WU-AX contents, old wheats possessed the highest mean WE-AX value (0.81%), followed by modern wheats (0.79%) and landraces (0.67%). WU-AX mean values were 7.24, 6.79, and 5.74% for modern wheats, old wheats, and landraces, respectively (Cetiner et al. 2020). Finnie et al. (2006) reported that WE-AX content was more greatly affected by genotype, while WU-AX content was primarily influenced by the environment.

Along with arabinoxylans, β-glucan is one of the most important dietary fibre components in wheat (Marotti et al. 2012). High β-glucan content is desirable to increase the health benefits by lowering blood cholesterol levels (Lia et al. 1997). β-Glucans are mostly located in the cell wall of the endosperm (Laroche and Michaud 2007). In a study carried out by Biel et al. (2016), the levels of β-glucans were found dependent on cultivar only. However, the contents of β-glucans in wheat grain varied between species, cultivars, and environmental conditions, ranging from 0.25% to 1.40% of dry weight (Marconi et al. 1999; Løje et al. 2003; Gebruers et al. 2008; Biel et al. 2016). In a comprehensive study, Gebruers et al. (2008) pointed out significant variations in β-glucan content of five wheat species. On average, the species of einkorn, emmer, and durum wheat possessed half of the β-glucan amount, noted for the spelt and bread wheat species. The ranges of variation were 0.25–0.35%, 0.30–0.40%, 0.55–0.70%, 0.50–0.95%, and 0.25–0.45% of dry weight in einkorn, emmer, spelt, bread, and durum whole meals, respectively. Marotti et al. (2012) reported significant differences in β-glucan contents of durum wheat genotypes, including two landraces and an old durum wheat cultivar. However, β-glucan content, ranging from 2.4 to 4.1 g/kg dmb, in all wheat genotypes was lower than those determined in other cereal grains such as oat and barley.

6.8 Rheological Properties

Rheology of dough plays an important role in determining the quality of wheat-based products (Kundu et al. 2017). Physical dough analyses such as Farinograph, Alveograph, and Mixograph are used to predict the dough mixing features in the world (Marchylo and Dexter 2001). Farinograph analysis is used widely to determine particularly water absorption of flour. Water absorption has a key role in baked products, affecting each step of the process, yield, and end-product quality. Water absorption is the amount of water which is needed to produce a dough of suitable consistency. Amount of protein, damaged starch, and non-starch polysaccharide (in particular pentosans) contents affect water absorption of flour. Protein can absorb water about twice its weight when the dough is mixing. Some polysaccharides can absorb even more water. Water absorptions of flour can vary between 50% and 70%, depending on grain hardness, milling, and the desired flour specifications (Miskelly et al. 2010).

Comparing breeding periods, the lowest Farinograph water absorption value (WA) was 50.4% at initial cultivars (released from the mid-1940s until the Green Revolution), followed by landraces (released before 1940) with 51.5% WA, and modern wheat cultivars (released from 1970 to 2001) with 53.1% WA (Sanchez-Garcia et al. 2015). Evaluating 330 Chinese wheat cultivars as four groups based on released year, development time, stability, and Farinograph quality number of cultivars released after 2000 were 17.9, 71.1, and 44.3% higher than those of cultivars released between 1949 and 1976. The results showed that these Farinograph characteristics increased significantly over time (Yang et al. 2014). Farinograph water absorption was associated positively with protein content (Corbellini et al. 1999). Sanchez-Garcia et al. (2015) reported that water absorption was strongly under genotype effects which accounted for 73.8% of the total variation.

Parameters obtained from Alveograph, one of the rheological analyses, are P (the pressure, related to the height of the curve), L (the length, extensibility), P/L (tenacity/extensibility ratio), and W (the work, related to the area of the curve). Especially, the Alveograph W value determines the strength of gluten by measuring the force needed to blow the bubble of dough until it ruptures. De Vita et al. (2007) compared the Alveograph parameters of Italian landraces and durum wheat genotypes. Landraces possessed the lowest baking strength (W values: 30–99 10^{-4} J) and dough-gluten properties. In another study, modern cultivars showed about three times and twice more Alveograph W and P values of the landraces, respectively (Sanchez-Garcia et al. 2015). Similarly, 35 lines, derived from landraces, possessed lower Alveograph W values (ranging from 37 to 253 10^{-4} J) than bread wheat checks (ranging from 135 to 431 10^{-4} J) (Guzmán et al. 2014). Significant differences were observed between landraces, ranging from 58 to 161 10^{-4} J, for Alveograph W values (Migliorini et al. 2016). Comparing between 37 Iranian and 42 Mexican durum wheat landraces, Hernández-Espinosa et al. (2019) found that Mexican landraces had a higher average Alveograph W value than the Iranian group. Although genotype, environment (year), and their interaction had significant effects on the

Alveograph W (Migliorini et al. 2016), the genotype effect was the strongest on the Alveograph W and P parameters (Sanchez-Garcia et al. 2015). A relatively high heritability value (0.61) was determined for dough strength (W), while Alveograph tenacity (P), extensibility (L), and P/L ratio possessed lower heritability values (ranging from 0.14 to 0.31) (Igrejas et al. 2002).

Mixograph is also a useful instrument for determining the gluten strength of wheat. For instance, midline peak time, a parameter of Mixograph, shows the highest correlation with gluten strength. A high midline peak time value shows strong gluten strength (Beta et al. 2019). The landraces and old cultivars showed a wider range of variation than modern wheat cultivars in terms of almost all traits of Alveograph and Mixograph (Bordes et al. 2008). Similarly, a large variation was determined among the Mixograph scores of landraces, ranging from 2 to 9. However, modern wheat cultivars possessed a higher mean Mixograph score than landraces (Blum et al. 1987). Guzmán et al. (2014) also reported that bread wheat check samples possessed about twice higher Mixograph dough development time and dough strength values on average than the lines derived from landraces. Likewise, the modern durum wheat cultivars showed the strongest dough properties, having higher work input to peak (WIP) and time-bandwidth (ETBW) values, compared to older cultivars. Sowing time affected WIP but not ETBW (Fois et al. 2011). Although both genotype and year had significant effects on almost all 11 Mixograph parameters, belonging to 150 lines of a landrace population, quite low heritability values ranging from 0.08 to 0.40 were reported by Igrejas et al. (2002).

The use of Rapid Visco Analyser (RVA) has been increased as an instrument in wheat breeding in the past two decades. The RVA instrument is used to determine various parameters associated with the starch pasting characteristics of wheat (e.g. viscosity, pasting temperature) (Cozzolino 2016). The range of peak viscosity was 175–295 Rapid Visco Analyser units (RVU) with an average of 254 RVU among the wheat landraces ($n = 133$) (Black et al. 2000). Bhattacharya et al. (1997) reported that the pasting characteristics of 242 hexaploid wheat landraces showed wide diversity in all RVA parameters. The average peak viscosity of modern wheats was 260 RVU and varied between 185 and 355 RVU, while it ranged from 139 to 305 RVU for landraces, and positively correlated ($r = 0.73$, $p < 0.001$) with flour swelling volume. High peak viscosity, low setback, high breakdown, and low final viscosity are the properties concerned with high eating quality of Japanese white salted noodles (Black et al. 2000).

6.9 Wheat Landrace-Based Foodstuffs

Limited studies have been carried out to determine the end-product quality of wheat landraces. In a study comparing bread volumes of 37 Iranian and 42 Mexican durum wheat landraces, bread volumes of both landrace groups were on average lower than bread wheat control; however, 14 Mexican and 6 Iranian landraces possessed similar or higher bread volumes than check (Hernández-Espinosa et al. 2019). Bread volumes of 35 lines derived from landraces, ranging from 495 to 745 ml with a mean of 698.9 ± 42.9 ml, were lower than bread wheat checks ranging between 670 and 900 ml (Guzmán et al. 2014). A landrace named Tir has moderate baking quality. Although it is not preferred by the bread wheat industry, local people mostly use it for making the lavash (flatbread) baked in tandoor (Ülker et al. 2019). Regarding the sensory analysis, breads produced from five landraces were preferred by consumers (Migliorini et al. 2016). In terms of aroma profile, landraces and old bread wheat cultivars had a softer aroma, while modern cultivars possessed a much stronger aroma (Starr et al. 2013).

Bulgur is a whole grain product that is commonly produced from *Triticum durum*. However, bulgur in different cities or regions of Turkey is produced with specific landraces such as einkorn and emmer to achieve regionally desired end-use product characteristics. On average, 55.7% of wheat landrace production is used in bread-making such as lavash, 35.8% is used in bulgur-making, and 2.6% is in pasta-making such as erişte in Turkey. However, about half of wheat landrace production is used in bulgur-making in the Mediterranean and Southeastern Anatolian regions of Turkey (Kan et al. 2015).

Protein content, rather than gluten strength, is the main determinant of high temperature dried pasta cooking quality (Dexter and Matsuo 1977; D'Egidio et al. 1990). Some old durum wheat cultivars such as Senatore Cappelli possessed good pasta texture compared to modern durum wheat cultivars (Fois et al. 2011). Protein quality and quantity, as well as starch gel properties, are the most important factors responsible for the oil content of instant noodles. Instant noodles made from Iranian wheat landraces possessed intermediate oil contents, compared to commercial wheat flours and US/Canadian samples (Wu et al. 2006).

Low protein content and hardness are the crucial parameters for predicting biscuit-making quality. Beside these parameters, some particular storage proteins such as HMW-GS 13 + 16 glutenin subunits are particularly useful in evaluating biscuit quality. In a comprehensive study comparing 98 lines from a landrace population named Barbela, biscuit mass, cooking time, length, width, thickness, density, and surface appearance of lines varied between 10.9 and 14.3 g, 6.58 and 10.12 min, 6.4 and 6.9 cm, 4.6 and 6.0 cm, 9.2 and 13.0 mm, 0.13 and 0.38 g/cm^3, and 1.0 and 4.5, respectively (Igrejas et al. 2002).

6.10 Conclusions

Wheat is the second most important grain after maize in the world. In terms of wheat production, it is not appropriate to make comparisons of the modern wheat cultivars and the landraces with their negative aspects, making each other useless. Instead of negative approaches, it should be emphasized which one should be preferred according to the conditions. Under conditions where wheat has the potential to provide high grain yield, choosing landraces instead of modern wheat cultivars will be risky in terms of food security. On the other hand, landraces have the opportunity to be produced under low yield, low input, and high-stress conditions. The landraces are naturally suitable for organic agriculture and environmental friendly practices. The landraces also provide some opportunities for breeders since they generally represent considerably wider genetic diversity than modern cultivars. Therefore, they could lead to extending the genetic base of modern wheat cultivars. The landraces are also suitable for regions with high local demands. The use of landraces could represent a strategy for local communities in the production of traditional niche food products.

Wheat is an important source of health-promoting components, particularly phytochemicals and antioxidants as well as the main components of protein, carbohydrate, and lipid. There are no extreme differences between landraces and modern wheat cultivars in terms of various nutritional components. Most of the components which are beneficial for health are mainly concentrated in the germ and aleurone layers of the wheat kernel. Maximum benefit from these components of wheat could be possible by consuming it as whole grain products. For this reason, consuming wheat as whole grain instead of separating it as a landrace or modern wheat will be more beneficial for health.

From the technological point of view, limited studies were carried out to screen the end-product quality characteristics of wheat landraces. Therefore, further detailed studies are required for the determination of the end-product quality of wheat landraces.

References

Abdel-Aal ES, Hucl P, Sosulski FW (1995) Compositional and nutritional characteristics of spring einkorn and spelt wheats. Cereal Chem 72:621–624

Abdel-Aal ESM, Young JC, Rabalski I et al (2007) Identification and quantification of seed carotenoids in selected wheat species. J Agric Food Chem 55:787–794

Akar T, Mert Z, Yazar S et al (2009) Sustainable use of winter durum wheat landraces under Mediterranean conditions. Afr J Biotechnol 8:4108–4116

Akçura M (2011) The relationships of some traits in Turkish winter bread wheat landraces. Turk J Agric For 35:115–125

Akcura M, Kokten K (2017) Variations in grain mineral concentrations of Turkish wheat landraces germplasm. Qual Assur Saf Crop Foods 9:153–159

Alfeo V, Jaskula-Goiris B, Venora G et al (2018) Screening of durum wheat landraces (*Triticum turgidum* subsp. *durum*) for the malting suitability. J Cereal Sci 83:101–109

Andersson AAM, Kamal-Eldin A, Fras A et al (2008) Alkylresorcinols in wheat varieties in the HEALTHGRAIN diversity screen. J Agric Food Chem 56:9722–9725

Arzani A (2019) Emmer (*Triticum turgidum* ssp. *dicoccum*) flour and bread. In: Preedy V, Watson R (eds) Flour and breads and their fortification in health and disease prevention, 2nd edn. Elsevier, Amsterdam, pp 89–98

Arzani A, Ashraf A (2016) Smart engineering of genetic resources for enhanced salinity tolerance in crop plants. Crit Rev Plant Sci 35:146–189

Azeez MA, Adubi AO, Durodola FA (2018) Landraces and crop genetic improvement. In: Grillo O (ed) Rediscovery of landraces as a resource for the future. InTechOpen. https://doi.org/10.5772/intechopen.75944

Bao J, Shen Y, Jin L (2006) Determination of thermal and retrogradation properties of rice starch using near-infrared spectroscopy. J Cereal Sci 46:75–81

Bedõ Z, Láng L, Rakszegi M (2010) Breeding for grain-quality traits. In: Wrigley CW, Batey IL (eds) Cereal grains assessing and managing quality, 1st edn. Woodhead Publishing, Oxford, pp 369–392

Bellato S, Ciccoritti R, Del Frate V et al (2013) Influence of genotype and environment on the content of 5-n alkylresorcinols, total phenols and on the antiradical activity of whole durum wheat grains. J Cereal Sci 57:162–169

Bertolini AC (2009) Trends in starch applications. In: Bertolini AC (ed) Starches characterization, properties, and applications, 1st edn. CRC Press, Boca Raton, pp 1–19

Beta T, Li W, Apea-Bah FB (2019) Flour and bread from black, purple, and blue-colored wheats. In: Preedy V, Watson R (eds) Flour and breads and their fortification in health and disease prevention, 2nd edn. Elsevier, pp 75–88

Bhattacharya M, Jafari-Shabestari J, Qualset CO, Corke H (1997) Diversity of starch pasting properties in Iranian hexaploid wheat landraces. Cereal Chem 74:417–423

Biel W, Stankowski S, Jaroszewska A et al (2016) The influence of selected agronomic factors on the chemical composition of spelt wheat (*Triticum aestivum* ssp. *spelta* L.) grain. J Integr Agric 15:1763–1769

Black CK, Panozzo JF, Wright CL, Lim PC (2000) Survey of white salted noodle quality characteristics in wheat landraces. Cereal Chem 77:468–472

Blum A, Sinmena B, Golan G, Mayer J (1987) The grain quality of landraces of wheat as compared with modern cultivars. Plant Breed 99:226–233

Bonilla JC, Schaber JA, Bhunia AK, Kokini JL (2019) Mixing dynamics and molecular interactions of HMW glutenins, LMW glutenins, and gliadins analyzed by fluorescent co-localization and protein network quantification. J Cereal Sci 89:102792

Bordes J, Branlard G, Oury FX et al (2008) Agronomic characteristics, grain quality and flour rheology of 372 bread wheats in a worldwide core collection. J Cereal Sci 48:569–579

Bouis HE, Welch RM (2010) Biofortification: a sustainable agricultural strategy for reducing micronutrient malnutrition in the global south. Crop Sci 50:20–32

Boukid F, Vittadini E, Prandi B et al (2018) Insights into a century of breeding of durum wheat in Tunisia: the properties of flours and starches isolated from landraces, old and modern genotypes. LWT 97:743–751

Brandolini A, Hidalgo A, Moscaritoloa S (2008) Chemical composition and pasting properties of einkorn (*Triticum monococcum* L. subsp. *monococcum*) whole meal flour. J Cereal Sci 47:599–609

Brandolini A, Hidalgo A, Plizzari L et al (2011) Impact of genetic and environmental factors on einkorn wheat (*Triticum monococcum* L. subsp. *monococcum*) polysaccharides. J Cereal Sci 53:65–72

Brown KH, Wuehler SE, Peerson JM (2001) The importance of zinc in human nutrition and estimation of the global prevalence of zinc deficiency. Food Nutr Bull 22:113–125

Cakmak I (2008) Enrichment of cereal grains with zinc: agronomic or genetic biofortification. Plant Soil 302:1–17

Cakmak I, Torun A, Millet E et al (2004) Triticum dicoccoides: an important genetic resource for increasing zinc and iron concentration in modern cultivated wheat. Soil Sci Plant Nutr 50:1047–1054

Carson GR, Edwards NM (2009) Criteria of wheat and flour quality. In: Khan K, Shewry PR (eds) Wheat: chemistry and technology, 4th edn. American Association of Cereal Chemists, Minnesota, pp 97–118

Çetiner B, Tömösközi S, Schall E et al (2018) Comparison of some landraces and a modern wheat in terms of quality, functional properties and gluten/gliadin content. Paper presented at the LACC4 – 4th ICC Latin American Cereals Conference, Mexico City, Mexico, 11–14 March 2018

Cetiner B, Tömösközi S, Török K et al (2020) Comparison of the arabinoxylan composition and physical properties of old and modern bread wheat (*Triticum aestivum* L.) and landraces genotypes. Cereal Chem 97:505–514

Ciccoritti R, Carbone K, Bellato S et al (2013) Content and relative composition of some phytochemicals in diploid, tetraploid and hexaploid *Triticum* species with potential nutraceutical properties. J Cereal Sci 57:200–206

Copeland L, Blazek J, Salman H, Tang MC (2009) Form and functionality of starch. Food Hydrocoll 23:1527–1534

Corbellini M, Empilli S, Vaccino P et al (1999) Einkorn characterization for bread and cookie production in relation to protein subunit composition. Cereal Chem 76:727–733

Cozzolino D (2016) The use of the rapid visco analyser (RVA) in breeding and selection of cereals. J Cereal Sci 70:282–290

Cubadda R, Marconi E (1996) Technological and nutritional aspects in emmer and spelt. In: Padulosi S, Hammer K, Heller J (eds) Hulled wheat, proceedings of the first international workshop on hulled wheats, 21–22 July 1995, Pascoli, Tuscany, Italy, p 203–211

Cubadda F, Aureli F, Raggi A et al (2009) Effect of milling, pasta making and cooking on minerals in durum wheat. J Cereal Sci 49:92–97

Čurná V, Lacko-Bartošová M (2017) Chemical composition and nutritional value of emmer wheat (*Triticum dicoccon* Schrank): a review. J Cent Eur Agric 18:117–134

D'Egidio MG, Mariani BM, Nardi S et al (1990) Chemical and technological variables and their relationships: a predictive equation for pasta cooking quality. Cereal Chem 67:275–281

De Vita P, Nicosia OLD, Nigro F et al (2007) Breeding progress in morpho-physiological, agronomical and qualitative traits of durum wheat cultivars released in Italy during the 20th century. Eur J Agron 26:39–53

De Vita P, Platani C, Fragasso M et al (2017) Selenium-enriched durum wheat improves the nutritional profile of pasta without altering its organoleptic properties. Food Chem 214:374–382

De Wals P, Tairou F, Van Allen MI et al (2007) Reduction in neural-tube defects after folic acid fortification in Canada. N Engl J Med 357:135–142

Delcour JA, Hoseney RC (2010) Principles of cereal science and technology. American Association of Cereal Chemists, Minnesota

Desheva G, Valchinova E, Kyosev B et al (2014) Grain physical characteristics of and breadmaking quality of alternative wheat cereals towards common and durum wheat. Emirates J Food Agric 26:418–424

Dexter JE, Matsuo RR (1977) Influence of protein content on some durum wheat quality parameters. Can J Plant Sci 57:717–727

Di Silvestro R, Marotti I, Bosi S et al (2012) Health-promoting phytochemicals of Italian common wheat varieties grown under low-input agricultural management. J Sci Food Agric 92:2800–2810

Dick JW, Quick JS (1983) A modified screening test for rapid estimation of gluten strength in early-generation durum wheat breeding lines. Cereal Chem 60:315–318

Dietrich O, Heun M, Notroff J et al (2012) The role of cult and feasting in the emergence of Neolithic communities. New evidence from Göbekli Tepe, south-eastern Turkey. Antiquity 86:674–695

Digesù AM, Platani C, Cattivelli L et al (2009) Genetic variability in yellow pigment components in cultivated and wild tetraploid wheats. J Cereal Sci 50:210–218

Dinelli G, Segura Carretero A, Di Silvestro R et al (2009) Determination of phenolic compounds in modern and old varieties of durum wheat using liquid chromatography coupled with time-of-flight mass spectrometry. J Chromatogr A 1216:7229–7240

Dinelli G, Marotti I, Di Silvestro R et al (2013) Agronomic, nutritional and nutraceutical aspects of durum wheat (*Triticum durum* Desf.) cultivars under low input agricultural management. Ital J Agron 8:85–93

Dotlačil L, Hermuth J, Stehno Z et al (2010) How can wheat landraces contribute to present breeding? Czech J Genet Plant Breed 46:70–74

Emeksizoğlu B (2016) Kastamonu yöresinde yetiştirilen siyez (*Triticum monococcum* L.) buğdayının bazı kalite özellikleri ile bazlama ve erişte yapımında kullanımının araştırılması. Dissertation, University of Ondokuz Mayıs (in Türkish)
Fan MS, Zhao FJ, Fairweather-Tait SJ et al (2008) Evidence of decreasing mineral density in wheat grain over the last 160 years. J Trace Elem Med Biol 22:315–324
Ficco DBM, Riefolo C, Nicastro G et al (2009) Phytate and mineral elements concentration in a collection of Italian durum wheat cultivars. Field Crops Res 111:235–242
Finnie SM, Bettge AD, Morris CF (2006) Influence of cultivar and environment on water-soluble and water-insoluble arabinoxylans in soft wheat. Cereal Chem 83:617–623
Fois S, Schlichting L, Marchylo B et al (2011) Environmental conditions affect semolina quality in durum wheat (*Triticum turgidum* ssp. *durum* L.) cultivars with different gluten strength and gluten protein composition. J Sci Food Agric 91:2664–2673
Frank J, Chin XWD, Schrader C et al (2012) Do tocotrienols have potential as neuroprotective dietary factors? Ageing Res Rev 11:163–180
Gebruers K, Dornez E, Boros D et al (2008) Variation in the content of dietary fiber and components thereof in wheats in the HEALTHGRAIN diversity screen. J Agric Food Chem 56:9740–9749
Gebruers K, Dornez E, Bedõ Z et al (2010) Environment and genotype effects on the content of dietary fiber and its components in wheat in the HEALTHGRAIN diversity screen. J Agric Food Chem 58:9353–9361
Giambanelli E, Ferioli F, Koçaoglu B et al (2013) A comparative study of bioactive compounds in primitive wheat populations from Italy, Turkey, Georgia, Bulgaria and Armenia. J Sci Food Agric 93:3490–3501
Giambanelli E, Ferioli F, D'Antuono LF (2016) Assessing the effect of traditional hulled wheat processing on bioactive compounds retention. J Cereal Sci 72:60–68
Gibson RS (2006) Zinc: the missing link in combating micronutrient malnutrition in developing countries. Proc Nutr Soc 65:51–60
Giunta F, Pruneddu G, Motzo R (2019) Grain yield and grain protein of old and modern durum wheat cultivars grown under different cropping systems. Field Crops Res 230:107–120
Gupta RB, MacRitchie F (1994) Allelic variation at glutenin subunit and gliadin loci, glu-1, glu-3 and gli-1 of common wheats. II. Biochemical basis of the allelic effects on dough properties. J Cereal Sci 19:19–29
Gupta RB, MacRitchie F, Shepherd KW, Ellison F (1991) Relative contributions of LMW and HMW glutenin subunits to dough strength and dough stickiness of bread wheat. In: Bushuk W, Tkachuk R (eds) Proc 4th Int workshop gluten proteins. St. Paul, Minnesota, pp 71–80
Guzmán C, Medina-Larqué AS, Velu G et al (2014) Use of wheat genetic resources to develop biofortified wheat with enhanced grain zinc and iron concentrations and desirable processing quality. J Cereal Sci 60:617–622
Hare R (2017) Durum wheat: grain-quality characteristics and management of quality requirements. In: Wrigley C, Batey I, Miskelly D (eds) Cereal grains: assessing and managing quality, 2nd edn. Elsevier, Duxford, pp 135–152
Hawkesford MJ, Zhao FJ (2007) Strategies for increasing the selenium content of wheat. J Cereal Sci 46:282–292
Hazard B, Zhang X, Naemeh M, Dubcovsky J (2014) Registration of durum wheat germplasm lines with combined mutations in SBEIIa and SBEIIb genes conferring increased amylose and resistant starch. J Plant Regist 8:334–338
Heidari B, Padash S, Dadkhodaie A (2016) Variations in micronutrients, bread quality and agronomic traits of wheat landrace varieties and commercial cultivars. Aust J Crop Sci 10:377–384
Hernández-Espinosa N, Payne T, Huerta-Espino J et al (2019) Preliminary characterization for grain quality traits and high and low molecular weight glutenins subunits composition of durum wheat landraces from Iran and Mexico. J Cereal Sci 88:47–56
Hidalgo A, Brandolini A (2017) Nitrogen fertilization effects on technological parameters and carotenoid, tocol and phenolic acid content of einkorn (*Triticum monococcum* L. subsp. *monococcum*): a two-year evaluation. J Cereal Sci 73:18–24

Hidalgo A, Brandolini A, Pompei C et al (2006) Carotenoids and tocols of einkorn wheat (*Triticum monococcum* ssp. *monococcum* L.). J Cereal Sci 44:182–193

Hoseney RC, Finney KF, Shogren MD et al (1969) Functional (breadmaking) and biochemical properties of wheat flour components. III. Characterisation of gluten protein fractions obtained by ultracentrifugation. Cereal Chem 46:126–135

Hussain S, Maqsood MA, Rahmatullah M (2010) Increasing grain zinc and yield of wheat for the developing world: a review. Emir J Food Agric 22:326–339

Hussain S, Maqsood MA, Rengel Z, Khan MK (2012a) Mineral bioavailability in grains of Pakistani bread wheat declines from old to current cultivars. Euphytica 186:153–163

Hussain A, Larsson H, Olsson ME et al (2012b) Is organically produced wheat a source of tocopherols and tocotrienols for health food? Food Chem 132:1789–1795

Igrejas G, Guedes-Pinto H, Carnide V et al (2002) Genetical, biochemical and technological parameters associated with biscuit quality. II. Prediction using storage proteins and quality characteristics in a soft wheat population. J Cereal Sci 36:187–197

Kan M, Küçükçongar M, Keser M et al (2015) Wheat landraces in farmers' fields in Turkey: national survey, collection and conservation, 2009-2014. Food and Agriculture Organization of the United Nations (FAO), Ankara

Kantor LS, Variyam JN, Allshouse JE et al (2001) Choose a variety of grain daily, especially whole grains: a challenge for consumers. J Nutr 131:473–486

Kiszonas AM, Morris CF (2018) Wheat breeding for quality: a historical review. Cereal Chem 95:17–34

Kondou Y, Manickavelu A, Komatsu K et al (2016) Analysis of grain elements and identification of best genotypes for Fe and P in Afghan wheat landraces. Breed Sci 66:676–682

Konvalina P, Moudrý J (2008) Quality parameters of emmer wheat landraces. J Cent Eur Agric 9:539–546

Konvalina P, Grausgruber H, Dang KT et al (2017) Rheological and technological quality of minor wheat species and common wheat. In: Wanyera R, Owuoche J (eds) Wheat improvement, management and utilization, 1st edn. InTechOpen, Rijeka, pp 255–273

Kundu M, Khatkar BS, Gulia N (2017) Assessment of chapatti quality of wheat varieties based on physicochemical, rheological and sensory traits. Food Chem 226:95–101

Lachman J, Orsák M, Pivec V et al (2012) Antioxidant activity of grain of einkorn (*Triticum monococcum* L.), emmer (*Triticum dicoccum* Schuebl [Schrank]) and spring wheat (*Triticum aestivum* L.) varieties. Plant Soil Environ 58:15–21

Lachman J, Hejtmánková K, Kotíková Z (2013) Tocols and carotenoids of einkorn, emmer and spring wheat varieties: selection for breeding and production. J Cereal Sci 57:207–214

Lampi AM, Nurmi T, Ollilainen V et al (2008) Tocopherols and tocotrienols in wheat genotypes in the HEALTHGRAIN diversity screen. J Agric Food Chem 56:9716–9721

Landberg R, Kamal-Eldin A, Salmenkallio-Marttila M et al (2008) Localization of alkylresorcinols in wheat, rye and barley kernels. J Cereal Sci 48:401–406

Laroche C, Michaud P (2007) New developments and prospective for β (1, 3) glucans. Recent Pat Biotechnol 1:59–73

Li G, He Z, Lillemo M et al (2008) Molecular characterization of allelic variations at Pina and Pinb loci in Shandong wheat landraces, historical and current cultivars. J Cereal Sci 47:510–517

Li S, Morris CF, Bettge AD (2009) Genotype and environment variation for arabinoxylans in hard winter and spring wheats of the U.S. Pacific Northwest. Cereal Chem 86:88–95

Lia A, Anderson H, Mekki N et al (1997) Postprandial lipidemia in relation to sterol and fat excretion in ileostomy subjects given oat bran and wheat test meals. Am J Clin Nutr 66:357–365

Løje H, Møller B, Laustsen AM et al (2003) Chemical composition, functional propreties and sensory profiling of einkorn (*Triticum monococcum* L.). J Cereal Sci 37:231–240

Lott JNA, Spitzer E (1980) X-ray analysis studies of elements stored in protein body globid crystals of Triticum grains. Plant Physiol 66:494–499

Lyons G, Ortiz-Monasterio I, Stangoulis J, Graham R (2005) Selenium concentration in wheat grain: is there sufficient genotypic variation to use in breeding? Plant Soil 269:269–380

Manickavelu A, Hattori T, Yamaoka S et al (2017) Genetic nature of elemental contents in wheat grains and its genomic prediction: toward the effective use of wheat landraces from Afghanistan. PLoS One 12:0169416

Maningat CC, Seib PA, Bassi SD et al (2009) Wheat starch: production, properties, modification and uses. In: BeMiller J, Whistler R (eds) Starch: chemistry and technology, a volume in food science and technology, 3rd edn. Elsevier, New York, pp 441–510

Marchylo BA, Dexter JE (2001) Pasta production. In: Owens G (ed) Cereals processing technology, 1st edn. Woodhead Publishing, pp 109–130

Marconi M, Cubadda R (2005) Emmer wheat. In: Abdel-Aal ESM, Wood PJ (eds) Speciality grains for food and feed, 1st edn. American Association of Cereal Chemists, St. Paul, pp 63–108

Marconi E, Carcea M, Graziano M et al (1999) Kernel properties and pasta-making quality of five European spelt wheat (*Triticum spelta* L.) cultivars. Cereal Chem 76:25–29

Marotti I, Bregola V, Aloisio I et al (2012) Prebiotic effect of soluble fibres from modern and old durum-type wheat varieties on Lactobacillus and Bifidobacterium strains. J Sci Food Agric 92:2133–2140

McKevith B (2004) Nutritional aspects of cereals. Nutr Bull 29:111–142

Mefleh M, Conte P, Fadda C et al (2019) From seed to bread: variation in quality in a set of old durum wheat cultivars. J Sci Food Agric. https://doi.org/10.1002/jsfa.9745

Melash AA, Mengistu DK, Aberra DA, Tsegay A (2019) The influence of seeding rate and micronutrients foliar application on grain yield and quality traits and micronutrients of durum wheat. J Cereal Sci 85:221–227

Mendis M, Simsek S (2014) Arabinoxylans and human health. Food Hydrocoll 42:239–243

Meuser F, Suckow P (1986) Nonstarch polysaccharides. In: Blanshard JMV, Frazier PJ, Galliard T (eds) Chemistry and physics of baking materials, processes, and product, 1st edn. The Royal Society of Chemistry, London, pp 42–61

Migliorini P, Spagnolo S, Torri L et al (2016) Agronomic and quality characteristics of old, modern and mixture wheat varieties and landraces for organic bread chain in diverse environments of northern Italy. Eur J Agron 79:131–141

Miskelly D, Batey IL, Suter DAI (2010) Processing wheat to optimize product quality. In: Wrigley CW, Batey IL (eds) Cereal grains assessing and managing quality, 1st edn. Woodhead Publishing, Duxford, pp 431–457

Morris CF, Bhave M (2008) Reconciliation of D-genome puroindoline allele designations with current DNA sequence data. J Cereal Sci 48:277–287

Motzo R, Fois S, Giunta F (2004) Relationship between grain yield and quality of durum wheats from different eras of breeding. Euphytica 140:147–154

Mujeeb-Kazi A, Kazi AG, Dundas I et al (2013) Genetic diversity for wheat improvement as a conduit to food security. In: Sparks DL (ed) Advances in agronomy. Elsevier, Burlington, pp 179–258

Nachit MM, Baum M, Impiglia A (1995) Studies on some grain quality traits in durum wheat grown in Mediterranean environments. CIHEAM, Options Méditerranéennes: Série A Séminaires Méditerranéens 2:181–187

Nazco R, Villegas D, Ammar K et al (2012) Can Mediterranean durum wheat landraces contribute to improved grain quality attributes in modern cultivars? Euphytica 185:1–17

Nurmi T, Nyström L, Edelmann M et al (2008) Phytosterols in wheat genotypes in the HEALTHGRAIN diversity screen. J Agric Food Chem 56:9710–9715

Okarter N, Liu CS, Sorrells ME et al (2010) Phytochemical content and antioxidant activity of six diverse varieties of whole wheat. Food Chem 119:249–257

Ottman MJ, Doerge TA, Martin EC (2000) Durum grain quality as affected by nitrogen fertilization near anthesis and irrigation during grain fill. Agron J 92:1035–1041

Ozkan H, Brandolini A, Torun A et al (2007) Natural variation and identification of microelements content in seeds of einkorn wheat (*Triticum monococcum*). In: Buck HT, Nisi JE, Salomón N (eds) Wheat production in stressed environments, 1st edn. Springer, Dordrecht, pp 455–462

Packer L (1995) Nutrition and biochemistry of the lipophilic antioxidants vitamin E and carotenoids. In: Ong ASH, Niki E, Packer L (eds) Nutrition, lipids, health, and disease, 1st edn. AOCS Press, Champaign, pp 8–35

Pasha I, Anjum FM, Morris CF (2010) Grain hardness: a major determinant of wheat quality. Food Sci Technol Int 16:511–522

Pechanek U, Karger A, Gröger S et al (1997) Effect of nitrogen fertilization on quantity of flour protein components, dough properties, and breadmaking quality of wheat. Cereal Chem 74:800–805

Piergiovanni AR (2013) Evaluation of genetic variation and grain quality of old bread wheat varieties introduced in north-western Italian environments. Genet Resour Crop Evol 60:325–333

Piironen V, Lindsay DG, Miettinen TA et al (2000) Plant sterols: biosynthesis, biological function and their importance to human nutrition. J Sci Food Agric 80:939–966

Piironen V, Edelmann M, Kariluoto S et al (2008) Folate in wheat genotypes in the HEALTHGRAIN diversity screen. J Agric Food Chem 56:9726–9731

Raciti CN, Doust MA, Lombardo GM et al (2003) Characterization of durum wheat Mediterranean germplasm for high and low molecular weight glutenin subunits in relation with quality. Eur J Agron 19:373–382

Rakszegi M, Batey IL, Vida GY et al (2003) Starch properties in different lines of an old Hungarian wheat variety, Bánkúti 1201. Starch 55:397–402

Rao AV, Janezic SA (1992) The role of dietary phytosterols in colon carcinogenesis. Nutr Cancer 18:43–52

Rizzello CG, Cavoski I, Turk J et al (2015) Organic cultivation of *Triticum turgidum* subsp. *durum* is reflected in the flour-sourdough fermentation-bread axis. Appl Environ Microbiol 81:3192–3204

Sanchez-Garcia M, Álvaro F, Peremarti A et al (2015) Changes in bread-making quality attributes of bread wheat varieties cultivated in Spain during the 20th century. Eur J Agron 63:79–88

Schmidt K (2007) Göbekli Tepe en eski tapınağı yapanlar/Taş çağı avcılarının gizemli anıtsal alanı (trans: Aslan R). Arkeoloji ve Sanat Yayınları, İstanbul

Schneider C (2005) Chemistry and biology of vitamin E. Mol Nutr Food Res 49:7–30

Scott J, Rébeillé F, Fletcher J (2000) Folic acid and folates: the feasibility for nutritional enhancement in plant foods. J Sci Food Agric 80:795–824

Serpen A, Gökmen V, Karagöz A et al (2008) Phytochemical quantification and total antioxidant capacities of emmer (*Triticum dicoccon* Schrank) and einkorn (*Triticum monococcum* L.) wheat landraces. J Agric Food Chem 56:7285–7292

Sezer B, Morgounov A, Ozdemir F et al (2019) Chemical composition of wheat landraces in Turkey. Paper presented at the international conference on wheat diversity & human health, İstanbul, 22–24 October 2019

Shewry RP, Hey S (2015) Do "ancient" wheat species differ from modern bread wheat in their contents of bioactive components? J Cereal Sci 65:236–243

Shewry PR, D'ovidio R, Lafiandra D et al (2009) Wheat grain proteins. In: Khan K, Shewry PR (eds) Wheat: chemistry and technology, 4th edn. American Association of Cereal Chemists, Minnesota, pp 223–298

Shewry PR, Piironen V, Lampi AM et al (2010) The HEALTHGRAIN wheat diversity screen: effects of genotype and environment on phytochemicals and dietary fiber components. J Agric Food Chem 58:9291–9298

Shewry PR, Hawkesford MJ, Piironen V et al (2013) Natural variation in grain composition of wheat and related cereals. J Agric Food Chem 61:8295–8303

Simsek S, Whitney K, Ohm JB, Mergoum M (2011) Refrigerated dough quality of hard red spring wheat: effect of genotype and environment on dough syruping and arabinoxylan production. Cereal Chem 88:445–450

Sofi BA, Wani IA, Masoodi FA et al (2013) Effect of gamma irradiation on physicochemical properties of broad bean (*Vicia faba* L.) starch. LWT 54:63–72

Starr G, Bredie WLP, Hansen AS (2013) Sensory profiles of cooked grains from wheat species and varieties. J Cereal Sci 57:295–303
Starr G, Petersen MA, Jespersen BM, Hansen AS (2015) Variation of volatile compounds among wheat varieties and landraces. Food Chem 174:527–537
Stehno Z, Paulickova I, Bradová J et al (2011) Evaluation of emmer wheat genetics resources aimed at dietary food production. J Life Sci 5:207–212
Swan G (2004) Findings from the latest national diet and nutrition survey. Proc Nutr Soc 63:505–512
Tiwari U, Cummins E (2009) Nutritional importance and effect of processing on tocols in cereals. Trends Food Sci Technol 20:511–520
Ülker M, Altuner F, Oral E et al (2019) Tir as a landrace and a sowing method. Paper presented at the international conference on wheat diversity & human health, İstanbul, 22–24 October 2019
Uthayakumaran S, Wrigley CW (2010) Wheat: characteristics and quality requirements. In: Wrigley CW, Batey IL (eds) Cereal grains assessing and managing quality, 1st edn. Woodhead Publishing, Duxford, pp 59–111
Vázquez D, Berger AG, Cuniberti M et al (2012) Influence of cultivar and environment on quality of Latin American wheats. J Cereal Sci 56:196–203
Velu G, Ortiz-Monasterio I, Cakmak I et al (2014) Biofortification strategies to increase grain zinc and iron concentrations in wheat. J Cereal Sci 59:365–372
Velu G, Crespo Herrera L, Guzman C et al (2019) Assessing genetic diversity to breed competitive biofortified wheat with enhanced grain Zn and Fe concentrations. Front Plant Sci. https://doi.org/10.3389/fpls.2018.01971
Wang ZH, Li SX, Malhi S (2008) Effects of fertilization and other agronomic measures on nutritional quality of crops. J Sci Food Agric 88:7–23
Wang S, Li C, Copeland L et al (2015) Starch retrogradation: a comprehensive review. Compr Rev Food Sci Food Saf 14:568–585
Waters BM, Uauy C, Dubcovsky J, Grusak MA (2009) Wheat (*Triticum aestivum*) NAM proteins regulate the translocation of iron, zinc, and nitrogen compounds from vegetative tissues to grain. J Exp Bot 60:4263–4274
Weaver CM, Kannan S (2002) Phytate and mineral bioavailability. In: Reddy NR, Sathe SK (eds) Food phytates, 1st edn. CRC Press, Boca Raton, pp 211–223
Welch RM, Graham RD (2004) Breeding for micronutrients in staple food crops from a human nutrition perspective. J Exp Bot 55:353–364
White PJ, Broadley MR (2009) Biofortification of crops with seven mineral elements often lacking in human diets-iron, zinc, copper, calcium, magnesium, selenium and iodine. New Phytol 182:49–84
Wieser H, Seilmeier W (1998) The influence of nitrogen fertilization on quantities and proportions of different protein types in wheat flour. J Sci Food Agric 76:49–55
Wrigley CW (2010) Cereal-grain morphology and composition. In: Wrigley CW, Batey IL (eds) Cereal grains assessing and managing quality, 1st edn. Woodhead Publishing, Duxford, pp 24–44
Wu J, Aluko RE, Corke H (2006) Partial least-squares regression study of the effects of wheat flour composition, protein and starch quality characteristics on oil content of steamed-and-fried instant noodles. J Cereal Sci 44:117–126
Yang X, Wu L, Zhu Z et al (2014) Variation and trends in dough rheological properties and flour quality in 330 Chinese wheat varieties. Crop J 2:195–200
Yilmaz VA, Brandolini A, Hidalgo A (2015) Phenolic acids and antioxidant activity of wild, feral and domesticated diploid wheats. J Cereal Sci 64:168–175
Zeng M, Morris CF, Batey IL, Wrigley CW (1997) Sources of variation for starch gelatinization, pasting, and gelation properties in wheat. Cereal Chem 74:63–71
Ziegler JU, Schweiggert RM, Würschum T et al (2016) Lipophilic antioxidants in wheat (*Triticum* spp.): a target for breeding new varieties for future functional cereal products. J Funct Foods 20:594–605

Chapter 7
Total Economic Value of Wheat Landraces

Sevinç Karabak and Mustafa Kan

7.1 Introduction

Ecosystem, biodiversity, food and nutrition, economy, development, and the most important part of this integrity human are the basic rings of an inseparable chain. Each ring is an indispensable guarantee of the sustainability of another. Therefore, for the continuation of life, human beings have to make maximum use of biological diversity in a sustainable manner. Wheat, which has wider biological diversity, is the main food source of many societies. Wheat has not only a nutritional and economic value but also a cultural and social value. Approximately 10,000–11,000 years ago, it was cultivated in the Fertile Crescent and played a leading role in the transition to settled life. Thus, wheat has been seen as a product with a sacred value in all areas of human life such as history, culture, health, food, and economy from the depths of history to the present day. It is a strategic product that is seen as a source of power and guarantee of the future for many countries in the world. It can be said that food safety, which is a part of food security, gains importance on the basis of healthy and balanced nutrition in developing countries as well as in developed countries. People tend to benefit more from biodiversity and tend to local and natural products. This situation has begun to add increasing value to wild forms and local landraces. Wheat landraces are also an important source of gene and one of the rural mainstays. Stuck in local areas, identified with that region, and some are lost, the rest are at risk of

S. Karabak (✉)
The Central Research Institute for Field Crops (CRIFIC) Gayret Mah. Biyoteknoloji Araştırma Merkezi Gayret, Şehit Cem Ersever Cd. No: 220, Yenimahalle, Ankara, Turkey
e-mail: sevinc.karabak@tarimorman.gov.tr

M. Kan
Kirsehir Ahi Evran University, Faculty of Agriculture, Department of Agricultural Economics Bağbaşı Mah. Sahir Kurutluoğlu Cad. No: 100 Merkez, Kırşehir, Turkey
e-mail: mustafa.kan@ahievran.edu.tr

N. Zencirci et al. (eds.), *Wheat Landraces*,
https://doi.org/10.1007/978-3-030-77388-5_7

disappearance. For the protection and sustainability of these populations, the benefit it provides to people must be made visible and maintained at an increasing level. From all these aspects, it is better understood the necessity of revealing its relationship with people and society in order to be able to recognize wheat landraces and provide sustainable maximum benefit. In order to examine a product, it is necessary to know very well the total economic value it provides to people in its journey from its historical past to today and the sociocultural value that brings the phenomenon of social objects to wheat landraces. In the process from the depths of history to the present, the journey that human and wheat landraces have carried out together with a social, cultural, and historical connection is very difficult to measure financially, although it is an economic value.

Landraces are produced, named, and preserved by manufacturers with traditional understanding to meet their social, economic, cultural, and environmental needs. In addition, they are referred to as landraces, farmer varieties, or people's village varieties (Jaradat 2011) in order to highlight the innovative role of these producer communities in their development and livelihoods. Due to the fact that the producers of wheat landraces are mostly subsistence family businesses living in rural areas, they have played an important but invisible role in rural development. They are a potential source of income in terms of both ecotourism, gastronomic tourism, and the evaluation of marginal areas. Local products have become an important tool for rural development, especially with the increasing popularity of the geographical indication system in recent years. The fact that the developed modern varieties (improved varieties) are on the market and the producers prefer higher productive products and landraces caused the local landraces to get stuck in local areas where they are compatible with the cultural structure. Rural development strategies must be determined by considering the socioeconomic and sociocultural structures of the producers living in these regions and producing local landraces and the social bond they establish with local landraces, because the policies to be carried out without knowing the human resources that carry out the activity and knowing the values will not be successful. Many socioeconomic, ethnobotanical, and archeobotanical researches have been carried out on wheat and wheat landraces. Combining these studies with the economic botanical concept in order to make an economic evaluation will guide us in determining the importance of these species and sustaining their production. There are different approaches and classifications on goods and services, which are the components of the "total economic value" provided by the species in the ecosystem. However, each class is known as the research subject of a different discipline. Although the economy gives the impression of a mathematical appearance, it is still a social science. Therefore, it requires a multidisciplinary understanding. There are very fine lines between the total economic value components, and they need to be analyzed with a multidisciplinary integrated approach that takes into account the differences between the components. Analysis of biodiversity and local landraces that have an important place in biodiversity concerns environmental and social sciences. Defining goods and services related to landraces requires examining ecosystem dynamics, human factor, economic activities, and the relationships between them. In the study, considering the relationship between wheat and society, socioeconomic, ethnobotanical, and archeobotanical studies were examined, and the economic, social, cultural, and historical values of wheat landraces were discussed. All these values are indicators of the total economic value of wheat landraces. In terms of this value, the importance of the use of landraces as a rural development tool was emphasized. In this study, the direct use value, which is one of the total economic value components, was examined in detail.

7.2 Total Economic Value Approach for Valuation of Wheat Landraces

The concept of value in terms of economy is used as a measure to determine the necessity, importance, and value of something (Genç Yılmaz and Çağlayan 2017) and is one of the basic concepts of microeconomic analysis (Aydın and Aydınlar 2011). There are three value theories in economic discipline. It is the theory of labor-value, benefit-value, and cost of production. The labor-value theory argues that the value of an object depends only on the amount of labor spent on its production. The utility-value theory takes into account the marginal utility of the object. Production cost theory explains the value by considering the production factors used in the production of the product. Some economists use the value of market-oriented change. These theories often fail to explain the value of biodiversity. Alternative suggestions are given on the methods used in many of the conducted researches. The most important reason for this is the difficulties experienced in evaluating many services in monetary terms and the search for solutions. One of these difficulties in monetary value estimates is that current economic valuation procedures and methods are insufficient in measuring passive or unknown values (Nunes and van den Bergh 2001; Gowdy 1998; Gowdy and Salman 2010). Another is that the cost of loss cannot be adequately measured, and in addition there are benefits that are not noticed or yet unknown. A number of new methods have been developed to determine the monetary value of biodiversity, and this is mostly done at the cost of loss. It is also used frequently today. Monetary economic estimates provide us only a perspective. It may not reflect the true value of the product. Therefore, it is necessary to accept that it is much higher than the estimated value. It is possible to explain the production value, exchange value, or preference value of wheat landraces in monetary terms. However, it is difficult to determine the total economic value and express it in monetary terms, because wheat landraces have many social values such as relative, objective, abstract, concrete, religious, philosophical, historical, traditional, environmental belief. Social values: the object itself, which is a value, should be considered as the object's capacity to meet social needs, and the appreciation of people for its ability to satisfy (Fichter 1990). Although it is not possible to express all of these with money, it is not accepted as a correct approach. To what extent life-supporting functions of biodiversity can be subjected to economic valuation is an important ethical issue, but it is argued that it may not be appropriate to subject spiritual values to economic valuation. It is more important to create a little idea and awareness by examining the quantitative and physical indicators of these values (Çelik 2010). Knowing the components that make up the economic value of all limited resources is also important for sustainable development (Karabak 2017). Wheat landraces are not just commodities that are subject to an economic activity. Although they are considered abstract, they are a social object and part of the ecosystem with their existence. In order to study economically, it is necessary to know the benefits it has achieved since the first day it was used for economic purposes and today. This brings together multiple research disciplines and interactions. While some usage value that constitutes the total economic value of wheat is common, some differences are seen when it is divided into local and commercial landraces. However, it is not always possible to explain these differences with very clear separations. It is possible to divide the total economic value into two as direct use and indirect use value (De Groot et al. 2002). These values are tried to be explained with examples below.

7.2.1 Direct Use Value (Directly Consumed or Available on the Market)

The direct use value is based on the fact that the goods are placed directly in the markets in line with supply and demand for a certain value. Benefits arising from the marketed goods, medicinal drugs that are produced or effective in their production, their contribution to the historical and cultural process, use as genetic resources, food and raw materials, production as agricultural products, benefits for tourism, and use for activities such as recreation can be shown as example for direct use value (Atasoy et al. 2014). This value of landraces should be examined under two subtitles "Production Function Value" and "Information Function Value."

7.2.1.1 Production Function Value

Commercial use value Although wheat landraces were widely cultivated in ancient times, some of these local landraces continue to be traditionally grown in marginal areas, generally mountain villages, rugged and barren lands, and small areas in Europe and Asia (Tan 2018; Cooper 2015; Kan et al. 2017). Since the production amounts are low, consumption primarily at home needs is aimed. After the producers meet their own needs as processed products or seeds, they trade the remaining amount. The producers use the seed they put aside and keep from the previous harvested crop. Seed change is mostly done by exchanging with other producers around. The sale of seeds started with the increase of market value of these landraces as processed products. Among the wheat landraces, emmer and einkorn, which are hulled wheat, are the most subject to trade. They are offered for sale after making bread, bulgur, flour, and noodles, processed in general. However, the amount sold is quite low commercially. In studies conducted in Turkey, it has been determined that the wheat landraces producers still continue to produce them because they are suitable for their own taste, they are used in the production of local products, and also the straw and chaff of them are high and they are suitable for animal feed. The wheat landrace producers generally do not seek a market to sell them because they are producing generally for their home consumption (Özdemir et al. 2018; Yaman et al. 2019).

The increase in the interest in local landraces in recent years has created a demand for wheat landraces, but it is not possible to meet this demand in the current situation due to the limited production. The most important disadvantage of wheat landraces is that they are insufficient to meet the market demand compared to modern wheat varieties that are more subject to trade (Karabak et al. 2019). This also triggers the rise in product prices. The increase in demand and the fact that they started to get more shares in the trade causes increase in the number of actors and product prices in the value chain. The increase in product prices will have an advantage in terms of actors such as producers, processors, wholesalers/traders, retailers, and big market operators in the value chain of wheat landraces and will also have negative consequences for the consumer. The fact that there is more subject to trade may affect the amount of consumption even if it increases the usage value, but it may cause the low-income producer to be unable to purchase the product. Wheat landraces find more value than other commercial modern varieties processed in the market. Therefore, it is also an important source of income for the actors who trade in production and processed products and operate in the countryside. Considering the supply and demand situations, it can be said that their commercial use is lower compared to other modern commercial varieties.

Use value in human and animal nutrition as a food source Cereals have high levels of antioxidants (flavonoids, phenolic acid, phytic acid, tocopherols, and carotenoid) and nutritional fibers (Mpofu et al. 2006; Serpen et al. 2008; Zengin 2015). An important part of 4.5 billion people's protein needs and 20 percent of their energy needs in 94 developing countries in the world are met by wheat products. The main foodstuff of approximately 40 countries, which constitute 35% of the world's population, is wheat (Şanal 2018).).

Although the purpose of use of wheat landraces varies depending on the cultural differences, they are used in both human and animal nutrition. Businesses that produce these landraces prefer it as a cheap source of feed in the livestock feeding. The long stems of emmer and einkorn are also one of the reasons they are used as animal feed. Grains are also used in the feeding of poultry. The increased interest in organic products in human nutrition, functional food searches, and health concerns have increased the tendency to wheat landraces, which is rich in vitamins, minerals, and dietary fiber. Understanding the health benefits of high nutritional products has increased the demand for consumption of products made from whole grain and whole-wheat flour. The source of increased demand for wheat landraces is thought to be findings that they have high protein, low allergic properties, and high antioxidant content (Şanal 2018). Many studies show that wheat landraces have higher protein content than other modern wheat varieties under the same growing technique conditions (Konvalina et al. 2013).

Considering the increasing demand for today's conditions and natural products, it has the feature of being a healthy and economical alternative in daily nutrition with its nourishment and naturalness (Yaman and Zencirci 2018).

Raw material value Wheat landraces are widely used in traditional bread making. At the same time, while bulgur, biscuits, flour, noodles, and beer are made from local landraces, their stems are also used for making reeds as baskets, hats, roofing materials, filling mattresses, mattresses, saddles, and harnesses (Peña-Chocarro et al. 2009a; Cooper 2015; Mann 2018; Karabak et al. 2019). Hulled wheats (Emmer, eincorn and spelt) can be named differently in different countries. Just as eincorn and emmer wheat are called "siyez" in Turkey. In Italy, it is known as "farro" (Karagöz 1996; Ertuğ 2004; Giuliani et al. 2009; Troccoli and Codianni 2005).

Today, bread making still continues in Switzerland, Italy, and Turkey. The bread called "pane di farro" in Italy is produced in some bakeries in local regions. The Siyez wheat in Turkey is used for making bulgur. Bulgur and flour are made by forging in traditional stone mills. It is planted especially in Kastamonu province in Turkey, İhsangazi district. There is also a bulgur factory in Seydiler district in Kastamonu. Different products such as flour and noodles are also produced in the factory (Karabak et al. 2019). Emmer is the raw material of Emmerbier at the Riedenburger ecological brewery in Bavaria, Germany (Cooper 2015).

Wheat landraces are considered as raw materials not only in food but also in different areas other than food in low-income areas. It was used as a cushion in Morocco (Peña-Chocarro et al. 2009b). It was also used in the construction of the roof of the houses in Morocco. It has been determined that the reason for the continuation of einkorn production in limited areas in the Chefchaouen region is still due to the low-income families continuing this practice in this region. It is preferred for roof construction because einkorn stalks are long, hard, and resistant to rain, and thatched roofs are hot in winter, cool in summer, and easily separable in case of fire. It has been documented that its stem and straw are also used to fill mattresses, cushions, saddles, and harnesses (Mann 2018). It was noted that einkorn was used in Hungary to make hats and tie vine branches and cornstalks (Gunda 1983) and also used in vine gardens in Romania.

Emmer is used to make "panchón" specific to the Aller region (Asturias) in Spain. Panchón is a pastry that is cooked slowly on the stove and consumed with milk and sugar (Peña-Chocarro and Zapata 1998; Borza 1945). Emmer flour is also used to produce a kind of porridge, pancakes, and pancake-like pastries that are traditionally consumed with milk (Peña-Chocarro 1996).

The disadvantage of other wheat landraces is that they are winnowed from their husks in the use of wheat landraces, which have mostly a husky structure. When wet, it easily separates from its husks. However, performing this process in modern facilities may reduce the use of labor.

Use value in traditional medicine The oldest information about the medicinal uses of plants in history comes from the history of China, Egypt, and Greece. Use value in traditional medicine: It is known that some drugs were produced and exported in Anatolia during the Hittites period (Sarı et al. 2006). People who live in rural areas and have difficulties in accessing modern medicine use the landraces that exist in their environment in the treatment of simple diseases. This tradition, which has been going on for centuries, continues in many countries of the world. Hittite tablets show that wheat was used for the treatment of diseases, medicine and magic. Ertuğ (2014) stated that in a pharmacological text, among the 33 plants that are understood to be prepared as porridge in a container with many types of bread and wine with red wheat as well. Today, no specific ethnobotanical study for the use of wheat landraces in traditional medicine has been encountered. In some publications, it was indicated that they were used for the treatment of a number of diseases, e.g., sore throat, boils, hemorrhoids, inflammations, digestive system diseases, ulcers, diarrhea, hoarseness, itching, hardness, callus, and cough. They are also used as wheat grain, stalk, wheat oil, and wheatgrass (Özdemir and Alpınar 2015; Korkmaz and Karakurt 2015).

Medical resource value Whole-wheat products are very rich in many important components such as dietary fiber, starch, fat, minerals, vitamins, and phytochemicals (Johnson and Gary 2003; Sidhu et al. 2007; Mpofu et al. 2006; Serpen et al. 2008; Lachman et al. 2011). It is known that cereals have cholesterol-lowering effect, reduce the effect caused by obesity because of their low glycemic index and satiety, and significantly reduce the effect caused by some types of cancer, cardiovascular diseases, diabetes, Alzheimer's disease, and cataracts when consumed together with fruits and vegetables (Slavin 2004; Mpofu et al. 2006; Serpen et al. 2008; Brouns et al. 2013; Yaman and Zencirci 2018). Natural antioxidants replace the radicals in oxidation reactions and prevent chain oxidation reactions that may occur by being oxidized themselves (Velioglu et al. 1998; Sidhu et al. 2007). It is seen that the wheat landraces have more nutritional value as a medical resource because they have richer components than other modern varieties. In a study conducted in Catalonia region of Spain, it was mentioned that the bread wheat is used as an antiseptic and infection reliever in animals (Bonet and Vallès 2007).

Use value in biotechnology Plant biotechnology enables further evaluation of genetic resources with molecular studies.

- Freezing and storing DNA in laboratory conditions in order to protect genetic resources and maintain their continuity.
- Molecular-assisted studies in addition to classical methods to identify genetic resources through modernized biotechnological methods (molecular marker-assisted selection (MAS), use of marker techniques, tissue culture studies, gene cloning and gene transfer techniques, and provision of transgenic lines).
- Creation of clone lines and their cultivation and development in order to increase the genetic resources.

- It can be expressed as developing the local and wild genetic resources with new in vitro studies as well as classical breeding methods due to their high allelic richness (İlhan 2017).

Using plant gene resources and gene pools is of critical importance for sustaining genetic diversity in plants for generations. For this purpose, they should be identified and put under protection in their geographies. Wild, local, and modern forms that are cultivated in terms of various features of the plant genetic resources developed to adapt to different climatic conditions are known as very valuable gene resources in order to sustain the rich genetic heritage (Jarvis and Hodgkin 1998; Maxted et al. 2000). Therefore, it is very important to preserve the gene structures of these genetic resources by protecting them through various new methods.

New clones and generations are created by using many new biotechnological methods such as tissue and cell culture, gene transfer, molecular marker-assisted selection (MAS), quantitative character locus (QTL) mapping, etc. in increasing yield and quality, obtaining lines that are resistant to various diseases or environmental pathogens, and providing the desired characters (İlhan 2017).

The local landraces have, until recently, been a dynamic and essential component of general agricultural biological landraces, which can be used almost exclusively in scientific breeding programs, and are considered to be the source of features that can increase the productivity of new crops (Jaradat 2013). It is mentioned that the genetic diversity has reduced in culture landraces of wheat, and this is mainly the result of selection processes in modern breeding programs (Cavanagh et al. 2013; Demir 2015). Although some of the desired features of selection processes come to the fore, it is thought that wheat, which is the focus of these breeding studies, increases its susceptibility to new diseases, pests, and adverse environmental conditions (Karcıcıo 2006; Demir 2015). Also, it is not possible to determine biotic and abiotic stress conditions by classical breeding methods since resistance to them is controlled by more than one gene. Therefore, use of the modern biotechnological methods in breeding comes to the fore. Among these, genetic engineering is a method with the highest level of hope and debate. However, genetic engineering techniques will only provide great convenience to plant breeders for the development of high-quality and productive new landraces resistant to environmental conditions, such as diseases and pests, drought, and salinity, with improved plant nutrient contents when used in conjunction with other molecular breeding methods (Çetiner 2005).

Local landraces are arguably seen as a key component in sustaining a safer food supply as they can increase productivity and make a much larger contribution in agricultural biodiversity (Jaradat 2013).

Determination of genetic diversity and population genetic structures in plants is very important for their more efficient use. The molecular markers are widely used for both breeding studies and identification and preservation of endangered species. However, plants are affected by environmental conditions and can show variations in both hereditary and non-hereditary forms even if molecular markers are not affected by environmental conditions. Therefore, knowing how much of the genetic diversity shown by populations is in the genotype of the plant and how much due to environmental factors will be effective in conducting correct studies (Demir 2015).

Genetic resources value The genetic resources of plants can be classified as wild species, close relative species, local landraces and improved materials, and commercial or modern varieties (improved varieties). Local landraces have a broad genetic basis (Özberk et al. 2016) and are locally acquired populations that people have bred themselves by natural selection before the development of modern varieties (improved varieties). Local landraces, which are seen as an important source of gene, are only present in some farms and gene banks in limited areas. The characteristic features that clearly reveal the importance of local landraces as genetic resources are that they have characteristics of different geographies, reflect the characteristics of different ecological environments, have a high level of allelic richness, enable development of new landraces with agricultural activities, and most importantly, preserve the rich genetic features that they have since the past (İlhan 2017).

They play a key role in combating major challenges such as use of plant genetic resources in agriculture, food safety, climate change, limited use of water, and periodic long-term drought, salinity, and desertification (Tan 2010).

In the evolutionary development process, many local landraces, which have different genetic structures and are especially suitable for arid and semiarid conditions, have emerged. The local landraces that continued to be produced until the 1950s were replaced by the cultivated landraces developed by breeding. As of the last century, 75% of genetic diversity, including the Fertile Crescent, has been lost (Harlan 1975; Jaradat 1992; Brown 2000; Jaradat 2013; Özberk 2018).

Genes that show allelic variation against the global warming threat can be transferred from local landraces back to modern varieties. In breeding, wheat landraces have the opportunity to be used more in development of drought resistance, tolerance to diseases, adaptation to low input environments, and cultivation of modern varieties (improved varieties) through hybridization (Srivastava and Damania 1989; Kyzeridis et al. 1995; Zaharieva et al. 2010; Talas et al. 2011; Özberk 2018).

7.2.1.2 Information Function Value (Values Without Direct Consumption)

Scientific and educational use value Wheat is a subject of very wide and diverse disciplines. It is extensively used in natural, social, and applied sciences both in scientific research and educational applications. We can see that there is a subject of scientific work in biology, physics, chemistry, mathematics, statistics, medicine, sociology, history, archeology, economics, geography, psychology, and anthropology, and even political sciences and sub-disciplines in these disciplines. In this respect, it can be said that wheat has a common scientific value without classifying it as wild, local landraces, and modern varieties. However, wheat landraces are the subject of study for more branches than the modern varieties. They are evaluated in a much wider range for historical, social, and biodiversity.

Cultural and historical value Sociocultural relations can be defined as beliefs, ceremonies, traditions, habits, and rules, which are formed by the interaction of people in a community, and the relations that arise with the effect of this structure on people's life. We can see the effects of traditional sociocultural values in our personal values and behavior, because the values that we learn, internalize and give various meanings, affect the behaviors of people.

People determine, produce, and meet their needs according to their economic status in line with their psychological, social, traditional, and cultural thoughts. When analyzing the history of local landraces, we seek answers to questions such as "what stages has it passed from past to present," "what are its social and economic contributions," "what is its cultural value and how can it be sustained in the future," and "can it be used as a tool for rural development?".

Wheat landraces constitute the history of agriculture and have a very old history. They appear as traditions in the areas where they are grown, and these traditions characterize the rural life in areas where wheat landraces are grown.

In order to understand the cultural value of wheat landraces, it would be a correct approach to firstly examine its historical past. Archaeological, archeobotanical, ethnobotanical, and ethnographic researches on wheat can give us this information. The information provided also allows us to generate ideas about the transition from hunting and gathering to settled life, the start of food production in line with the needs, and determining how it has been shaped up to now and what can be done next. Some research results on this subject have been used in order to give an idea about cultural and historical value.

Wheat and human have 10,500 years of history and culture cooperation. In the researches, it is reported that the Neolithic age is important for humanity, and agriculture was initiated by people living in the Mediterranean settlement called Fertile Crescent (Vavilov 1926; Braidwood and Braidwood 1950; Heun et al. 1997; Diamond 2002). Genetic studies show that these wheat landraces are found in Turkey's southeastern part, Karacadağ (Kimber and Sears 1983), and they have spread to the world from here. Einkorn (*Triticum monococcum* ssp. *monococcum*) and emmer (*Triticum dicoccon*) are wheat landraces that have been cultivated in the early period and einkorn cultivated form of wild wheat species from *Triticum boeoticum*. The journey of bread wheat, which started from Anatolia to Greece 8000 years ago, has reached the Central Europe (mainly Italy and France), Scandinavia, England, Central Asia, China, Iran, Egypt, Africa, Mexico, Spain, and Australia (Keser 2018).

While the wild landraces were collected from the nature and consumed as food in ancient times, they were later cultivated and produced. It is thought that the first criterion for wheat cultivation is to be suitable for making bread (Hammer and Perrino 1984; Salamini et al. 2002). Communities engaged in hunting and gathering have been replaced by communities that settle and produce. It is said that the first villages in which the people were engaged in agriculture were in Southeast Anatolia and Northern Syria, and (Nesbitt and Samuel 1996) Abu Hurairah in Syria, and archaeological sites such as Cafer Höyük, Çayönü, and Nevali Çöri in Turkey are among the first agricultural villages.

The beginning of agriculture is considered as a key to the human history, and this historical journey of local landraces also sheds light on the journey of human history. Human history is a sociological process. This process has turned into a socio-economic process with the introduction of the economy over time and has become our current values by integrating with the sociocultural structure.

The first foundations of food supply have been laid in the process that has survived with the start of agriculture, and the adventure of plants has now been linked to the adventure of people who have started to apply their own laws (Pelt et al. 2002). Transition to food production in 8500 BC and effective use of scarce resources, which constitute the main subject of economic science, against unlimited needs, have constituted the first behaviors. When we examine the historical process of wheat landraces, we can say that no consumption behavior is independent from each other and the factors that shape the point of view of today's consumption arise from the behavioral patterns that have emerged in the past (Özüşen and Yıldız 2012).

Grain stocks, pots, and wheat forging containers found in the excavations in Çatalhöyük provide very important information from that period. In particular, it is stated that the grain stocks provide cultural data as the largest, richest, and best stored among those discovered (Özüşen and Yıldız 2012). Wheat residues found have shown that contain hexaploid wheat that is similar to the contemporary hexaploid wheat landraces including both hulled (*T. aestivum*) and unhulled (*T. spelta*) wheat. Unlike the grain-based Neolithic food systems in some parts of the world, bread is claimed to be a feature of the cultural traditions of the Neolithic people in the Near East (Haaland 2007; Fuller and Rowlands 2011). The social importance of bread as a cultural food item has contributed to the importance of grain consumption and is thought to be one of the factors that supports cultivation.

Different disciplined researches show us that cereals and the products produced from these, especially bread, have an important place in Anatolian, Hittite, Ancient Greek, Roman, and Egyptian nutrition culture. One of the cultural elements is the eating habits. Social, economic, and health are the factors that determine eating habits. These habits have continued as a traditional process for many years but have entered into an important change process due to sociocultural environment and economic reasons. The consumption habits of wheat landraces are also tucked in local areas. But the desire to eat healthy has begun to increase the popularity of wheat landraces.

Many products that have an important economic value in the world have actually reached this level due to their traditional sociocultural importance, and it is their social and cultural values that brought them the economic power. The fact that many wheat landraces are limited in certain areas, depending on not only the needs but also the traditional sociocultural environment, and that, although the consumption patterns differ, it is called with traditional names proves this. Traditional sociocultural factors have a determining feature in the process from the production of local landraces to their consumption. Although the demand for local landraces has increased today, we can also see that the planting areas have not spread far beyond the areas where they have been grown for many years.

Aesthetic value All living things in nature have a physical appeal and beauty. Aesthetic values refer to the value every person gives this beauty through his/her eyes. Wheat ear is an indicator of elegance not only for people living in rural areas but also for everybody. This is evidenced by the aesthetic value turned into art.

Artistic and spiritual value Ears of wheat symbolize flexibility and harmony in the symbol language. The common feature of wheat in societies with different beliefs and cultural structures is that it is considered sacred. This meaning that people place on wheat has turned into rituals and reflected on art.

With the symbolizing ability of humans, they unconsciously transform objects and forms into symbols. They place psychological significance on these symbols and expose them both within belief and visual arts (Jung 2009). Wheat has also become a universal symbol and it represents fertility, productivity, seed, and rebirth in many cultures. For example, it is said that the blonde girl who holds a wheat spike in her right hand and a torch in her left hand is the goddess of harvest and fertility in Greek mythology (Çakır 2019). Since wheat has a dominant role in the Roman Empire, the nation was then called the "Wheat Empire." It is mentioned that one of the three products that are important for Pharaoh in Egypt is emmer (*Triticum dicoccon*) and bread made from it (Mayerson 2002).

In Hattuşaş, the capital of the Hittites, who established the oldest and first empire in Anatolia, near Çorum, wheat silos with a capacity of 4200–5900 tons dating to the thirteenth century BC were found (Seeher 2001). Ivriz Rock Relief near the Hittites, Konya, indicates the social and religious importance of wheat. The grain silos and the wheat remains (Balkan 1964) found near the Urartu temple and palace dating back to the 800–700s show that similar traditions have continued in Anatolia for thousands of years (Zengin 2015). It is the symbol of fertility in Mesopotamia.

Art is the expression of aesthetic and spiritual feelings with different tools. This can be exampled as using wheat motifs in traditional arts, handicrafts and embroidery, stone and metalwork, and ceramics (Tan 2018; Sezgin Ceyhun and Bülbül 2017) and its being an inspiration for folklore, poetry, literature, idioms, proverbs, and folk songs.

Recreation and tourism (ecotourism and gastronomy tourism) value Wheat landraces, such as emmer and einkorn, which are considered as cultural heritages today, are still produced by traditional methods. Products such as bread, flour, bulgur, noodles, tarhana, beer, biscuits, basket, hat, and saddle produced from these wheats are important opportunities for tourism.

In our developing world, the consumption of traditional and cultural values along with consumption habits has created a trigger force for gastronomy tourism. Sociocultural interaction and orientation to the natural products and the fact that wheat landraces have a unique taste increases the interest in local products and dishes produced from the wheat. It is known that wheat landraces have touristic attraction with its historical and traditional aspects. Only these elements are needed to be enriched with art and food, to increase the product range for demand and to be used as a tool for developing regional tourism. Wheat landraces are of a great potential for tourism policies aiming to protect the cultural heritage and values of local people. Thus, they can be used as tools for local development (Kan et al. 2016a, b).

7.2.2 *Indirect Use Value*

7.2.2.1 The Value of Service in Functioning, Order, and Protection of the Ecosystem

Since the wheat landraces adapt to harsh conditions by cultivation areas, their production can be grown as close to natural, naturally or organically. Use of weed, pesticides, and fertilizer is very low. Therefore, habitat and regulatory service value can be increased more than the modern wheat varieties. There are two methods in biodiversity and conservation. These are in gene banks (ex situ) and in the hands of the producers (in situ). Producers are considered more important in the preservation of wheat landraces, because they have brought them together with the traditional structure until today. Their indirect use values with their assets and services to the habitat and the environment are given below:

- Habitat value: Shelter function value
- Biodiversity value
- Protection value

- Regulatory service value: Contributions to the regulation of atmospheric gases
- Contribution to the climate regulation
- Contribution to the food chain
- Contribution to water supply and conservation of resources
- Contribution to soil formation
- Contribution to erosion prevention
- Contribution to biological control
- Contribution to pollination

Preference usage value It expresses the value that it gave up as a result of preferring wheat landraces. Choosing wheat landraces means giving up a modern variety with the same usage value. The value we lose turns into opportunity cost.

7.3 Out-of-Use Value

Heritage value It is the value of payment willingness to protect for future generations to use. Producers show great sacrifice in protecting wheat landraces. There is also a willingness to protect behind this behavior. The value here is how much more you can sacrifice to continue your protection. The losses it will experience and the extra expenses it will make constitute the inheritance value.

Existence value Every creature has a value from its existence. Even if they do not offer any service or benefit, all living things have value with their own existence. It is a difficult value to explain in monetary terms.

In order to ensure the sustainable use of wheat landraces, it is necessary to create awareness in order to transfer these values not only to people who research, produce, and consume but to all people. It should not be forgotten that awareness is the most important solution to maintain some habits in humans or to give up some behavior. Sustainable use of biodiversity depends on the benefits that people receive from it and the value it generates from its perspective. Economic value is not just a commercial value. It contains the benefits of all services and benefits in it. Although wheat landraces are evaluated with their price, it should be considered that they are a social object and add more economic value to them. Awareness is a key to be included in all plans and policies. The appropriateness of the economic plans to be made at local, national, and global levels to the social and economic structure is the trigger of awareness. Awareness can be raised if wheat landraces are included in health, education, and local development plans and in all environmental policies.

7.4 Use of Wheat Landraces as a Rural Development Tool

The aim of rural development is to provide sustainable, economic, social, cultural, and political development of rural residents. Rural areas could not achieve the targeted development as a result of supporting economic development approaches and developments in modernization in agriculture (Ellis and Biggs 2001) and experienced significant economic, social, and physical changes (Kan et al. 2020). It has shown itself especially in developing countries. Political revenue-oriented policies have caused rural people to lose their cultural values after a while and try to survive. The orientation has started from the producing societies and continued in consuming societies, while the agricultural production sector was predominantly in rural areas, the service sector came to the fore. With this change, besides rural areas, agriculture, and forestry, other economic activities such as tourism, small-scale industry, and handicrafts have started to develop. Thus, the need for different sectors and the need for physical change in rural areas have been opened to discussion (Davoudi and Stead 2003; Costis 2003; Noronha Vaz et al. 2006; OECD 2006; Yenigül 2017; Kan et al. 2020).

As the world changes, people and their needs change, and people's behaviors change to reach these needs. Communities living and producing in rural areas should also meet their needs arising from this change. Shifting agriculture-based industry to these areas instead of moving away from agriculture will surely create a pressure on nature. On the other hand, the most important issue to be emphasized is to prepare a living space where the sector that produces it can be developed on-site, where they can meet their needs in their fields and ensure sustainability in production.

Rural areas have begun to lose these features with changes and orientations. The terms of rural area and rural communities need to be renegotiated. Who should be the target audience in development? Are the people living in rural areas and continuing their agricultural production activities? Do they live in the countryside and continue to work in the service sector? Do they live in the city and continue production in rural areas? In fact, a new definition is needed to answer all these questions. Development plans need multidisciplinary, extensive, and rigorous research. First of all, identification and selection of target audience must be made correctly.

Worldwide, policies focusing on local areas and local issues have begun to be developed. Approaches aimed at improving the agricultural structure and eliminating the negativities have been replaced by the approaches aiming to spread the service offered to the local people with an understanding that respects social and cultural values and protects nature. Especially in rural development programs, sociocultural structures of rural people are taken into consideration as well as socioeconomic status. The target audience we consider becomes more important in this respect.

As result of the changes in rural areas, policy approaches, and all areas, a stronger interaction started between rural and urban areas. This interaction triggered the change of the sociocultural structure and revealed the interest of the people living in the rural areas to the cities, the desire or longing of the people living in the urban area to natural life.

The negativity of technology age, diseases, epidemics, and people's desire for the nature gradually increases. Local landraces are capable of bringing together the bilateral demands of the people living in rural and urban areas. From a cultural approach perspective in rural areas identified with agriculture, local landraces should be considered as a good alternative for development. However, there should be an approach in which other sectors such as processing industry can be included for wheat landraces.

Although more income requests, nature conservation practices, and sociological factors create a harmony and balance today, they are also seen as elements that destroy each other in rural areas. It is important that the policies developed are sustainable as well as include measures to eliminate this risk. At this point, the historical process and socioeconomic and sociocultural structure of the rural people come into play. As a matter of fact, considering the history of humanity as a sociological process, it becomes more important. It is the common meeting point of agriculture, economics, sociology, and psychology. The survival of wheat landraces to this date is the best example of this.

Wheat landraces are produced mostly in small areas and family businesses. They are an important source of income for the people living in the regions where they are produced. The areas where these farms are located are often called rural areas. Since the lands of these family farms, which are mostly low-income and small, have low productivity, the number of alternative products is very low. The landraces they prefer are suitable for hard conditions and unproductive soil. In addition, it is seen that the average age of landrace producers is over 50 in most countries (Negri 2003; Tsegaye and Berg 2007; Hajnalová and Dreslerová 2010; Montesano et al. 2012; Baboev et al. 2015; Husenov et al. 2015; Kan et al. 2016a, b, 2019a, b; Karabak et al. 2019). It makes it compulsory to use the product that does not require much workforce and meets the consumption and the financial needs of the family. They do not prefer to take risks. Producers compromise some social values, produce local landraces in line with their needs, and continue to protect according to their needs. However, the proportion of these self-sacrificing producers is also decreasing.

While wheat landraces are mostly used as animal feed, they also serve different purposes in many countries, and these areas of use have played an important role in maintaining their production. Unfortunately, local landraces have begun to disappear as their use also disappeared. Food need and health factor created an opportunity to revive local landraces. After many years, direct use value of wheat landraces has created a change value in line with the quality features suitable for consumption demands.

Two important issues should be carefully considered for the sustainability of wheat landraces. The first is to ensure that they are protected in the areas where they are grown, and the second is to ensure that those who produce these landraces earn more. It is known by the name of the region where many products are produced in the world. Wheat landraces are known as local products produced by traditional methods. Local products are economic values resulting from an economic activity at the end of a production process. Their special features are based on natural conditions specific to the region or the knowledge, skills, methods, and techniques developed by the producers for a very long time. It should not be forgotten that people living and producing in that region reflect their traditions and cultures with their local products besides defining local products as economic values (Çandır 2010; Altuntaş and Gülçubuk 2014; Kan et al. 2016a, b).

The issue of using local dynamics from rural development as a means of development comes to the forefront all over the world. Local economic development envisages a participatory approach that supports private and public cooperation by using local resources on-site and using these resources for the economic and social well-being of the local community and supporting competitive advantage (Altuntaş and Gülçubuk 2014; Çetin 2007). Among the local sources, especially the local products come to the forefront and make an important contribution to tourism (Kan et al. 2012), because today, it is seen that touristic preferences are directed toward countries and regions that protect their local values (Babcock and Clemens 2004; Yenipınar et al. 2014; Kan et al. 2016a, b). In order to improve tourism in local governments, they have attempted to get geographical indicators for local products in their regions. Usage of wheat landraces in geographical indication system will contribute to the development of local people with its contributions to ecotourism and gastronomy tourism.

Nature-friendly, human-oriented economic planning should help preserve cultural heritage, traditional knowledge, and biodiversity. Planning should be made to ensure vertical and horizontal integration among all stakeholders serving development with local, regional, national, and global cooperation. Wheat landraces will be very valuable in human nutrition in the future as in the past. The main target should be to increase the diversity in nutrition for social, economic, environmental, and health and to ensure their sustainable use, by considering the local people first and then the country and the whole humanity at the global level.

7.5 Use of Wheat Landraces in Geographical Indication System

Although it is not a definitive description, the limits of the local products are tried to be determined with the perception that it differs from its peers in terms of quality, taste, flavor, and aroma on the producer's side, and with the thought that traditions, skills, and human and environmental factors are brought together with the product on the consumer's side. Because of the negative effects of the results of R&D and innovation studies regarding the environmental pollution that started with the industrialization process in the world and the increase in the use of technology in the industry, biotechnological products, mass production, and long-lasting foodstuffs, people have begun to return to the past especially in terms of healthy life. Organic product, additive-free product, natural product, local/traditional product, and local landraces/populations emerge as terminologies with increasing popularity in this process. People's interest and demand for such products are increasing day by day. Accordingly, a new sector has begun to emerge in the economy of many countries. This new sector is described as an alternative food economy, and quality and healthy food criteria come to the fore rather than price within this structure (Marsden et al. 2000; Holt 2005).

Local food, local landraces/populations, and handcrafts, especially food, are important elements describing the traditions, customs, cultures, past, people, geography, and climate of a region. Therefore, such elements are important tools for development and can be described as local development dynamics. If we take the Industrial Revolution as a beginning, the failure of the classical development theories that continued until the 1970s showed that the concept of development cannot be addressed with holistic approaches (Stamer 2003; Başkaya 2005). The existence of different structures in a country or even a region brought up the issue that local dynamics should be mobilized in development, and local/regional development models begun to come to the fore especially after the 1980s. For this reason, it is recommended to define local dynamics in the development rhetoric instead of gravitational regions and to develop area-specific development strategies and plans accordingly (Doğan 2011; Kan et al. 2016a, b). Among these local dynamics, genetic resources and production of value-added products based on these resources are an important starting point for the economic pillar of development.

Turkey is home to the genetic resources of many plants and animals with its geography, climate, and natural resources (Şehirali et al. 2005). This natural treasure has evolved into culture with the unification of the human element, and this culture has been sustained for generations to become a world heritage site. Even today, the role of plant and animal genetic resources is great for Turkey to come to this point in terms of local knowledge, particularly in gastronomy. Many European countries that are aware of this situation have protected their existing resources by legal means in terms of protecting these resources and ensuring their sustainability and have succeeded in turning this into a commercial advantage. Although the homeland of grape is the field called Asia Minor covering Anatolia and Caucasus, France is famous with its wines; although the homeland of wheat is shown as the area known as the Fertile Crescent which covers Turkey too, Italy has become the center of pasta and bread landraces; the Mediterranean basin is the homeland of olives and it has made Greece, Spain, Italy, and Turkey important brands in olives and olive oil; and although Turkey has over 160 types of cheese, France, Italy, and the Netherlands have become brands. As it can be understood from here, the local resources (including genetic resources) are only understood and owned locally, gaining value when legal, economic, and even social measures are taken to ensure their sustainability. This treasure, combined with other natural and human resources, disappears and is forgotten day by day.

Turkey is the homeland of many plant species and wheat, which is one of humanity's most important foods, is among these (Van-Slageren 1994; Harlan 1998; Şehirali et al. 2005; Blood et al. 2015). Turkey is the homeland of 23 species of wild wheat and more than 400 breeding wheat landraces (WWF-Turkey 2016). The Fertile Crescent, which is shown as the homeland of wheat, is known as the region where the Western and Near East/Middle East/Pre-Asian civilizations were born. The first agriculture, domestication of animals, and the first villages have emerged in this region. This region covers Turkey, Syria, Lebanon, Israel, Iran, and Iraq today. The Fertile Crescent is the natural homeland of eight products (wheat (emmer), einkorn wheat (Siyez), barley, flax, chickpea, lentil, pea, and Burçak), which are the founding products of Neolithic culture (Lev-Yadun et al. 2000; Kahyaoğlu 2018).

Being located in the region of such an important product in human history and human development is an important reason for this product to be the local development dynamics. It can be said that wheat and wheat products have an important place both in the culture and trade of our geography. But the most important question to be asked is "How important is such an important gene source in the development of its homeland?" Local development dynamics have an important place in the new development theory, and these dynamics need to be determined. For this purpose, the Geographical Indication System, which was firstly developed in France and then spread across all EU countries, is an important initiative. The geographically indicated products, called as the new food chain, include quality, food safety, and commitment to origin, culture reflection, and all of the human contribution. Geographical Indication System, which has four different registration structures as PDO (Protected Designation of Origin), PGI (Protected Geographical Indication), Geographical Indication (GI), and TSG (Traditional Speciality Guaranteed), serves important purposes such as protecting the region-specific products, preventing unfair competition, preventing the use of names unfairly, making connections with its origin, knowing the production standards, and having the right information about the product. It is stated that these products contribute to local economic development with effects such as bringing extra revenue, contributing to tourism (especially gastronomic tourism), and providing additional employment (Kan 2011; Kan and Kan 2020).

When the Geographical Indication System's development in Turkey and the EU is monitored, it is seen that 493 products are registered, in which 185 of them have origin indicators and 304 have geographical indicators as of June 30, 2020, in Turkey (TURKPATENT 2020). In the European Union geographical indicators registration system, 1836 of the 3336 registered geographical indications are PDO, 1211 are PGI, 247 are GI, and 63 are TSG (eAmbrosia 2020). The most important one is the group that is based on wheat and wheat products, and a total of 70 products in two groups are registered in the system of Turkey (14.20%). In the EU, 504 products (13.65%) are registered in five groups (Class 1.6-2.24-2.26-2.3-2.5).

As can be seen, the group of wheat-based products is important in Turkey. However, we have one registered geographical indicator (Kastamonu Siyez Bulgur-PDO) and three geographical indicators (Kastamonu Siyez Wheat-PDO, Kastamonu Siyez Flour-PDO, Bolu Seben Iza Wheat-PDO) in which the products are directly related to the wheat landraces. The reason why Turkey has very few registered geographical indicators based on local populations although it is the homeland of wheat is that its local populations have been lost or not given enough importance.

There are countries in the world that use the geographical indicators and register and turn them into an economic activity. For example, Spain and Italy are the leaders of these countries. Pa de Pagès Català in Spain and Pane Di Altamura and Pagnotta del Dittaino in Italy are examples of traditional bread with geographical indicators (eAmbrosia 2020). Especially Pane Di Altamura and Pagnotta del Dittaino in Italy are products of origin indicators and are produced and marketed only in certain regions. In addition, Farro della Garfagnana in Italy can be given as an example for PGI and Farro di Monteleone di Spoleto for emmer (*Triticum dicoccum*) wheat with PDO (Buerli 2006). Turkey is an important country for einkorn (*Triticum monococcum ssp. monococcum*) and emmer (*Triticum Dicoccum Schrank.*) wheat landraces. There are important attempts especially for registering einkorn wheat and the products made of it, and Siyez and Iza wheat are among these products.

An example of a local economic development initiative for wheat can be given in India. There is one local wheat with a geographical indication in India. The name of this local wheat is registered as "Bhalia Wheat," and it is produced in Bhal region of Gujarat state of India. It is stated that it has a wheat origin indicator, and it takes its unique feature from the region where it is registered. It is known as "Daudkhani Wheat," and it has high carotene, low water absorption, and high protein content, and approximately 5000 farmers grow it in the region. The producers of this wheat landrace in Bhal region sold of 25% more price than the other commercial wheat landraces and 40–50% more than the other bread wheat landraces (The Hindu Businessline 2011; Chaudhary et al. 2017).

As a result, the ability of wheat and wheat products to create local development model based on the geographical indication system can be seen in different examples. The fact that Turkey is the homeland of wheat, which is in the most front row in the food consumption of humans that there is a culture based on wheat and wheat products, makes it necessary for us to protect our genetic resources as well as to use it economically too. Economic factors are closely related to the protection of genetic resources (Kan et al. 2015; Kan 2018; Kan et al. 2019a, b; Kan 2019), and we need to support protection policies and economic use policies to combat genetic erosion.

7.6 General Evaluation and Conclusion

Wheat landraces, which have important social and economic values, also illuminate the common past of many countries. Although they have been forgotten in many regions in terms of culture in the modern world, they have begun to be remembered again with nutrition and health concerns brought by modern life. Conservation of these landraces mostly depends on the societies that produce it. However, the strategies and policies to be determined by the states will also guide the productive behaviors. With this study, the social and economic relationship between human and wheat landraces has been tried to be explained by researches in different disciplines. The results show that social and economic factors were effective in wheat landraces in the past, and the effects of environmental and nutritional and health factors are increasing. The diseases seen in the recent years, especially chronic diseases and global epidemics, have highlighted food safety, reliability, and self-sufficiency issues of the states. The priority has been nutrition and health. Especially, the Covid-19 pandemic has once again showed the importance of agriculture and biodiversity.

The main purpose of the economy is to ensure the highest sustainable use of scarce resources. Environmental, social, and economic criteria are considered as three separate criteria in sustainability. Both research results and current needs reveal that economic and social factors are intertwined, and the nutrition and health factors should be considered as sustainability criteria. The governments should take measures to ensure sustainability in terms of "environment," "society," and "economy" and "nutrition and health" in determining policies and strategies for sustainable use of these valuable landraces that provide nutrition to the people. The Green Deal recently announced by the European Commission and the "Farm to Fork Strategy" and "EU Biodiversity Strategy for 2030" declared on May 20, 2020, exactly show how important this issue will be in the future (European Commission (EC) (2020). Wheat landraces are products that have the potential to be used as development tools in local economic development with their nutritional content and social and economic value for these purposes. A sustainable use should be gained with the measures to be taken without destroying its traditional features and transforming it into a commercial commodity. Turkey is one of the most fortunate countries having this potential. Although the issues of biological diversity, conservation, and sustainable use of natural resources come to the forefront in both agricultural policies and rural development strategy documents, it is necessary to perform the act faster.

References

Altuntaş A, Gülçubuk B (2014) Yerel kalkınmada yaygınlaşan bir araç olarak geleneksel gıdalar ve geleneksel gıda mevzuatının yaygınlaştırılabilirliği. Gaziosman Paşa Üniversitesi Ziraat Fakültesi Dergisi 3(3):73–81

Atasoy Y, Selçukcan Erol IÇ, Ayvaz Reis Z et al (2014) A perspective of infonomics to economic value of biodiversity. Paper presented at akademik bilişim'14 - XVI. Mersin Üniversitesi, Mersin, 5–7 Şubat 2014. https://ab.org.tr/ab14/bildiri/327.pdf

Aydın M, Aydınlar K (2011) 'Emek-değer' teorisinden 'fayda-değer' teorisine. Bilgi Sosyal Bilimler Dergisi 1:1–12

Babcock B, Clemens R (2004) Geographical indications and property rights: Protecting value added agricultural products. MATRIC Briefing paper 04-MBP 7

Baboev S, Morgounov A, Muminjanov H (2015) Wheat landraces in farmers' fields in Uzbekistan. natıonal survey, collectıon, and conservatıon, 2013-2015. Food and Agriculture Organization of the United Nations, Ankara, 2015. pp. 18

Balkan K (1964) "Patnos'ta Keşfedilen Urartu Tapınağı ve Urartu Sarayı", TürkTarih Kurumu Yıllık Konferansları I: Atatürk Konferansları, Ankara: Türk Tarih Kurumu Yayınları, XVII. Seri-No I: p:235–243

Başkaya F (2005) Kalkınma İktisadının Yükselişi ve Düşüşü, Beşinci Baskı, Türkiye ve Ortadoğu Forumu Vakfı Yayınları, No: 44, Ankara

Bonet MA, Vallès J (2007) Ethnobotany of Montseny biosphere reserve (Catalonia, Iberian Peninsula): plants used in veterinary medicine. J Ethnopharmacol 110:130–147

Borza A (1945) L'alac (engrain, Triticum monococcum) chez les roumains. atude botanique et d'histoire culturelle. atude botanique et d'histoire culturelle. Sibiu, Roumania, Centru de studi si cercetäri privitoare la Transilvania

Braidwood RJ, Braidwood L (1950) Jarmo: a village of early farmers in Iraq. Antiquity 24:189–195

Brouns FJPH, Buul VJV, Shewry P (2013) Does wheat make us fat and sick? J Cereal Sci 58:209–215

Brown AHD (2000) The genetic structure of crop landraces and the challenge to conserve them in situ on farms. In: Maxted N, Ford-Lloyd BV, Hawkes JG (eds) Plant genetic conservation: the in situ approach. Chapman and Hall, London

Buerli M (2006) Farro in Italy: A desk study. The Global Facilitation Unit for Underutilized Species. http://www.underutilized-species.org/Documents/PUBLICATIONS/farro_in_italy.pdf

Çakır E (2019) Kitabın dışından iç dünyamıza mitolojik bir yolculuk: Yayınevi logoları. Mitoloji araştırmaları kitabı Hiperyayın, İstanbul, pp 201–232

Çandır A (2010) Anadolu'nun sahipsiz değerleri: yöresel ürünler. Borsanomi Antalya Ticaret Borsası Bülteni 25:30–41

Cavanagh CR, Chao S, Wang S et al (2013) Genome-wide comparative diversity uncovers multiple targets of selection for improvement in hexaploid wheat landraces and cultivars. Proc Natl Acad Sci U S A 110:8057–8062. https://doi.org/10.1073/pnas.1217133110

Çelik İ (2010) Biyoçeşitlilik ne kadar değerli. Bilim Teknik Dergisi 516:50–53

Çetin M (2007) Yerel ekonomik kalkınma yaklaşımı ve uluslararası organizasyonlar. Yönetim ve Ekonomi Dergisi 14(1):156

Çetiner S (2005) Türkiye ve Dünyada Tarımsal Biyoteknoloji ve Gıda Güvencesi: Sorunlar ve Öneriler. GDO Bilgi Platformu

Chaudhary RC, Yadav SK, Kumar S (2017) Geographical indications in Indian agriculture on the Anvil. J Bio Innov 6(5):790–816

Cooper R (2015) Re-discovering ancient wheat varieties as functional foods. J Tradit Complement Med 5(2015):138–143. https://doi.org/10.1016/j.jtcme.2015.02.004

Costis H (2003) Imagining rurality in the new Europe and dilemmas for spatial policy. Eur Plan Stud 11(2):103–113

Davoudi S, Stead D (2003) Urban and rural relationships: an introduction and brief history. Built Environ J 28(4):269–277

De Groot RS, Wilson M, Boumans R (2002) A typology for the description, classification and valuation of ecosystem functions, goods and services. Ecol Econ 41(3):393–408. https://doi.org/10.1016/S0921-8009(02)00089-7

Demir S (2015) Türkiye'de yetişen yerel emmer buğday [Triticum turgidum L. ssp. dicoccon (Schrank) Thell.] popülasyonlarında genetik çeşitliliğin moleküler yöntemlerle karakterizasyonu (Yüksek Lisans Tezi)

Diamond J (2002) Evolution, consequences and future of plant and animal domestication. Nature 418:700–707

Doğan BB (2011) Kalkınma iktisadının xx. yüzyıldaki gelişim süreci, iktisat politikalarına etkisi ve son on yıllık konjonktürün disiplinin geleceğine olası etkiler. Kocaeli Üniversitesi Sosyal Bilimler Enstitüsü Dergisi (22)2011/2:41–83

eAmbrosia (2020) European Commission eAmbrosia–the EU geographical indications register database. Available online at https://ec.europa.eu/info/food-farming-fisheries/food-safety-and-quality/certification/quality-labels/geographical-indications-register/#

Ellis F, Biggs S (2001) Evolving themes in rural development 1950s–2000s. Dev Policy Rev 19(4):437–449

Ertuğ F (2004) Wild edible plants of the Bodrum Area (Mugla, Turkey). Turk J Botany 28:161–174

Ertuğ F (2014) Etnobotanik Kaynakları. In: Güner A, ve Ekim T (eds) Resimli Türkiye Florası, Cilt 1. Ali Gökyiğit Vakfı, Flora Araştırmaları Derneği ve Türkiye İş Bankası Kültür Yayınları, İstanbul

European Commission (EC) (2020) European Green Deal. Available online at https://ec.europa.eu/info/strategy/priorities-2019-2024/european-green-deal_en

Fichter J (1990) Sosyoloji nedir? (Çev.,Nilgün Çelebi). Selçuk Üniversitesi Yayınları, Konya

Fuller DQ, Rowlands M (2011) "Ingestion and food technologies: maintaining differences over the long-term in West, South and East Asia". In Interweaving worlds: systematic interactions in Eurasia, 7th to 1st millennia BC, Wilkinson, T. C., Sherratt, S. and Bennett, J. 37–60. Oxford: Oxbow Books

Genç Yılmaz S, Çağlayan T (2017) İktisadi Düşüncede Değerin Kaynağı Sorunsalı. Çalışma ve Toplum Dergisi 53:667–680

Giuliani A, Karagöz A, Zencirci N (2009) Emmer (*Triticum dicoccon*) production and market potential in marginal mountainous areas of Turkey. Mount Res Dev (MRD) 29(3):220–229

Gowdy J (1998) Limited wants, unlimited means: a reader on hunter-gatherer economics and the environment. Island Press, Washington, D.C.. https://books.google.com.tr

Gowdy J, Salman A (2010) Institutions and ecosystem functions: the case of Keti Bunder, Pakistan. In: Kumar P, Wood M (eds) Valuation of regulating services of ecosystems. Routledge, London, pp 201–221

Gunda B (1983) Cultural ecology of old cultivated plants in the Carpathian area. Ethnol Eur 13(2):145–179

Haaland R (2007) Porridge and pot, bread and oven: food ways and symbolism in Africa and the Near East from the Neolithic to the present. Camb Archaeol J 17(2):167–183

Hajnalová M, Dreslerová D (2010) Ethnobotany of einkorn and emmer in Romania and Slovakia: towards interpretation of archaeological evidence. PamátkyArcheologické, 101:169–202

Hammer K, Perrino (1984. Further information on farro (Triticum monococcum L. & T. dicoccon Schrank) in South Italy. Kulturpflanze, 32, 143 151

Harlan JR (1975) Crops & man, 2nd edn. American Society of Agronomy, Crop Science Society of America, Inc., Madison

Harlan JR (1998) Distribution of agricultural origins: a global perspective. In: Damania AB, Valkoun J, Willcox G, Qualset CO (eds) The origins of agriculture and the domestication of crop plants in the Near East. The Harlan Symposium Book of abstracts. Icarda, Alepo

Heun M, Schafer-Pregl R, Klawan D et al (1997) Site of einkorn wheat domestication identified by DNA fingerprinting. Science 278:1321–1314

Holt G (2005) Local foods and local markets: strategies to grow the local sector in the UK. Anthropology of food, https://journals.openedition.org/aof/179

Husenov B, Otambekova M, Morgounov A et al (2015) Wheat landraces in farmers' fields in Tajikistan: national survey, collection, and conservation, 2013-2015. Food and Agriculture Organization of the United Nations, Ankara, p 11

İlhan D (2017) Bitki Biyoteknolojisinde Genetik Kaynakların Önemi, Kafkas University. J Nat Appl Sci 10(2):134–144

Jaradat AA (1992) Breeding potential of durum wheat landraces from Jordan. II Differential response to drought. Hereditas 116:305–309

Jaradat AA (2011) Wheat landraces: genetic resources for sustenance and sustainability. USDA-ARS, Morris. https://www.ars.usda.gov/ARSUserFiles/50600000/products-wheat/AAJ-Wheat%20Landraces.pdf

Jaradat AA (2013) Wheat landraces: a mini review. Emir J Food Agric 25(1):20–29

Jarvis D, Hodgkin T (1998) Strengthening the scientific basis of in situ conservation of agricultural biodiversity on-farm. Options for data collection and analysis. Proceedings of a workshop to develop tools and procedures for in situ conservation on-farm, 25–29 August 1997. IPGRI, Rome, Italy

Johnson I, Gary W (2003) Phytochemical functional foods. Woodhead Publishing Ltd and CRC Press LLC, Cambridge and Boca Raton

Jung CG (2009) İnsan ve Sembolleri, çev. Ali Nahit Babaoğlu. Okuyan Us Yayınevi, İstanbul

Kahyaoğlu D (2018) Bereketli Hilala (Fertile Crescent). History Education, https://tarihegitimi.blogspot.com/2018/03/bereketli-hilal.html

Kan M (2011) Yerel Düzeyde Ekonomik Kalkınmada Coğrafi İşaretlerin Kullanımı ve Etkisi: Akşehir Kirazı Araştırması. Ph.D. diss. Ankara Üniversity, Grad. Sch. Nat. and Appl. Sci. Dept. of Agr. Econ., Ankara-Turkey

Kan M (2018) The relationship production behaviours and perception of genetic resource conservation of the farmers producing dry bean landraces in Turkey. Fresenius Environ Bull 27(12-B):9913–9921

Kan A (2019) Role of dry bean landrace women producers on conservation of plant genetic resource in middle Kizilirmak Valley region of Turkey. Pak J Agric Sci 56(1):301–309. https://doi.org/10.21162/PAKJAS/19.7115

Kan M, Kan A (2020) Qualitative and quantitative analysis of the geographical indication system in Turkey. J Glob Innov Agric Soc Sci 8(2):114–123. https://doi.org/10.22194/JGIASS/8.899

Kan M, Gülçubuk B, Küçükçongar M (2012) Coğrafi işaretlerin kırsal turizmde kullanılma imkanları. KMU J Soc Econ Res 14(22):93–101

Kan M, Küçükçongar M, Keser M et al (2015). Wheat Landraces in Farmers' Fields in Turkey National Survey, Collection and Conservation, 2009-2014, FAO

Kan M, Kan A, Gülçubuk B et al (2016a) Türkiye'de yerel ürünlerin bölgesel kalkınma dinamikleri içindeki önemi. In: A.E. Peker, Bölgesel Kalkınma, Paradigma Akademi – Kasım 2016, Çanakkale-Turkey, pp. 231–280

Kan M, Küçükçongar M, Morgounov A et al (2016b) Wheat landraces production on farm level in Turkey; who is growing in where? Pak J Agri Sci 53(1):159–169. https://doi.org/10.21162/PAKJAS/16.3754

Kan M, Küçükçongar M, Morgounov A et al (2017) Türkiye'de Yerel Buğday Popülasyonlarının Durumu ve Yerel Buğday Üreten Üreticilerin Üretim Kararlarında Etkili Olan Faktörlerin Belirlenmesi. Gaziosmanpaşa Üniversitesi Ziraat Fakültesi Dergisi [Journal of Agricultural

Faculty of Gaziosmanpasa University-JAFAG], 34(2): 54–64. https://doi.org/10.13002/jafag4225

Kan M, Kan A, Sözen Ö et al (2019a) Evaluation of the farms producing dry bean landraces by capital approach in the Middle Kızılırmak Valley of Turkey. Plant Genet Resour 17(5):391–400. https://doi.org/10.1017/S1479262119000169

Kan M, Kan A, Bayraktar ÖV (2019b) Türkiye'de Coğrafi İşaret Sistemindeki Gelişmelerin Yansımaları. Şırnak Üniversitesi. Uluslararası İktisadi ve İdari Bilimler Kongresi. 02-04 Mayıs 2019, s:1312–1322, Şırnak

Kan M, Kan A, Nizam D et al (2020) Dünyada Ve Türkiye'de Kırsal Kalkınma Uygulamalarındaki Mevcut Durum ve Gelecek. TMMOB Ziraat Mühendisleri Odası, Türkiye Ziraat Mühendisliği IX. Teknik Kongresi, 13–17 Ocak 2020, Bildiriler Kitabı 2, s:687–710, Ankara

Karabak S (2017) Economic and SocioCultural importance of edible wild species Anadolu. J AARI 27(2):26–38

Karabak S, Taşcı R, Ceyhan V et al (2019) Cultural heritage siyez wheat from the İhsangazi farms to the cuisine. Toprak Su Dergisi:86–93

Karagöz A (1996) Agronomic practices and socioeconomic aspects of emmer and einkorn cultivation in Turkey. In: Padulosi S, Hammer K, Heller J (eds) Hulled wheat. Promoting the conservation and use of underutilized and neglected crops, 4, 172–177. Proceedings of the first international workshop on hulled wheats, Castelvecchio Pascoli, Tuscany, Italy 1995. International Plant Genetic Resources Institute, Rome

Karcıcıo M (2006) Yerel Durum Buğdayı Çeşitlerinde (Triticum durumDesf.) RAPD-PCR Tekniği ile Genetik Çeşitlilik Analizi. Doktora Tezi, Hacettepe Üniversitesi Fen Bilimleri Enstitüsü, Ankara. https://hdl.handle.net/11491/5324

Keser M (2018) Türkiye Yerel Buğdaylar Sempozyumu, 20–22 Aralık 2018, p: 43–44, Bolu

Kimber G, Sears ER (1983) Assignment of genome symbols in Triticeae. In: Sakamoto S (ed) Proc 6th Int Wheat Genet Symp. Maruzen, Kyoto, pp 1195–1196

Konvalina P, Capouchová I, Stehno Z et al (2013) Baking quality and high molecular weight glutenin subunit composition of emmer wheat, old and new varieties of bread wheat. Rom Agric Res 30:83–89

Korkmaz M, Karakurt E (2015) An ethnobotanical investigation to determine plants used as folk medicine in Kelkit (Gumushane/Turkey) district. Bio Divers Conserv 8(3):290–303

Kyzeridis N, Biesantz A, Limberg P (1995) Comparative trials with durum-wheat landraces and cultivars in different ecological environments in the Mediterranean region. J Agron Crop Sci 174:133–144

Lachman J, Miholová D, Pivec V, Jíru K, Janovská D (2011) Content of phenolic antioxidants and selenium in grain of einkorn (*Triticum monococcum*), emmer (*Triticum dicoccum*) and spring wheat (*Triticum aestivum*) varieties. Plant Soil Environ 57:2235–2243

Lev-Yadun S, Gopher A, Abbo S (2000) The cradle of agriculture. Science, New Series 288(5471):1602–1603

Mann S (2018) Socioeconomics of agriculture. Springer International Publishing, Cham

Marsden T, Banks J, Bristow G (2000) Food supply chain approaches: exploring their role in rural development. Sociol Rural 40(4):424–438

Maxted N, Tan A, Amri A et al (2000) In situ conservation. In: Maxted N, Bennet S (eds) Plant genetic resources of legumes in the Mediterranean. Kluver, Dordrecht, pp 1–386. ISBN 0-7923-6707-3. 292–308

Mayerson P (2002) Three pharaonic crops in Ptolemaic Egypt: Ólura (Emmer wheat) and maslins of krithópuron and olurókrithon. Z. Papyrologie Epigraphik 141:210–213

Montesano V, Negro D, Sarli G et al (2012) Landraces in inland areas of the Basilicata region, Italy: monitoring and perspectives for on farm conservation. Gen Res Crop Evol 59:701–716. https://doi.org/10.1007/s10722-011-9712-7

Mpofu A, Sapirstein HD, Beta T (2006) Genotype and environmental variation in phenolic content, phenolic acid composition, and antioxidant activity of hard spring wheat. J Agric Food Chem 54:1265–1270

Negri V (2003) Landraces in Central Italy: where and why they are conserved and perspectives for their on-farm conservation. Genet Resour Crop Evol 50:871–885
Nesbitt M, Samuel D (1996) From staple crop to extinction? The archaeology and history of the hulled wheats. In: Padulosi S, Hammer K, Heller J (eds) Hulled wheats. Promoting the conservation and use of underutilised and neglected crops 4. Proceedings of the first international workshop on hulled wheats. IPGRI, Rome, pp 4–41
Noronha Vaz T, Morgan EJ, Nijkamp P (2006) The new European rurality. Ashgate Publishing Ltd., Farnham
Nunes PALD, van den Bergh JCJM (2001) Economic valuation of biodiversity: sense or nonsense. Ecol Econ 39:203–222. https://doi.org/10.1016/S0921-8009(01)00233-6
OECD (2006) OECD Rural Policy Reviews, The New Rural Paradigm Policies and Governance, http://www.oecd.org/gov/regionalpolicy/thene-wruralparadigmpoliciesandgovernance
Özberk İ (2018) Türkiye Yerel Buğdaylar Sempozyumu, 20–22 Aralık 2018, pp 40–42, Bolu, ISBN 978-605-80871-0-1
Özberk F, Karagöz A, Özberk İ et al (2016) Buğday Genetik Kaynaklarından Yerel ve Kültür Çeşitlerine; Türkiye'de Buğday ve Ekmek. Tarla Bitkileri Merkez Araştırma Enstitüsü Dergisi 25(2):218–233
Özdemir E, Alpınar K (2015) An ethnobotanical survey of medicinal plants in western part of central Taurus Mountains: Aladaglar (Nigde–Turkey). J Ethnopharmacol 166:53–65
Özdemir F, Küçükçongar M, Morgounov A et al (2018) Anadolu'nun genetik hazinesi "yerel buğdaylar". Türkiye Yerel Buğdaylar Sempozyumu. 20–22 Aralık 2018. Bolu. s. 32–33
Özüşen B, Yıldız Z (2012) Buzul Çağı'ndan İlk Çağ'a tüketimin tarihi. Süleyman Demirel Üniversitesi Vizyoner Dergisi 4(7):1–16
Pelt JM, Mazoyer M, Monod T et al (2002) Bitkilerin En Güzel Tarihi. Kültür Yayınları, İstanbul
Peña-Chocarro L (1996) In-situ conservation of hulled-wheat species: the case of Spain. In: Padulosi S, Hammer K, Heller J (eds) Hulled wheats. Promoting the conservation and use of underutilised and neglected crops 4. Proceedings of the first international workshop on hulled wheats. IPGRI, Rome, pp 129–146
Peña-Chocarro L, Zapata L (1998) Hulled wheats in Spain: history of minor cereals. In: Jaradat AA (ed) Triticeae III. Oxford & NBH, New Delhi and Science Publishers, New Hampshire, pp 45–52
Peña-Chocarro L, Zapata Pena L, Gonzalez-Urquijo JE et al (2009a) Einkorn (Triticum monococcumL.) cultivation in mountain communities of the western Rif (Morocco): an ethnoarchaeological project. In: Fairbairn AS, Weiss E (eds) From foragers to farmers: papers in honour of Gordon C. Hillman. Oxbow Books, London
Peña-Chocarro L, Zapata Pena L, Gonzalez-Urquijo JE et al (2009b) Einkorn (Triticum monococcum L.) cultivation in mountain communities of the western Rif (Morocco): an ethnoarchaeological project. In: Fairnbairn AS, Weiss E (eds) From foragers to farmers. Gordon Hillman Festschrift, Oxbow, Oxford, pp 103–111
Salamini F, Özkan H, Brandolini A (2002) Genetics and geography of wild cereal domestication in the near east. Nat Rev Genet 3:429–441. https://doi.org/10.1038/nrg817
Şanal T (2018) Bazı yerel buğday çeşitlerinin kalite parametreleri. TÜRKTOB Dergisi:38–43
Sarı AO, Oğuz B, Bilgiç A et al (2006) Batı Anadolu'da Halk İlacı Olarak Kullanılan Lamiaceae Türleri. Anadolu. J AARI 16(2):50–67. MARA
Seeher J (2001) Hattuşa Kazı Sonuçları, T.C. Kültür Bakanlığı, Anıtlar ve Müzeler Genel Müdürlüğü 22. Kazı Sonuçları Toplantısı (22–26 Mayıs 2000, İzmir) Bildirileri Ankara, s. 303–314 http://acikerisimarsiv.selcuk.edu.tr:8080/xmlui/bitstream/handle/123456789/12355/393093.pdf?sequence=1&isAllowed=y
Şehirali S, Özgen M, Karagöz A et al (2005) Bitki Genetik Kaynaklarının Korunma Ve Kullanımı. TMMOB Ziraat Mühendisleri Odası VI. Teknik Kongresi. Cilt 1. Kozan Ofset, Ankara. 253–273
Serpen A, Gökmen V, Karagöz A et al (2008) Phytochemical quantification and total antioxidant capacities of emmer (T. dicoccon shrank) and einkorn (T. monococcum L.) wheat landraces. J Agric Food Chem 56:7285–7292

Sezgin Ceyhun A, Bülbül S (2017) Türk Sanatı ve Mutfak Kültüründe Buğday. Uluslararası Sosyal Araştırmalar Dergisi 10(54):1080–1091

Sidhu SJ, Kabir Y, Huffman GF (2007) Functional foods from cereal grains. Int J Food Prop 10(2):231–244

Slavin J (2004) Whole grains and human health. Nutr Res Rev 17:99–110

Srivastava JP, Damania AB (1989) Use of collections for cereal improvement in semi-arid areas. In: Brown AHD, Frankel OH, Marshall DR, Williams JT (eds) The use of plant genetic resources. Cambridge University Press, Cambridge, pp 88–104

Stamer JM (2003) Why is Local Economic Development so difficult, and What can we do to make it more effective? Mesopartner Working Paper, 4, Duisburg, 1–41

Talas F, Longin F, Miedaner T (2011) Sources of resistance to Fusarium head blight within Syrian durum wheat landraces. Plant Breed 130:398–400

Tan A (2010) Türkiye Bitki Genetik Kaynakları ve Muhafazası. Anadolu J AARI 20(1):7–25

Tan A (2018) Buğday yerel çeşitlerinin geleneksel bilgileri, Türkiye Yerel Buğdaylar Sempozyumu, 20-22 Aralık 2018, pp 25–26 Bolu, ISBN 978-605-80871-0-1

The Hindu Businessline (2011) Bhalia wheat gets GI tag. Erişim Tarihi 09(05):2019. https://www.thehindubusinessline.com/economy/agri-business/bhalia-wheat-gets-gi-tag/article20322702.ece1

Troccoli A, Codianni P (2005) Appropriate seeding rate for einkorn, emmer, and spelt grown under rainfed condition in southern Italy. Eur J Agron, 22 (2005), pp. 293–300. https://doi.org/10.1016/j.eja.2004.04.003

Tsegaye B, Berg T (2007) Genetic erosion of Ethiopian tetraploid wheat landraces in Eastern Shewa, Central Ethiopia. Genet Resour Crop Evol 54:715–726

Turk Patent (2020) Geographical Indication Database of Turkish Patent and Trademark Office. Available online at https://www.ci.gov.tr/

Van-Slageren MW (1994) Wild wheats: a monograph of Aegilops L. and Amblyopyrum (Jaub. & Spach) Eig (Poaceae). Veenman Drukkers, Wageningen, Hollanda

Vavilov NI (1926) The origin of the cultivation of 'primary' crops, in particular cultivated hemp. In: Studies on the origin of cultivated plants. Institute of Applied Botany and Plant Breeding, Leningrad, pp 221–233

Velioglu YS, Mazza G, Gao L et al (1998) Antioxidative activity and total phenolics in selected fruits, vegetables and grain products. J Agric Food Chem 46:4113–4117

WWF-Türkiye (Doğal Hayatı Koruma Vakfı) (2016) Türkiye'nin Buğday Atlası. http://awsassets.wwftr.panda.org/downloads/turkiye_nin_buday_atlas_web.pdf

Yaman HM, Zencirci N (2018) IZA buğdayı nedir, özellikleri, tarımı ve kullanımı, Türkiye Yerel Buğdaylar Sempozyumu, 20-22 Aralık 2018, Bolu 76

Yaman HM, Ordu B, Zencirci N et al (2019) Coupling socioeconomic factors and cultural practices in production of einkorn and emmer wheat species in Turkey. Environ Dev Sustain 2019. https://doi.org/10.1007/s10668-019-00562-7

Yenigül SB (2017) Kırsal Kalkınma Politikalarında Yeni Yaklaşımlar ve Bu Yaklaşımların Türkiye'nin Kırsal Kalkınma Politikalarına Etkisi. Plan Theory 27(1):16–25

Yenipınar U, Köşker H, Karacaoğlu S (2014) Turizmde Yerel Yiyeceklerin Önemi ve Coğrafi İşaretleme: Van Otlu Peyniri. J Tour Gastron Stud 2(2):13–23

Zaharieva M, Ayana NG, Al Hakimi A et al (2010) Cultivated emmer wheat (Triticum dicoccon Schrank), an old crop with promising future: a review. Genet Resour Crop Evol 57:937–962

Zengin G (2015) Bazı ilkel buğdaylarda kalite parametrelerinin belirlenmesi üzerine bir araştırma Selçuk Üniversitesi, Fen Bilimleri Enstitüsü (Yüks. Lisans Tezi) Konya http://acikerisimarsiv.selcuk.edu.tr:8080/xmlui/bitstream/handle/123456789/12355/393093.pdf?sequence=1&isAllowed=y

Chapter 8
Chemical Contents of Wheat Landraces and Their Contribution to Human Health

Cisem Nildem Keskin, Fatma Pehlivan Karakas, and Ferdi Ağıl

8.1 Introduction

Huge attention on Turkish wheat variation continued since the beginning of the twentieth century. Searching and collecting germplasms undertook as a duty, and these collected materials were evaluated in different countries (Gökgöl 1935, 1939; Harlan 1950; Zhukovskyi et al. 1951). According to the Gökgöl (1935), Turkey has almost all varieties of wheat in the world, and Turkish local wheat offers an endless treasure for wheat breeders. Having various climatic conditions provided a suitable environment to create natural hybridization and recombination, and therefore, it has created richness in wheat variety. As a result, 20,000 wheat varieties, including wheat varieties unique to the country, were examined (Gökgöl 1939). It is believed that Anatolia is the warehouse of wheat varieties, and it is an important region that distributes wheat varieties to neighboring countries (Zhukovskyi et al. 1951).

Anatolia is the place where wheat domestication occurred around 10,000 BP. The primary gene center of wheat diversification is in Southeast Anatolia (Harlan 1981; Diamond 1997; Heun et al. 1997; Nesbit and Samuel 1998; Lev-Yadun et al. 2000). Because different regions in Anatolia have different climatic conditions, different wheat varieties can grow in Turkey such as bread and durum wheat that are grown widely in Southeast Anatolia and bread wheat that is mostly grown in Central Anatolia. Moreover, cultivation of hulled wheats *Triticum monococcum* (cultivated einkorn) and *T. dicoccon* Schrank (cultivated emmer) are planted around the north transition zone (Karagöz and Zencirci 2005).

Wheat is a staple food around the world and *Triticum aestivum* (hexaploid, $2n = 6x = 42$, unhulled wheat) and *Triticum durum* (tetraploid, $2n = 4x = 28$, unhulled wheat) are the most traded wheat types. Besides these, *Triticum*

C. N. Keskin · F. Pehlivan Karakas (✉) · F. Ağıl
Department of Biology, Faculty of Science and Art, Bolu Abant Izzet Baysal University, Bolu, Turkey

N. Zencirci et al. (eds.), *Wheat Landraces*,
https://doi.org/10.1007/978-3-030-77388-5_8

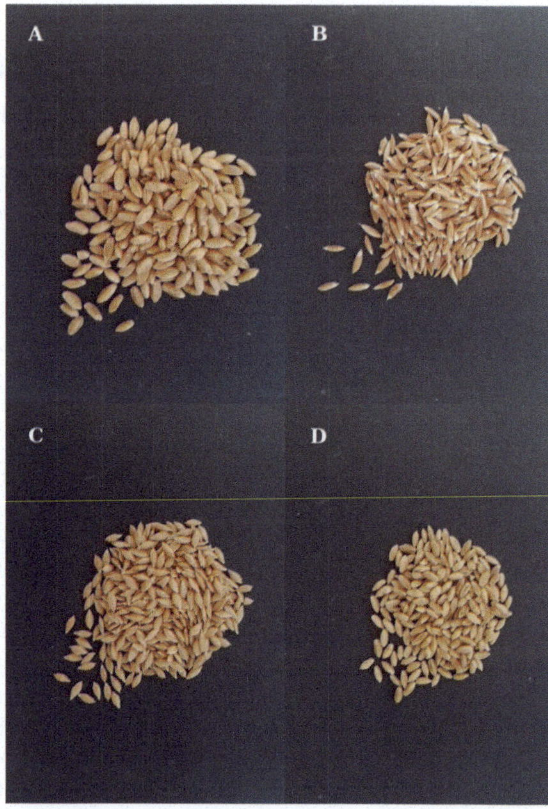

Fig. 8.1 (**a**) *Triticum durum* (Kunduru-1149), (**b**) *Triticum dicoccum* Schrank, (**c**) *Triticum monococcum* ssp. *monococcum* (IZA), (**d**) *Triticum aestivum* (Kıraç-66)

monococcum ssp. *monococcum* (diploid, 2n = 2x = 14, hulled wheat) and *Triticum dicoccum* Schrank (hexaploid, 2n = 4x = 28, hulled wheat) are also commercially important (Stevenson et al. 2012) (Fig. 8.1). On the one hand, einkorn (*Triticum monococcum* ssp. *monococcum*) has got low fiber content, high protein content, and mostly unsaturated fatty acids. Moreover, it contains high level of Zn and Fe elements. On the other hand, emmer (*Triticum dicoccum* Schrank) has also been known as one of the ancient wheat species, and ancient people of Egypt used it in bread making (Shewry 2009a, b). Emmer is rich in bioactive compounds and dietary fiber, but digestion of its starch is slow. Einkorn and emmer have been known as healthy cereals because of their higher protein, mineral, and carbohydrate, but they have poor fat contents. Therefore, nowadays, there is a big interest on raising emmer and einkorn in organic farming (Dhanavath and Prasada Rao 2017).

Seeds of wheat consist of different tissues such as germ, endosperm, aleurone layer, and pericarp. Every layer of wheat contains different concentrations of nutrients. Endosperm, for instance, contains higher concentration of starch (Stevenson et al. 2012). Moreover, the bran of wheat is very important for health since it is rich in phenolic acids, some phytochemicals, and a few types of vitamin B (such as thiamine, riboflavin, niacin, pyridoxine, and folates) and vitamin E (α-, β-, ϒ-, δ-tocotrienols and α-, β-, ϒ-, δ-tocopherols) (Shewry 2009a, b).

Bran fraction contains higher antioxidant capacity than other milled fractions (Liyana-Pathirana and Shahidi 2007). Consumption of whole grain wheat reduces the risk of cardiovascular diseases, LDL level, and the risk of some certain cancers such as colon cancer. Moreover, it reduces the risk of type 2 diabetes and certain chronic diseases and protects the cell against free radical-induced oxidative damage (Zhou et al. 2004; Yu et al. 2005; de Munter et al. 2007; Seal 2006; Schatzkin et al. 2007; Mellen et al. 2008).

8.2 Chemical Contents and Effects of Wheat Varieties on Human Health

8.2.1 B Vitamins

Vitamin B1 (thiamine) which has a small storage in the body before discharging plays a role in maintaining the blood level and energy production in the body. Cereals, whole grains, brown rice, pork, nuts, soybeans, peas, and beans are rich in thiamine. For humans, men need to consume 1.2 mg/day of thiamine in their daily diet, women need to consume 1.1 mg/day, and babies need to consume 0.2 mg/day. Thiamine, which is included in enzymatic activities, is essential for mitochondrial activity, so decreasing the level of consumption of thiamine leads to alteration of mitochondrial activities. Deficiency of thiamine decreases energy production. Neurons are the cells which need higher energy; therefore, thiamine is important for energy requirement of neurons. Deficiency of vitamin B1 can cause Wernicke-Korsakoff syndrome (WKS) and Beriberi disease. Both these diseases are neurological and have been mainly observed in a person who consumes alcohol (Julianna and Franklin 2019). Thiamine also takes part in postnatal brain development. Moreover, it is involved in the absorption of iron, in immune defense, and in the inhibition of carcinogen-induced DNA damage (Thakur et al. 2017).

B2 (riboflavin) is another important B vitamin found in wheat. Animal products (liver, kidney, and heart), almonds, mushrooms, and whole grains are rich in vitamin B2 as well. It is a cofactor that helps in the production of flavin adenine mononucleotide (FMN) and flavin adenine dinucleotide (FAD) which are placed in energy producing process. Riboflavin decreases the risk of cardiovascular disease, complications of pregnancy, skin lesion, anemia, and nerve degeneration.

Niacin (B3) has been found in wheat as well. Niacin is included in the enzymatic activities of peripheral and brain cells. It plays an important role as a precursor for nicotinamide adenine dinucleotide (NAD) and NAD phosphate (NADP). NAD and NADP participate in energy production processes in cells. Moreover, they are high antioxidant molecules and take part in redox reactions (Kennedy 2016). Niacin deficiency causes genomic instability. Besides this, cell cycle becomes arrested, DNA repair lagged, single and double strand of DNA breakage, and cancer cell occurrence (Kirkland 2012). The most obvious disorder resulting in the deficiency of niacin is pellagra (Prakash et al. 2008).

One of the other B vitamins of wheat is B9 (folate). It helps in protein and red blood formation (Arzani and Ashraf 2017). Its deficiency causes neural tube defects in infants (NTF) (US Preventive Services Task Force 2009) during pregnancy.

B6 (pyridoxal) is also found in wheat. Besides wheat, meat, fish, nuts, and grains are also rich in B6 vitamin. B6 is a cofactor in enzymatic reaction like places in amino acid, carbohydrate, and lipid metabolisms. It is also a cofactor of gluconeogenesis and glycogenolysis. Moreover, B6 is also very important for fetal brain development (Brown and Beier 2019).

These B vitamin types are more common in wheats. But they are variable in bread and durum wheats according to the controlled scientific studies. Also, the type of these vitamins is connected to the wheat variety, location (for thiamine and riboflavin), year, and soil type. The thiamine concentration in wheat increases with fertilizer usage and decreases with pesticide usage. On the other hand, riboflavin is not affected (Batifoulier et al. 2006). Some processes such as making bread and milling grains can change the concentration of B vitamins as well (Adrian and Petit 1970).

Consumption of carbohydrate is decreasing and consumption of lipid is increasing in Western populations (Batifoulier et al. 2006). But low cereal consumption is, on the other hand, associated with the deficiency of micronutrients such as thiamine (Hercberg et al. 1994; Bertrais et al. 2000). This diet can also induce other vitamin B type deficiencies. In return, this leads to obesity, cancer, and cardio-vascular diseases (Kaaks et al. 1998; Van den Berg et al. 2002; Bruce et al. 2003). Because of that, it is very important to consume wheat to get enough micronutrients for the body. Batifoulier et al. (2006) have observed that *Triticum spelta* contained higher thiamine than *Triticum durum* and *Triticum aestivum*. In other words, *T. spelta* can decrease Wernicke-Korsakoff syndrome (WKS) and Beriberi disease.

Moreover, according to the wheat's layer and the bread type, B vitamin concentrations may change. Even if the endosperm represents nearly 85% of grain, it doesn't contain B vitamins as much as the external layers. Pyridoxine and thiamine (80%) are higher in the outer layer, whereas endosperm contains 6% of pyridoxine and 3% of thiamine (Batifoulier et al. 2006). Coarse bran (non-endosperm tissue) has also higher concentration of thiamine and pyridoxine (Fig. 8.2) (Keagy et al. 1980). Because of that, white bread contains less concentration of B vitamins than whole wheat breads. Whole wheat breads have two and nine times higher B vitamins than white wheat breads (Batifoulier et al. 2006). It provides 20% of riboflavin in our daily life. Pyridoxine of whole wheat bread, on the other hand, provides approximately 16% of daily requirements of human being (Rivlin and Pinto 2001). The process of making bread leads to decrease in the concentration of B vitamins. Thiamine and pyridoxine decrease during the bread making process. Fermentation time and type of ferment can also reduce these B vitamins as well (Batifoulier et al. 2006).

In HEALTHGRAIN project, bread wheat (spring and winter wheat genotypes), durum wheat, emmer wheat, and spelt were observed for formic acid concentration.

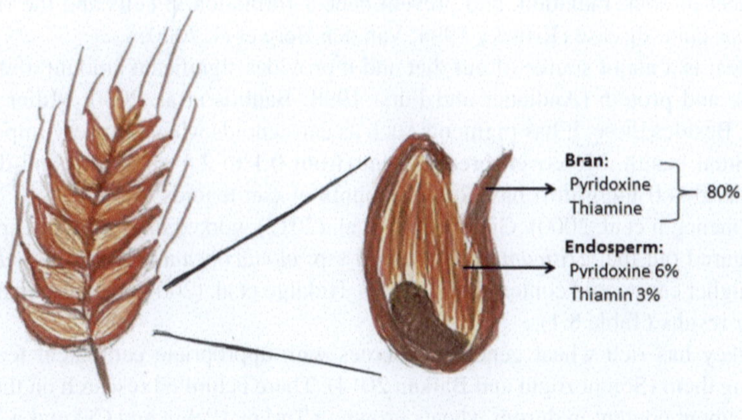

Fig. 8.2 Wheat layers where types of vitamin B are concentrated (Batifoulier et al. 2006)

The total formic acid was higher in winter wheats than spring wheats. On the other hand, durum wheat contained higher formic acid than emmer, bread wheat, and spelt. Moreover, emmer and spelt have higher B9 vitamin than bread wheat (Piironen et al. 2008a, b).

There are some reports that show spelt wheat has higher content of vitamins. For the niacin concentration, spelt cultivar, PGR8801, has higher content than einkorn, spelt SK0263, and common wheat (Abdel-Aal et al. 1995). Also, in another study, when thiamine, riboflavin, and niacin analyzed in spelt wheat and niacin found higher (5%) than red winter wheat (Ranhorta et al. 1995).

8.2.2 Carotenes

Yellow, orange, and red colors of flowers and birds' feathers come from the pigment called carotenoids. In addition to these features of carotenoids, they are lipid-soluble antioxidants, which means they reduce the risk of chronic and age-related diseases. There are two major classes of carotenoid which are carotenes and xanthophylls. Carotenes are hydrocarbon compounds; xanthophylls, on the other hand, contain oxygen atoms, which make them more polar than carotenes (Van den Berg et al. 2000). Antioxidant activities of these compounds are determined by their repeats of double and single bond in polyenoic chain (Britton 1995). As tocols, carotenoids cannot be synthesized by animals, so animals take these compounds from plants in their diet. α- and β-Carotenes play an important role in biosynthesis of vitamin A and take part in the embryo improvement and fetus development, cell replication, and visual functions (Zile 1998). With antioxidant activities of carotenoids, tissues and cells are protected from free radicals, and lutein and zeaxanthin carotenoids have big roles on health. They protect the macula region in the retina, inhibit cataract formation, exert protective effects on functions of the immune system, decrease the effect of solar radiation, and prevent cancer formation in cells and the risk of cardiovascular disease (Krinsky 1994; Van den Berg et al. 2000).

Wheat is a major source of our diet and it provides significant amount of carbohydrate and protein (Andlauer and Fürst 1998; Baublis et al. 2000; Miller et al. 2000). Besides these, it has pigments such as carotenoids which are very important for animal health. Moreover, bread wheat (from 0.1 to 2.4 mg/g dm) and durum wheat (1.5–4.0 mg/g dm) have high amounts of carotenoids (Panfili et al. 2004; Zandomeneghi et al. 2000). Giambanelli et al. (2013) worked on four wheat species and figured out that *Triticum monococcum* ssp. *monococcum* and *Triticum durum* have higher carotenoid contents than others. Hidalgo et al. (2006) have found nearly similar results (Table 8.1).

Turkey has rich wheat genetic resources with appropriate ecological features growing them (Sönmezoğlu and Balkan 2014). There is limited research on the yellow pigment content in durum wheats grown in Turkey (Pekin and Çakmakli 1987; Coşkun and Ercan 2003). Pigment values change in durum wheat depending on the genotype (Coşkun and Ercan 2003; Sayaslan et al. 2012). Yellow color in durum is very important for pasta making and preferred by pasta manufacturers, because after the flour process, losing the yellow color is not preferred. The presence of lipoxygenase (LOX) decreases the yellow pigment concentration because of oxidation. As the yellow pigment content changes in durum wheats, the LOX content also varies between genotypes (Coşkun and Ercan 2003). Among different durum varieties in Turkey, Gediz 75 has the lowest LOX concentration with higher yellow pigment concentration (Coşkun and Ercan 2003).

The ancient wheat, einkorn (*Triticum monococcum* ssp. *monococcum* L.), is sown abundantly in its motherland, Turkey. Consumption of einkorn is popular because of its high protein content. The yellow pigment of einkorn has higher carotene content than other wheat species. Lutein of whole einkorn flour is four times higher than the bread wheat (Abdel-Aal et al. 2002). Hidalgo et al. (2006) reported that lutein (on average, 91%) is the highest carotene in einkorn, and also, several einkorn varieties contained 25–33% α- + β-carotenes. *T. aestivum* has, on the other hand, acquired the lowest lutein content among einkorn and *Triticum turgidum* (Abdel-Aal et al. 2002; Adom et al. 2003; Konopka et al. 2004; Panfili et al. 2004,

Table 8.1 Lutein, α-carotene, β-carotene, and total carotenoid contents of different wheat species (Hidalgo et al. 2006)

	μg g^{-1} dry matter			
Wheat species	Lutein	α-Carotene	β- carotene	Total carotenoids
T. monococcum ssp. *monococcum*	8.20	0.33	0.29	8.82
T. turgidum ssp. *durum*	4.79	nd	0.21	5.00
T. turgidum ssp. *dicoccum*	1.90	nd	nd	1.90
T. aestivum ssp. *aestivum*	2.51	0.31	0.23	3.05
T. aestivum ssp. *spelta*	2.45	0.34	0.17	2.96

nd non-detectable

**T. monococcum* ssp. *monococcum* L. (Einkorn) species is from Turkey. The data with the highest values in the varieties were shown

Hidalgo et al. 2006). Since tocol content changes because of genotypes and environmental factors, carotenoid contents change as well (Hidalgo et al. 2006).

8.2.3 Macro- and Microelements

Very important minerals in human diet are divided into two groups (Martinez-Ballesta et al. 2009).

- Macro minerals are important and needed by the human body in high amounts. These are magnesium (Mg), calcium (Ca), potassium (K), sodium (Na), chloride (Cl), phosphorus (P), and sulfur (S) (Martinez-Ballesta et al. 2009).
- Micro minerals or trace elements including iron (Fe), zinc (Zn), iodine (I), selenium (Se), copper (Cu), manganese (Mn), fluoride (F −), chromium (Cr), and molybdenum (Mo) are required in small amount by animal or human bodies (Martinez-Ballesta et al. 2009).

Enzymes which perform metabolic functions become active with minerals. In other words, all body processes rely on minerals. Wheat is a very good nutrient-rich food across the world, because of its higher production and consumption. Therefore, receiving these minerals from wheat is very important for the health of human being (Galan et al. 1997).

Globally, more than three billion people are suffering from mineral deficiencies (Welch and Graham 2004). Some chronic diseases can arise when a person lacks specific minerals (Golden 1991; Branca and Ferrari 2002). Mineral deficiencies caused increased risk of premature death, high healthcare costs, and decreased working ability. People who are not earning enough money to buy mineral-rich foods in developing countries are at high risk of experiencing mineral deficiencies (Graham et al. 1999). Mostly Fe and Zn deficiencies are seen worldwide. Fe and Ze are unfortunately low in cereals (Hurrell 2001). Mineral concentration in wheat changes depending on organic and inorganic conditions. For instance, Mg and P in wheat have been higher under inorganic conditions than under organic conditions. On the other hand, Hussain et al. (2010) found higher mineral values in wheat by using specific organic system as well as specific genotypes.

Consumption of ancient wheat becomes popular nowadays, and it is pointed out that they have more health-supporting properties than modern wheat (Giambanelli et al. 2013). Einkorn (*T. monococcum* ssp. *monococcum*) and emmer (*T. dicoccum* Schrank) primitive wheats have higher nutritional values, such as protein and carotenoids (Grausgruber et al. 2004). Zn and Fe concentrations are higher in einkorn than the other species (Table 8.2). Worldwide, these primitive species have the potential to decrease mineral deficiencies.

Consumption of these minerals is really important for human health. For example, iron reduces drowsiness and insomnia problems; calcium reduces varicose veins, body swelling, and slow motions; potassium reduces hypertension, palpitation, and depression problems; zinc lowers prostate risk; sodium provides acid-base

Table 8.2 Mineral values in different wheat species

	mg kg^{-1} of dry weight								
Wheat species	Ca	Cu	Fe	K	Mg	Mn	Na	Zn	References
T. monococcum ssp. *monococcum*	na	6.2	51	na	na	56	na	59	Ozkan et al. (2007)
T. turgidum ssp. *durum*	na	3.31	40.54	na	na	40.53	na	52.94	Rachoń et al. (2015)
T. turgidum ssp. *dicoccum*	0.36 10^3	4.1 10^3	49	4.39 10^3	1.67 10^3	24	12	54	Suchowilska et al. (2012)
T. aestivum ssp. *aestivum*	0.45 10^3	10.19	26.21	0.81 10^3	0.78 10^3	34.90	1.76 10^3	19.69	Plaza et al. (2003)

na not analyzed

and water balance in the body; and magnesium is good for intestinal and muscle health (Kumar et al. 2016).

But there is a counterargument on bioavailability of these minerals, because of phytic availability in wheats. Pericarp and aleurone layer as well as germ of wheat contain phytic acid (Cheryan 1980). Almost all minerals exist as complex with phytic acid in wheat kernels. Phytase hydrolyzes phytate and makes the minerals nutritionally available. But phytate is an "anti-nutrient" in humans, thus affecting the bioavailability of magnesium, iron, zinc, and calcium. Phytate-mineral complexes are formed by binding phytic acid to the mineral cations. As a result, it leads to a disadvantage of absorption or hydrolysis of minerals in the human body (Stevenson et al. 2012). High phytic acid diets can also cause decreased absorption of calcium, iron, magnesium, and zinc. In some minerals, absorption of them, on the other hand, depends on the rate between minerals and phytic acid doses. For example, if the ratio of phytate/zinc is higher than 15–20, it increases the absorption of zinc in the human body (Navert et al. 1985).

In addition to the consumption of wheat seeds, wheat grass juice is consumed worldwide. Wheat grass juice or powders are usually produced from *Triticum aestivum* L. species. Chlorophyll constitutes 70% of the chemical content in wheat grass. This means that there are a lot of magnesium molecules in the grass. Instead of iron, chlorophyll contains magnesium in its structure. When we compare the structure of this molecule with hemoglobin, it is almost identical. Because of this similarity, wheat grass is known as "green blood" (Padalia et al. 2010). Consumption of wheatgrass can prevent Mg deficiency in the human body.

8.2.4 Tocols

α-, β-, ϒ-, δ-tocotrienols and α-, β-, ϒ-, δ-tocopherols are vitamin E vitamers. Recent studies show that α-tocopherols take part in vitamin E activity (European Food Safety Authority 2015). Because of antioxidant activities, tocols have serious health benefits (Lachman et al. 2018). They protect the structure of cell membrane by preserving its lipids and proteins (Wolf 2005; Dörmann 2007). Besides taking part in antioxidant activity, vitamin E also takes part in gene expression. Vitamin E is a gene regulator in the immune system with its inflammation process. Moreover, abundant tocopherols and tocotrienols in wheat and barley reduce the LDL cholesterol in blood by dropping the heptic enzyme activity (Baik and Ullrich 2008). Tocotrienols suppress breast cancer formation (Nesaretnam et al. 1998). Furthermore, tocotrienols, during tumor development, can help also in forming blood cells (Tonini et al. 2003).

Animals cannot produce tocols; thus, they acquired these components from plants in their diets. Tocols and carotenoids are lipophilic secondary metabolites and are abundant in cereals (Atanasova-Penichon et al. 2016). That is why consumption of wheats is good to inhibit tocol and vitamin E deficiency (Johansson et al. 2014). Daily consumption of 200 g of wheat provides 20% of vitamin E (Hussain et al. 2012).

Cereals contain tocopherols (α-, β-, δ-, and γ-tocopherols) and tocotrienols (α-, β-, δ-, and γ-tocotrienols), but mainly α-forms are predominant in cereals, and germ is mainly composed of tocopherols (Panfili et al. 2003; Gutierrez-Gonzalez et al. 2013). Whole wheat grain contains higher vitamin E than endosperm (Zielinski et al. 2018). Pericarp and endosperm of the cereals, on the other hand, mostly contain tocotrienols (Falk et al. 2004). In brief, tocotrienols are the major tocols in wheat, barley, and oat. Among all wheat species, einkorn and emmer have high tocol values. According to Tsao (2008) β-tocotrienol, α-tocotrienol, α-tocopherol, and β-tocopherol are abundant in einkorn, respectively. The same occurred in an experiment by Hidalgo et al. (2006) (Table 8.3).

Among 15 diploid, tetraploid, and hexaploid *Triticum* species, wild types of diploid wheats such as *T. thaoudar*, *T. aegilopoides*, *T. monococcum* ssp. *monococcum*, and *T. urartu* contain higher tocol contents (Brandolini et al. 2015; Hidalgo et al. 2006). Comparing durum, bread wheat, and triticale, bread wheat contained higher tocols than durum and triticale. According to HEALTGRAIN, outside of barley (46.2–68.8 mg/kg), tocols are abundant in spelt (40.2–50.6 mg/kg), durum wheat (40.1–62.7 mg/kg), einkorn (29.0–57.5 mg/kg), emmer (29.0–57.5 mg/kg), and bread wheat (27.6–79.7 mg/kg) (Shewry et al. 2013).

Table 8.3 Tocol values in wheat species (Hidalgo et al. 2006)

	$\mu g\ g^{-1}$ dry matter				
Wheat species	α-T	β-T	α-T3	β-T3	Total tocols
T. monococcum ssp. *monococcum*	17.35	5.19	17.83	43.43	83.80
T. turgidum ssp. *durum*	8.34	3.70	5.35	39.89	57.27
T. turgidum ssp. *dicoccum*	12.24	6.26	4.74	44.69	67.92
T. aestivum ssp. *aestivum*	18.15	11.89	6.42	38.48	74.94
T. aestivum ssp. *spelta*	16.05	10.16	5.46	37.52	69.18

α-T, α-Tocopherols; β-T, β-Tocopherols; α-T3, α-Tocotrienol; β-T3, β-Tocotrienol

**Triticum monococcum* ssp. *monococcum* L. (einkorn) species is from Turkey. The data of the varieties with the highest values were taken

Environmental conditions and genetic factors affect tocol content in wheat species. The cultivation conditions of the year can affect some wheat species' secondary metabolites such as tocols; on the other hand, some wheat species can show stable concentration of tocols even if the weather conditions change (Lachman et al. 2018). Moreover, spring and winter wheats can show different content of tocols (Lampi et al. 2008). But within all these researches, there is one thing similar with all wheat species, i.e., β-tocotrienol is a dominant tocols in all dehulled and hulled wheats (Lachman et al. 2018). On the other hand, barley, oat, rice, and corn generally have lower concentrations of β-tocotrienols than wheats (Hussain et al. 2012).

In developing countries, it is hard to consume vitamin E-rich foods, so there is a higher risk of vitamin E deficiency. The prevalence of some oxidative stressors such as HIV and malaria can also increase the depletion of vitamin E intake (Dror and Allen 2011). So, selecting vitamin E-rich wheat varieties decreases the risk of developing illnesses because of vitamin E deficiency.

8.2.5 Phenols

Eating whole grain decreases the risk of chronic diseases such as cancer, cardiovascular diseases, and diabetes. Besides containing nutrients such as proteins, carbohydrates, and dietary fiber, whole grain also contains healthy phenolic compounds (Dinelli et al. 2011; Karakas et al. 2017). The secondary metabolites of polyphenols are antioxidant, antimutagenic, anti-inflammatory, and anticancer (Fardet 2010; Karakas et al. 2017). The antioxidant compounds (i.e., phenolics) lower the damage to DNA, cell wall, and enzymes by neutralizing free radicals (ROS) (Carter et al. 2006). That is why they protect humans from the risk of chronic diseases such as cancer and cardiovascular disease (Anderson 2004).

Most phenolics are in bound form that attach to the cell wall. Because they take place in antioxidant action, they are health beneficial compounds. They are important since they cannot be digested by the upper gastrointestinal area. They are digested when they reach to the colon microflora, and, therefore, they reduce colon cancer risk and other gastrointestinal diseases (Liu 2007).

In whole grain, phenolic acid is the major form of phenolic compounds, and it consists of three forms which are soluble free, soluble conjugates, and insoluble-bound forms. Soluble conjugates make ester and ether linkages to polysaccharides in the cell wall. The insoluble-bound form of phenolic acids is the main form that makes cross-linkage with polymers in cell walls. Like other phytochemicals, phenolic acids are concentrated in the bran of cereals. White flour has, on the other hand, a lower concentration of phenolic acids than bran (Piironen et al. 2008a, b).

Phenolic acids are divided into two groups which are hydroxycinnamic acid and hydroxybenzoic acid. These two groups have acquired derivatives (Fig. 8.3). These derivatives exist in wheat species. Even though ferulic, *p*-coumaric, and vanillic acids are dominant in wheat, caffeic, *p*-hydroxybenzoic, and syringic acids have also been found. Phenolic acids in wheat are commonly seen in insoluble form and

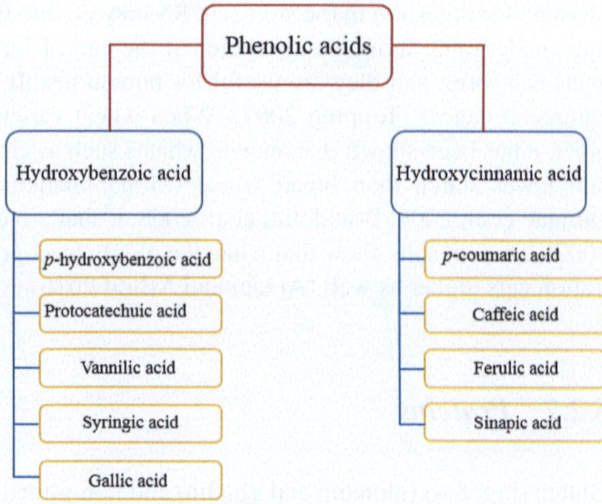

Fig. 8.3 Derivatives of hydroxybenzoic acid and hydroxycinnamic acid which are found in wheat (Li et al. 2008)

linked to the cell wall (Li et al. 2008). These phenolic acids reduce the risk of some type of cancers, cardiovascular disease, and type 2 diabetes. Also, they are decreasing the risk of age-related diseases (Serpen et al. 2008).

Phenolic acids of bread, durum, einkorn, and spelt have been compared in much research (Li et al. 2008). Emmer has got the highest concentration of total phenolic composition, while spelt has the lowest concentration of total phenolic acids. Again, in the same research, total bound phenolic acids (77%) presented the highest total phenolic acids (total free and conjugated phenolic acids). Moreover, einkorn had the lowest bound phenolic acids, whereas emmer genotypes contained the highest total bound phenolic acids. On the other hand, phenolic and ferulic acids are dominant in all wheat species.

Flavonoids include two-thirds of dietary phenols and are health beneficial ones (Robbins 2003). Lipid peroxidation which is associated with carcinogenesis, antherogenesis, and thrombosis is moderated by flavonoids. Flavonoids are free radical scavengers and antioxidants. They also hinder the oxidative and hydrolytic enzymes such as lipoxygenase, phospholipase A2, and cyclooxygenase (Dinelli et al. 2011).

8.2.6 *Starch*

Starch, which is present in white bread, is digested in the small intestine and increased the glucose level in the blood. Consumption of starch-based food increases obesity risk. People, especially Asian people, who are fed with higher levels of starch-based foods have an increased risk of having type 2 diabetes (Hu et al. 2012). Moreover, animal and human investigations have shown a relationship between type 1 diabetes and gluten. On the other hand, resistant starch (RS) has become resistant for digestion in the stomach. RS may go into the small intestine and colon; thus, undergoing fermentation there. At the end of fermentation, small-sized fatty acids can form, and they are useful for human health such as lowering the risk of colorectal cancer (Topping 2007). When wheat varieties are compared with each other, it has been shown that ancient wheats such as einkorn, emmer, and spelt contain lower starch than bread wheat (Mohammadkhani et al. 1998; Rodriguez-Quijano et al. 2004; Brandolini et al. 2008; Caballero et al. 2008; Haghayegh et al. 2007). These results show that when the ploidy level gets higher, the starch concentration gets higher as well (Arzani and Ashraf 2016).

8.2.7 *Proteins*

Gluten (Fig. 8.4) (glutenin and gliadin) and non-gluten (globulin and albumin) proteins do exist in all wheat species. Glutenin gives the elastic structure to the dough; on the other hand, gliadin gives viscosity (Arzani and Ashraf 2016). In the gastrointestinal digestion, proteins break down into peptides. Proline, which is an amino acid found abundant in gluten, causes difficulties in digestion (Arentz-Hansen et al. 2000). These kinds of peptides cause autoimmune response from the human body. For instance, celiac disease is an autoimmune disease which responds to gliadin 33-mer peptide of gluten (Shan et al. 2002). Protein content of wheat changes not only by different species but also by environment. According to the studies, the protein content of the ancient whole grains such as einkorn, emmer, and spelt is superior to bread wheat (Abdel-Aal et al. 1995; Ranhotra et al. 1996; Loje et al. 2003; Marconi and Cubadda 2005; Brandolini et al. 2008; Shewry et al. 2013). Some structural differences can give some clue about the composition of grains such as grain size and weight. Heavy and big grains have large endosperm and the

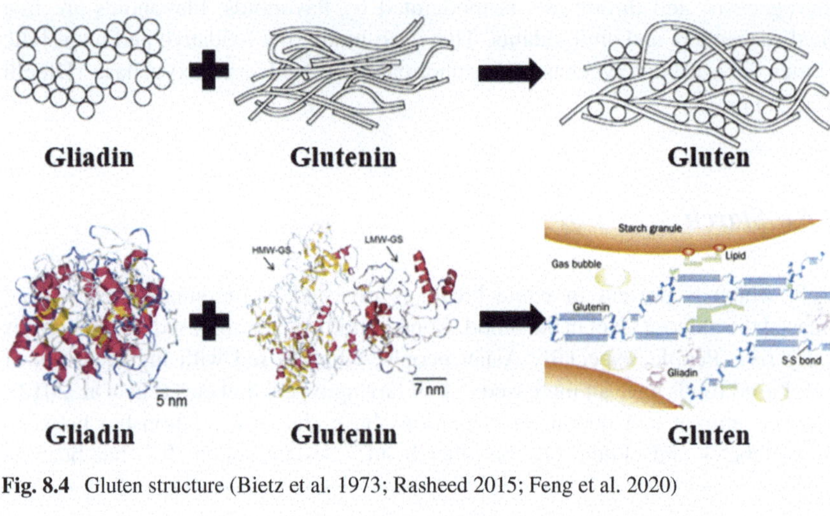

Fig. 8.4 Gluten structure (Bietz et al. 1973; Rasheed 2015; Feng et al. 2020)

endosperm contains nearly 90% of grain. Because it has bigger and heavier grain than ancient wheat, modern wheat contains lower protein content (Arzani and Ashraf 2017).

The protein of gluten causes celiac disease (CD), which is a T-cell-mediated disease. About 0.5–2% of the human population is suffering from this disease (Rewers 2005). CD is an autoimmune disease and a disease of gluten intolerance that causes inflammation on the intestine (Diosdado and Wijmenga 2005). α-Gliadin is a gluten component and has an immunodominant role in the stimulation of T cells in celiac patients (Ciccocioppo et al. 2005). Gli-2 gene codes α-gliadin. In hexaploid wheats, Gli-2 locus is placed on the short arms of three homologous chromosomes (6AS, 6BS and 6DS). Gli-A2, Gli-B2, and Gli-D2 are individual locus (Van Herpen et al. 2006) α-gliadin copies, and their distributions are different and complex in each Gli-2 locus. Considering T-cell toxicity based on α-gliadin epitopes, the Gli-2 locus in D genome is considered to be the most relevant (Van Herpen et al. 2006). Also, Spaenij-Dekking et al. (2005)'s study confirmed that epitopes of D genome species has got higher T-cell stimulation than A and B genome species. Same or equal fragment of αG-33 mer protein is encoded by the α-gliadin genes which are on chromosome 6D. Either einkorn or some pasta wheat does not have these genes, because einkorn is a diploid (A genome) and pasta wheat is a tetraploid (AB genome) (Molberg et al. 2005). Comparing Gli-D2 and Gli-B2 expression, Gli-B2 gene expression was found to be lower than Gli-D2. The expression level can be changed according to seed maturation. For example, most α-gliadin genes in Gli-A2 locus are expressed late in seed maturation (Kawaura et al. 2005).

8.2.8 Fiber

Wheat contains 11.6–12.7% of fiber in dry weight (Carson and Edwards 2009). Wheat bran contains higher amount of fiber compared to other layers. The fiber content of the bran is mostly (46%) non-starch polysaccharide (NSP). Arabinoxylan contains 70%, cellulose contains 24%, and beta-glucan contains 6% of NSP in wheat bran. Wheat has less soluble fiber than other cereals such as barley and oat (Maes and Delcour 2002).

Bound forms of polyphenols and carotenoids covalently linked to arabinoxylan. Fermentation process may release these compounds from arabinoxylan in colon (Vitaglione et al. 2008). Ferulic acid is mostly seen in bound form in wheat. They are usually bound to polysaccharides and arabinoxylans (Liu 2007; Mateo Anson et al. 2009). Even though these bound forms of compounds are released in the colon, some scientists have doubts about their absorption (Mateo Anson et al. 2009). Others suggest that these compounds have site-specific effects in the colon. Most of phenolics are in bound form in wheat, and they are difficult to digest not until they reach to the colon. Microflora in the colon digest these compounds. So, they have potential health benefit for the local area of the colon (Liu 2007).

Increasing interest in organic products in recent years has increased the inclination to rediscover and re-evaluate old wheats which are rich in vitamins, minerals, and nutritional fiber. *Triticum monococcum* ssp. *monococcum* has high fiber value that is easy to digest and is known to have a lowering effect on cholesterol (Şanal 2017).

8.3 Conclusion

Wheat is a staple crop around the world. The importance wheat cultivation is going to increase due to the threat in food security. Wheat also has an important place in the world trade (Karagöz and Zencirci 2005; Karakas et al. 2017). Because of its importance, knowledge of its contents and health support is also becoming important. It has got high antioxidant contents such as phenolics, carotenoids, and tocols. These antioxidants protect the cell membrane and ensure gene expression properly. They decrease the risk of chronic diseases such as cancer, cardiovascular disease, and type 2 diabetes (Krinsky 1994; Nesaretnam et al. 1998; Van den Berg et al. 2000; Anderson 2004; Wolf 2005; Carter et al. 2006; Dörmann 2007; Fardet 2010; Karakas et al. 2017). Wheat also contains minerals but not in high levels. Zn and Fe concentrations are higher than other elements. Especially, hemoglobin is built around Fe, which exists in red blood cells. That's why it is important to consume iron-based foods. On the other hand, Zn is effective in strengthening the immune system. The other content of wheat is starch. The perception of consuming starch-based food by humans is not positive. Even though high starch consumption can cause obesity and type 2 diabetes, resistant starch has health benefits such as decreasing the risk of colorectal cancer (Topping 2007). Gluten is the dominant protein type of wheat. In some people, consumption of gluten activates the autoimmune system. Celiac disease is an example of that autoimmune disorder (Shan et al. 2002). We cannot say that there is no gluten in any wheat species. For example, the expression degree of α-gluten is different depending on wheat species (Abdel-Aal et al. 1995; Ranhotra et al. 1996; Loje et al. 2003; Marconi and Cubadda 2005; Brandolini et al. 2008; Shewry et al. 2013). Wheat also contains fiber. These fibers are mostly concentrated in the bran layer (Maes and Delcour 2002). Consumption of whole wheat helps in digesting wheat easily. Because bound phenolics are abundant in wheat and they are bound on fibers, they are digested with the help of colon flora. Thus, they decrease the risk of colon cancer (Liu 2007). Types of B vitamins are seen in wheat as well. Riboflavin, pyridoxine, formic acid, and niacin are main vitamins of wheat. As a result, wheat, which is an essential nutrient, has many health benefits.

References

Abdel-Aal ES, Hucl P, Sosulski FW (1995) Compositional and nutritional characteristics of spring einkorn and spelt wheats. Cereal Chem 72:621–624

Abdel-Aal ESM, Young JC, Wood PJ et al (2002) Einkorn: a potential candidate for developing high lutein wheat. Cereal Chem 79:455–457. https://doi.org/10.1094/CCHEM.2002.79.3.455

Adom KK, Sorrells ME, Liu RH (2003) Phytochemical profiles and antioxidant activity of wheat varieties. J Agric Food Chem 51:7825–7834. https://doi.org/10.1021/jf030404l

Adrian J, Petit L (1970) Vitamins of cereals and their evolution during technologic treatment. Ann Nutr Aliment 24:B131–B168

Anderson JW (2004) Whole grains and coronary heart disease: the whole kernel of truth. Am J Clin Nutr 80:1459–1460. https://doi.org/10.1093/ajcn/80.6.1459

Andlauer W, Fürst P (1998) Antioxidative power of phytochemicals with special reference to cereals. Cereal Foods World 43:356–360

Arentz-Hansen H, Körner R, Molberg Ø et al (2000) The intestinal T cell response to α-gliadin in adult celiac disease is focused on a single deamidated glutamine targeted by tissue transglutaminase. J Exp Med 191:603–612. https://doi.org/10.1084/jem.191.4.603

Arzani A, Ashraf M (2016) Smart engineering of genetic resources for enhanced salinity tolerance in crop plants. CRC Crit Rev Plant Sci 35:146–189. https://doi.org/10.1080/07352689.2016.1245056

Arzani A, Ashraf M (2017) Cultivated ancient wheats (*Triticum* spp.): a potential source of health-beneficial food products. Compr Rev Food Sci Food Saf 16:477–488. https://doi.org/10.1111/1541-4337.12262

Atanasova-Penichon V, Barreau C, Richard-Forget F (2016) Antioxidant secondary metabolites in cereals: potential involvement in resistance to Fusarium and mycotoxin accumulation. Front Microbiol 7:1–16

Baik BK, Ullrich SE (2008) Barley for food: characteristics, improvement, and renewed interest. J Cereal Sci 48:233–242

Batifoulier F, Verny MA, Chanliaud E et al (2006) Variability of B vitamin concentrations in wheat grain, milling fractions and bread products. Eur J Agron 25:163–169. https://doi.org/10.1016/j.eja.2006.04.009

Baublis AJ, Lu C, Clydesdale FM et al (2000) Potential of wheat-based breakfast cereals as a source of dietary antioxidants. J Am Coll Nutr 19:308–311. https://doi.org/10.1080/07315724.2000.10718965

Bertrais S, Luque MLP, Preziosi P et al (2000) Contribution of ready-to-eat cereals to nutrition intakes in French adults and relations with corpulence. Ann Nutr Metab 44:249–255. https://doi.org/10.1159/000046692

Bietz JA, Huebner FR, Wall JS (1973) Glutenin—the strength protein of wheat flour. Bakers Digest 47(1):26–34

Branca F, Ferrari M (2002) Impact of micronutrient deficiencies on growth: the stunting syndrome. Ann Nutr Metab 46:8–17. https://doi.org/10.1159/000066397

Brandolini A, Hidalgo A, Moscaritolo S (2008) Chemical composition and pasting properties of einkorn (*Triticum monococcum* L. subsp. *monococcum*) whole meal flour. J Cereal Sci 47:599–609. https://doi.org/10.1016/j.jcs.2007.07.005

Brandolini A, Hidalgo A, Gabriele S, Heun M (2015) Chemical composition of wild and feral diploid wheats and their bearing on domesticated wheats. J Cereal Sci 63:122–127. https://doi.org/10.1016/j.jcs.2015.03.005

Britton G (1995) Structure and properties of carotenoids in relation to function. FASEB J 9:1551–1558

Brown MJ, Beier K (2019) Vitamin B6 Deficiency (Pyridoxine). In: StatPearls. StatPearls Publishing, Treasure Island (FL)

Bruce WR, Furrer R, Shangari N et al (2003) Marginal dietary thiamin deficiency induces the formation of colonic aberrant crypt foci (ACF) in rats. Cancer Lett 202:125–129. https://doi.org/10.1016/j.canlet.2003.08.005

Caballero L, Bancel E, Debiton C, Branlard G (2008) Granule-bound starch synthase (GBSS) diversity of ancient wheat and related species. Plant Breed 127:548–553. https://doi.org/10.1111/j.1439-0523.2008.01512.x

Carson GR, Edwards NM (2009) Criteria of wheat and flour quality. In: Khan K, Shewry P (eds) Wheat chemistry and technology, 4th edn. American Association of Cereal Chemists, St Paul, pp 97–118

Carter JW, Madl R, Padula F (2006) Wheat antioxidants suppress intestinal tumor activity in Min mice. Nutr Res 26:33–38. https://doi.org/10.1016/j.nutres.2005.10.005

Cheryan M (1980) Phytic acid interactions in food systems. Crit Rev Food Sci Nutr 13(4):297–335

Ciccocioppo R, Di Sabatino A, Corazza GR (2005) The immune recognition of gluten in coeliac disease. Clin Exp Immunol 140(3):408–416

Coşkun E, Ercan R (2003) Determination of lipoxygenase activity in *durum* wheats. Gıda 28(3):221–226

De Munter JSL, Hu FB, Spiegelman D et al (2007) Whole grain, bran, and germ intake and risk of type 2 diabetes: a prospective cohort study and systematic review. PLoS Med 4:1385–1395. https://doi.org/10.1371/journal.pmed.0040261

Dhanavath S, Prasada Rao UJS (2017) Nutritional and nutraceutical properties of *Triticum dicoccum* wheat and its health benefits: an overview. J Food Sci 82:2243–2250

Diamond J (1997) Location, location, location: the first farmers. Science 278:1243–1244

Dinelli G, Segura-Carretero A, Di Silvestro R et al (2011) Profiles of phenolic compounds in modern and old common wheat varieties determined by liquid chromatography coupled with time-of-flight mass spectrometry. J Chromatogr A 1218:7670–7681

Diosdado B, Wijmenga C (2005) Molecular mechanisms of the adaptive, innate and regulatory immune responses in the intestinal mucosa of celiac disease patients. Expert Rev Mol Diagn 5(5):681–700

Dörmann P (2007) Functional diversity of tocochromanols in plants. Planta 225:269–276. https://doi.org/10.1007/s00425-006-0438-2

Dror DK, Allen LH (2011) Vitamin e deficiency in developing countries. Food Nutr Bull 32:124–143. https://doi.org/10.1177/156482651103200206

European Food Safety Authority (2015) Scientific opinion on dietary reference values for vitamin E as tocopherol. EFSA Panel on dietetic products, nutrition and allergies. EFSA J 13:4149

Falk J, Krahnstöver A, Van Der Kooij TAW et al (2004) Tocopherol and tocotrienol accumulation during development of caryopses from barley (Hordeum vulgare L.). Phytochemistry 65:2977–2985. https://doi.org/10.1016/j.phytochem.2004.08.047

Fardet A (2010) New hypotheses for the health-protective mechanisms of whole-grain cereals: what is beyond fibre? Nutr Res Rev 23:65–134. https://doi.org/10.1017/S0954422410000041

Feng W, Ma S, Wang X (2020) Quality deterioration and improvement of wheat gluten protein in frozen dough. Grain Oil Sci Technol 3(1):29–37

Galan P, Preziosi P, Durlach V et al (1997) Dietary magnesium intake in French adult population. In: Theophanides T, Anastassopoulou J (eds) Magnesium: current status and new developments. Springer, Dordrecht, pp 147–149. https://doi.org/10.1007/978-94-009-0057-8_36

Giambanelli E, Ferioli F, Koçaoglu B et al (2013) A comparative study of bioactive compounds in primitive wheat populations from Italy, Turkey, Georgia, Bulgaria and Armenia. J Sci Food Agric 93:3490–3501. https://doi.org/10.1002/jsfa.6326

Gökgöl M (1935) Turkish wheats, Vol. I. Ministry of Agriculture, Yesilköy seed breeding institute publications. No: 7. Devlet Press, Istanbul. (In Turkish), 436 pp

Gökgöl M (1939) Turkish wheats, Vol. II, Yesilköy seed breeding institute publications. No. 14. Tan Press, Istanbul. (In Turkish), 955 pp

Golden MHN (1991) The nature of nutritional deficiency in relation to growth failure and poverty. Acta Paediatr Scand Suppl 80:95–110

Graham R, Senadhira D, Beebe S et al (1999) Breeding for micronutrient density in edible portions of staple food crops: conventional approaches. Field Crop Res 60:57–80. https://doi.org/10.1016/S0378-4290(98)00133-6

Grausgruber H, Scheiblauer J, Schönlechner R et al (2004) Variability in chemical composition and biologically active constituents of cereals. Genet Var Plant Breed 26:23–26

Gutierrez-Gonzalez JJ, Wise ML, Garvin DF (2013) A developmental profile of tocol accumulation in oat seeds. J Cereal Sci 57:79–83. https://doi.org/10.1016/j.jcs.2012.10.001

Haghayegh GH, Schönlechner R, Berghofer E (2007) Comparison of functional properties of isolated emmer and einkorn wheat starches. Paper presented at the Proceedings of 4th international Congress 'Flour-Bread 07' and 6th Croatian Congress of Cereal Technologists, Opatija, 24–27 October 2007

Harlan JR (1950) Collection of crop plants in Turkey. Agron J 42:258–259

Harlan JR (1981) The early history of wheat: earliest traces to the sack of Rome. In: Evans LT, Peacock WJ (eds) Wheat science – today and to-morrow. Cambridge Press, pp 1–19

Hercberg S, Preziosi P, Galan P (1994) Vitamin status of a healthy French population: dietary intakes and biochemical markers. Int J Vitam Nutr Res 64:220–232

Heun M, Schafer-Pregl R, Klawan D, Castagna R, Accerbi M, Borghi B, Sala-mini F (1997) Site of Einkorn domestication identified by DNA finger printing. Science 278:1312–1314

Hidalgo A, Brandolini A, Pompei C, Piscozzi R (2006) Carotenoids and tocols of einkorn wheat (*Triticum monococcum* ssp. *monococcum* L.). J Cereal Sci 44:182–193. https://doi.org/10.1016/j.jcs.2006.06.002

Hu EA, Pan A, Malik V, Sun Q (2012) White rice consumption and risk of type 2 diabetes: meta-analysis and systematic review. BMJ 344:1–9. https://doi.org/10.1136/bmj.e1454

Hurrell RF (2001) Modifying the composition of plant foods for better human health. In: Nösberger J, Geiger HH, Struik PC (eds) Crop science: progress and prospects. CABI Publishing, New York, pp 53–64

Hussain A, Larsson H, Kuktaite R, Johansson E (2010) Mineral composition of organically grown wheat genotypes: contribution to daily minerals intake. Int J Environ Res Public Health 7:3442–3456. https://doi.org/10.3390/ijerph7093442

Hussain A, Larsson H, Olsson ME et al (2012) Is organically produced wheat a source of tocopherols and tocotrienols for health food? Food Chem 132:1789–1795. https://doi.org/10.1016/j.foodchem.2011.11.141

Johansson E, Hussain A, Kuktaite R et al (2014) Contribution of organically grown crops to human health. Int J Environ Res Public Health 11:3870–3893

Julianna LM, Franklin DS. Last Update: January 26, 2019. Vita-min B1 (Thiamine). https://www.ncbi.nlm.nih.gov/books/NBK482360/

Kaaks R, Tuyns AJ, Haelterman M, Riboli E (1998) Nutrient intake patterns and gastric cancer risk: a case-control study in Belgium. Int J Cancer 78:415–420. https://doi.org/10.1002/(SICI)1097-0215(19981109)78:4<415:AID-IJC4>3.0.CO;2-X

Karagöz A, Zencirci N (2005) Variation in wheat (Triticum spp.) landraces from different altitudes of three regions of Turkey. Genet Resour Crop Evol 52(6):775–785

Karakaş FP, Bozat BG, Aslan D, Zencirci N (2017) Diversity exists in development parameters and enhancement of antioxidant mechanisms of some einkorn and bread wheats under combined water deficits and salt stress. Prog Nutr 19:460–471. https://doi.org/10.23751/pn.v19i4.6203

Kawaura K, Mochida K, Ogihara Y (2005) Expression profile of two storage-protein gene families in hexaploid wheat revealed by large-scale analysis of ex-pressed sequence tags. Plant Physiol 139(4):1870–1880

Keagy PM, Borenstein B, Ranum P et al (1980) Natural levels of nutrients in commercially milled wheat flours. II: vitamin analysis. Cereal Chem 57:59–65

Kennedy DO (2016) B vitamins and the brain: mechanisms, dose and efficacy—a review. Nutrients 8. https://doi.org/10.3390/nu8020068

Kirkland JB (2012) Niacin requirements for genomic stability. Mutat Res Fundam Mol Mech Mutagen 733:14–20. https://doi.org/10.1016/j.mrfmmm.2011.11.008

Konopka I, Kozirok W, Rotkiewicz D (2004) Lipids and carotenoids of wheat grain and flour and attempt of correlating them with digital image analysis of kernel surface and cross-sections. Food Res Int 37:429–438. https://doi.org/10.1016/j.foodres.2003.12.009

Krinsky NI (1994) The biological properties of carotenoids. Pure Appl Chem 66:1003–1010

Kumar NS, Murali M, Nair AM, Nair AS (2016) Green blood therapy of wheat grass – nature' s finest medicine' – a literature review. J Pharm Biol Sci 11:57–64. https://doi.org/10.9790/3008-1102045764

Lachman J, Hejtmánková A, Orsák M et al (2018) Tocotrienols and tocopherols in colored-grain wheat, tritordeum and barley. Food Chem 240:725–735. https://doi.org/10.1016/j.foodchem.2017.07.123

Lampi AM, Nurmi T, Ollilainen V, Piironen V (2008) Tocopherols and tocotrienols in wheat genotypes in the healthgrain diversity screen. J Agric Food Chem 56:9716–9721

Lev-Yadun S, Gopher A, Abbo S (2000) The cradle of agriculture. Science 288:1602–1603

Li L, Shewry PR, Ward JL (2008) Phenolic acids in wheat varieties in the healthgrain diversity screen. J Agric Food Chem 56:9732–9739. https://doi.org/10.1021/jf801069s

Liu RH (2007) Whole grain phytochemicals and health. J Cereal Sci 46:207–219. https://doi.org/10.1016/j.jcs.2007.06.010

Liyana-Pathirana CM, Shahidi F (2007) The antioxidant potential of milling fractions from breadwheat and *durum*. J Cereal Sci 45:238–247. https://doi.org/10.1016/j.jcs.2006.08.007

Loje H, Møller B, Laustsen AM, Hansen Å (2003) Chemical composition, functional properties and sensory profiling of einkorn (*Triticum monococcum* L.). J Cereal Sci 37:231–240. https://doi.org/10.1006/jcrs.2002.0498

Maes C, Delcour JA (2002) Structural characterization of water-extractable and water-unextractable arabinoxylans in wheat bran. J Cereal Sci 35:315–326. https://doi.org/10.1006/jcrs.2001.0439

Marconi M, Cubadda R (2005) Emmer wheat. In: Abdel-Aal ESM, Wood P (eds) Speciality grains for food and feed. American Association of Cereal Chemists, Inc., St. Paul, pp 63–108

Martinez-Ballesta MC, Dominguez-Perles R, Moreno DA et al (2009) Minerals in plant food: effect of agricultural practices and role in human health. Sustain Agric 2:111–128. https://doi.org/10.1007/978-94-007-0394-0_8

Mateo Anson N, van den Berg R, Havenaar R et al (2009) Bioavailability of ferulic acid is determined by its bio accessibility. J Cereal Sci 49:296–300. https://doi.org/10.1016/j.jcs.2008.12.001

Mellen PB, Walsh TF, Herrington DM (2008) Whole grain intake and cardiovascular disease: a meta-analysis. Nutr Metab Cardiovasc Dis 18:283–290. https://doi.org/10.1016/j.numecd.2006.12.008

Miller HE, Rigelhof F, Marquart L et al (2000) Antioxidant content of whole grain breakfast cereals, fruits and vegetables. J Am Coll Nutr 19:312S–319S. https://doi.org/10.1080/07315724.2000.10718966

Mohammadkhani A, Stoddard FL, Marshall DR (1998) Survey of amylose content in Secale cereale, *Triticum monococcum*, *T. turgidum* and *T.* tauschii. J Cereal Sci 28:273–280. https://doi.org/10.1016/S0733-5210(98)90007-8

Molberg Ø, Uhlen AK, Jensen T et al (2005) Mapping of gluten T-cell epitopes in the bread wheat ancestors: implications for celiac disease. Gastroenterology 128:393–401

Navert B, Sandstron B, Cederblad A (1985) Reduction of the phytate con-tent of bran by leavening in bread and its effect on zinc absorption in man. Br J Nutr 53:47–53

Nesaretnam K, Stephen R, Dils R, Darbre P (1998) Tocotrienols inhibit the growth of human breast cancer cells irrespective of estrogen receptor status. Lipids 33:461–469

Nesbit M, Samuel D (1998) Wheat domestication: archaebotanical evidence. Science 279:1433–1435

Özkan H, Brandolini A, Torun A et al (2007) Natural variation and identification of microelements content in seeds of einkorn wheat (*Triticum monococcum*). In: Wheat production in stressed environments. Springer, Dordrecht, pp 455–462. https://doi.org/10.1007/1-4020-5497-1_55

Padalia S, Drabu S, Raheja I et al (2010) Multitude potential of wheatgrass juice (green blood): an overview. Chronicles Young Sci 1:23–28
Panfili G, Fratianni A, Irano M (2003) Normal phase high-performance liquid chromatography method for the determination of tocopherols and tocotrienols in cereals. J Agric Food Chem 51:3940–3944. https://doi.org/10.1021/jf030009v
Panfili G, Fratianni A, Irano M (2004) Improved normal-phase high-performance liquid chromatography procedure for the determination of carotenoids in cereals. J Agric Food Chem 52:6373–6377. https://doi.org/10.1021/jf0402025
Pekin F, Çakmakli Ü (1987) Research on technological colour features of some turkish breed wheat varieties. Turkey Cereal Symposium, 06–09 October 1987, Bursa, Turkey, pp 18–20
Piironen V, Edelmann M, Kariluoto S, Bedo Z (2008a) Folate in wheat genotypes in the HEALTHGRAIN diversity screen. J Agric Food Chem 56:9726–9731. https://doi.org/10.1021/jf801066j
Piironen V, Lampi AM, Ekholm P et al (2008b) Micronutrients and phytochemicals in wheat grain. In: Khan K, Shewry PR (eds) Wheat: chemistry and technology, 4th edn. AACC, St. Paul, in press, pp 179–222
Plaza L, De Ancos B, Cano MP (2003) Nutritional and health-related compounds in sprouts and seeds of soybean (Glycine max), wheat (*Triticum aestivum*.L) and alfalfa (Medicago sativa) treated by a new drying method. Eur Food Res Technol 216:138–144. https://doi.org/10.1007/s00217-002-0640-9
Prakash R, Gandotra S, Singh LK et al (2008) Rapid resolution of delusional parasitosis in pellagra with niacin augmentation therapy. Gen Hosp Psychiatry 30:581–584. https://doi.org/10.1016/j.genhosppsych.2008.04.011
Rachon L, Szumiło G, Brodowska M, Woźniak A (2015) Nutritional value and mineral composition of grain of selected wheat species depending on the intensity of a production technology. J Elem 20:705–715. https://doi.org/10.5601/jelem.2014.19.4.640
Ranhotra GS, Gelroth JA, Glaser BK, Lorenz KJ (1995) Baking and nutritional qualities of a spelt wheat sample. LWT Food Sci Technol 28:118–122. https://doi.org/10.1016/S0023-6438(95)80022-0
Ranhotra GS, Gelroth JA, Glaser BK, Stallknecht GF (1996) Nutritional profile of three spelt wheat cultivars grown at five different locations. Cereal Chem 73:533–535
Rasheed F (2015) Tailoring the structure-function relationship in wheat Gluten. Doctoral thesis, Swedish University of Agricultural Sciences, Alnarp 2015:13
Rewers M (2005) Epidemiology of celiac disease: what are the prevalence, in-cidence, and progression of celiac disease? Gastroenterology 128(4):47–51
Rivlin RS, Pinto JT (2001) Riboflavin. In: Ruker RB, Suttie JW (eds) The handbook of vitamins, 3rd edn. Dekker, New York, pp 255–273
Robbins RJ (2003) Phenolic acids in foods: an overview of analytical methodology. J Agric Food Chem 51:2866–2887. https://doi.org/10.1021/jf026182t
Rodriguez-Quijano M, Vázquez JF, Carrillo JM (2004) Waxy proteins and amylose content in diploid Triticeae species with genomes A, S and D. Plant Breed 123:294–296. https://doi.org/10.1111/j.1439-0523.2004.00883.x
Şanal T (2017) Bazi Yerel Buğday Çeşitlerínín Kalíte Parametrelerí. Türkiye Tohumcular Birliği Derg 24:27–31
Sayaslan A, Koyuncu M, Yildirim A et al (2012) Some quality characteristics of selected *durum* wheat (*Triticum durum*) landraces. Turk J Agric For 36:749–756. https://doi.org/10.3906/tar-1201-69
Schatzkin A, Mouw T, Park Y et al (2007) Dietary fiber and whole-grain consumption in relation to colorectal cancer in the NIH-AARP Diet and Health Study. Am J Clin Nutr 85:1353–1360. https://doi.org/10.1093/ajcn/85.5.1353
Seal CJ (2006) Whole grains and CVD risk. Proc Nutr Soc 65:24–34. https://doi.org/10.1079/pns2005482

Serpen A, Gökmen V, Pellegrini N, Fogliano V (2008) Direct measurement of the total antioxidant capacity of cereal products. J Cereal Sci 48:816–820. https://doi.org/10.1016/j.jcs.2008.06.002

Shan L, Molberg Ø, Parrot I et al (2002) Structural basis for gluten intolerance in Celiac Sprue. Science (80-) 297:2275–2279. https://doi.org/10.1126/science.1074129

Shewry PR (2009a) The HEALTHGRAIN programme opens new opportunities for improving wheat for nutrition and health. Nutr Bull 34:225–231. https://doi.org/10.1111/j.1467-3010.2009.01747.x

Shewry PR (2009b) Wheat. J Exp Bot 60:1537–1553. https://doi.org/10.1093/jxb/erp058

Shewry PR, Hawkesford MJ, Piironen V et al (2013) Natural variation in grain composition of wheat and related cereals. J Agric Food Chem 61:8295–8303. https://doi.org/10.1021/jf3054092

Sönmezoğlu ÖA, Balkan AS (2014) Molecular and biochemical analysis of *durum* wheat genotypes to examine carotenoid pigment content and lipoxygenase enzyme activity. Cereal Res Commun 42:218–228. https://doi.org/10.1556/CRC.2013.0070

Spaenij-Dekking L, Kooy-Winkelaar Y, Van Veelen P et al (2005) Natural variation in toxicity of wheat: potential for selection of nontoxic varieties for celiac disease patients. Gastroenterology 129:797–806

Stevenson L, Phillips F, O'sullivan K, Walton J (2012) Wheat bran: its composition and benefits to health, a European perspective. Int J Food Sci Nutr 63:1001–1013. https://doi.org/10.3109/09637486.2012.687366

Suchowilska E, Wiwart M, Kandler W, Krska R (2012) A comparison of macro- and microelement concentrations in the whole grain of four *Triticum* species. Plant Soil Environ 58:141–147. https://doi.org/10.17221/688/2011-pse

Thakur K, Tomar SK, Singh AK et al (2017) Riboflavin and health: a review of recent human research. Crit Rev Food Sci Nutr 57:3650–3660. https://doi.org/10.1080/10408398.2016.1145104

Tonini T, Rossi F, Claudio PP (2003) Molecular basis of angiogenesis and cancer. Oncogene 22:6549–6556. https://doi.org/10.1038/sj.onc.1206816

Topping D (2007) Cereal complex carbohydrates and their contribution to human health. J Cereal Sci 46:220–229. https://doi.org/10.1016/j.jcs.2007.06.004

Tsao R (2008) Carotenoid, tocopherol, lignin, flavonoid, and phytosterol compositions of wheat grain and its fractions. In: Yu L (ed) Wheat antioxidants. Wiley, Hoboken, pp 42–53

U.S. Preventive Services Task Force (2009) Folic acid for the prevention of neural tube defects: U.S. preventive services task force recommendation statement. Ann Intern Med 150(9):626–631. https://doi.org/10.7326/0003-4819-150-9-200905050-00009. PMID: 19414842

Van Den Berg H, Faulks R, Granado HF et al (2000) The potential for the improvement of carotenoid levels in foods and the likely systemic effects. J Sci Food Agric 80:880–912. https://doi.org/10.1002/(SICI)1097-0010(20000515)80:7<880:AID-JSFA646>3.0.CO;2-1

Van den Berg H, Van der Gaag M, Hendriks H (2002) Influence of lifestyle on vitamin bioavailability. Int J Vitam Nutr Res 72:53–59. https://doi.org/10.1024/0300-9831.72.1.53

van Herpen TWJM, Goryunova SV, van der Schoot J et al (2006) Alpha-gliadin genes from the A, B, and D genomes of wheat contain different sets of celiac disease epitopes. BMC Genomics 7:1–13. https://doi.org/10.1186/1471-2164-7-1

Vitaglione P, Napolitano A, Fogliano V (2008) Cereal dietary fibre: a natural functional ingredient to deliver phenolic compounds into the gut. Trends Food Sci Technol 19:451–463. https://doi.org/10.1016/j.tifs.2008.02.005

Welch RM, Graham RD (2004) Breeding for micronutrients in staple food crops from a human nutrition perspective. J Exp Bot 55:353–364. https://doi.org/10.1093/jxb/erh064

Wolf G (2005) The discovery of the antioxidant function of vitamin E: the contribution of Henry A. Mattill. J Nutr 135:363–366

Yu L, Zhou K, Parry JW (2005) Inhibitory effects of wheat bran extracts on human LDL oxidation and free radicals. LWT Food Sci Technol 38:463–470. https://doi.org/10.1016/j.lwt.2004.07.005

Zandomeneghi M, Festa C, Carbonaro L et al (2000) Front-surface absorbance spectra of wheat flour: determination of carotenoids. J Agric Food Chem 48:2216–2221. https://doi.org/10.1021/jf9909003
Zhou K, Su L, Yu L (2004) Phytochemicals and antioxidant properties in wheat bran. J Agric Food Chem 52:6108–6114
Zhukovskyi PM, Kipcak C, Nouruzhan H, Türkistanli S (1951) Ecological types and economic importance of Turkish durum wheat, agricultural structure of Turkey. Turk Sugar Beet Plants Publ 20:158–214
Zielinski H, Ciska E, Kozlowska H (2018) The cereal grains: focus on vitamin E. Czech J Food Sci 19:182–188. https://doi.org/10.17221/6605-cjfs
Zile MH (1998) Vitamin A and embryonic development: an overview. J Nutr 128:455–458

Chapter 9
Climate Change and Global Warming Effect(s) on Wheat Landraces: A General Approach

Hakan Ulukan

9.1 Introduction

The world's human population is predicted to reach over 12 billion by the year 2050 (Anonymous 2018a) according to UN, and population projection and shifts in diets toward animal products, oils, and other resource-intensive foodstuffs are placing even more pressure on agricultural systems to increase production (Kastner et al. 2012). Changes in temperatures and precipitation are known as climate change (CC). Nevertheless, global warming (GW) is a different phenomenon. Shortly, their impacts depend on their size and frequency/frequencies (Semenov et al. 2014). As known, CC affects many sectors, particularly the agricultural sector (Mengü et al. 2008), with its amount and time (Valizadeh et al. 2014). The main factor, due to the human activity, is an increase in the greenhouse gas concentration (CO_2, CH_4, N_2O, and types of halocarbons (CFC)), and the gases that regulate the climate system and absorb the sun's light rays (Tubiello et al. 2000).

While *climate* is important for agricultural production due to its parameters such as temperature, precipitation, humidity, etc., global warming threatens the agricultural production because of GHG accumulation in the atmosphere. Due to GHG accumulation, sunlight is not reflected back to space, thereby increasing the Earth's temperature (IPCC 2007, 2014). Researcher Fuhrer (2003) reported that global warming would lead to an increase in world temperature by 2100 (1.4–5.8) °C, leading to significant agricultural losses. The amount of losses is related to the increase in CO_2 and CH_4 and other greenhouse gas concentrations in the atmosphere (Zavarzin 2001). The CO_2 gas content was 270 ppm before the Industrial Revolution, reaching up to 355 ppm in the modern age. It is expected to reach 600 ppm in the twenty-first century (Rogers et al. 1994) (Fig. 9.1).

H. Ulukan (✉)
Ankara University, Agricultural Faculty, Field Crops Department, Ankara, Turkey
e-mail: ulukan@ankara.edu.tr

N. Zencirci et al. (eds.), *Wheat Landraces*,
https://doi.org/10.1007/978-3-030-77388-5_9

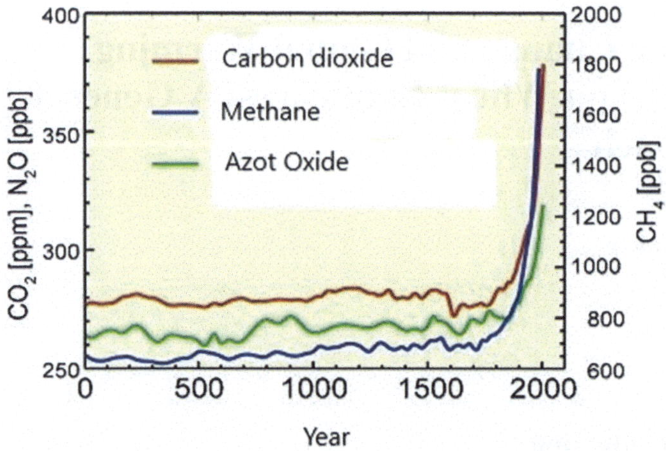

Fig. 9.1 Some greenhouse gas concentrations and global temperature change over the last 2000 years (Di Norcia 2008)

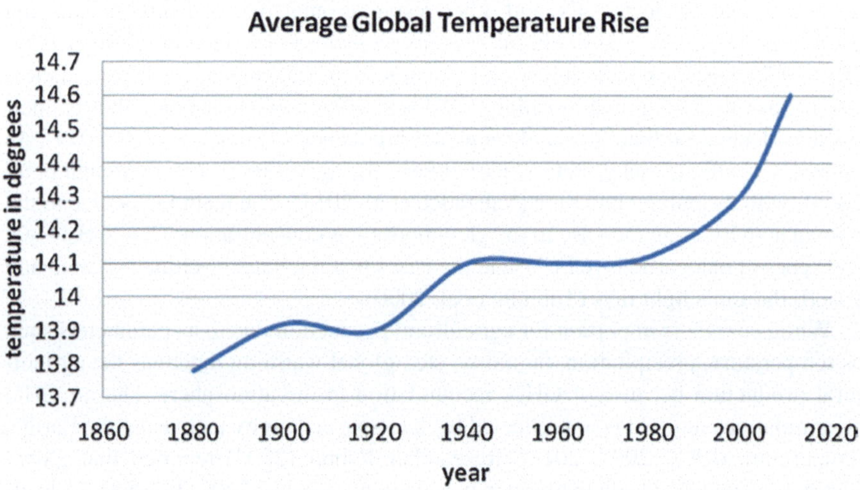

Fig. 9.2 The graph of mean global temperature rise in the measured period (Nema et al. 2012)

The most recent value of CO_2 gas is 411.97 ppm according to the records in March 2019 (https://www.co2.earth/). Greenhouse gases CH_4 and N_2O have annual growth rates of 1% and 0.3%, respectively. All GHG gases protect the Earth by acting as a shield against harmful rays of the sun and negatively affect O_3 gas in the troposphere (Krupa 1997); they are (GHGs) spreading from the refineries, rice paddies, and various elements such as the atmosphere (Mei et al. 2007) (Fig. 9.2).

The effect(s) of CC and GW can be illustrated as follows (Fig. 9.3):

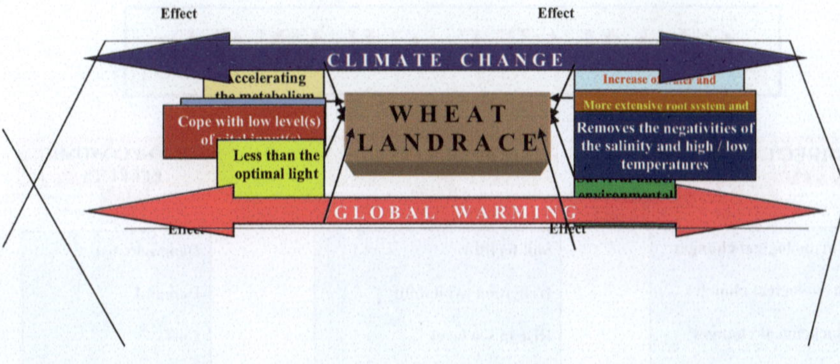

Fig. 9.3 Schematic showing CC and GW effects on wheat landraces (original)

Future climate change scenarios suggest that abiotic stress may occur at unexpected stages of plant development, thus decreasing yield consistency, and various global warming scenarios could reduce wheat productivity in zones where optimal temperature already exists, potentially increasing food insecurity and poverty (Elissavet et al. 2014), and consequently all the landrace's genotypes, including the wheat, have a very important/crucial/vital place in plant breeding for the elimination of many agricultural deficiencies against abiotic and biotic stresses. However, in predominantly self-pollinated species like wheat, no long-term investments are attractive when farmers use their own seeds (Stamp and Visser 2012). There is an inverse relationship between the ability to compete with yield and adverse conditions in the wheat lands, and they are valuable gene pools due to many other superior properties (such as high protein content) (Lopes et al. 2015) (Fig. 9.4).

On the other hand, CC and GW have some both positive and negative effects. All of them can be illustrated as follows (Fig. 9.5):

Climate change refers to changes in climate measures over a long period of time, say approximately 100 years, but global warming is a natural phenomenon that affects all living and nonliving things arising from greenhouse gases. The information obtained in all these processes is very important for the sustainability of the agricultural sector. In fact, the agricultural sector is extremely vital to various inputs (such as biodiversity, soil, water, etc.). However, in any case, CC and its natural consequence, the GW, are the factors that threaten our planet, and the effect is getting felt more and more every day. In another study, it was found that wheat yield decreases by 4.1% to 6.4% in each crop due to global climate change. Consumption is estimated to be more than 30% in 40 years of production (Tricker et al. 2018). As known, all culture plants are classified as C3, C4, and Crassulacean acid metabolism (CAM) according to the number of carbons they bind to the nutrients they form by photosynthesis. C3 plants are trees, edible legumes, rice (*Oryza* spp.), wheat (*Triticum* spp.), barley (*Hordeum* spp.), soybean (*Glycine max*), potato (*Solanum tuberosum*), vegetables, citrus (*Citrus* spp.), grape (*Vitis vinifera*), coffee (*Coffea arabica*), tea (*Camellia sinensis*), peanut (*Arachis hypogea*), lemon (*Citrus limon*), peach (*Prunus persica*), mango (*Mangifera indica*), carrot (*Daucus carota*), etc.

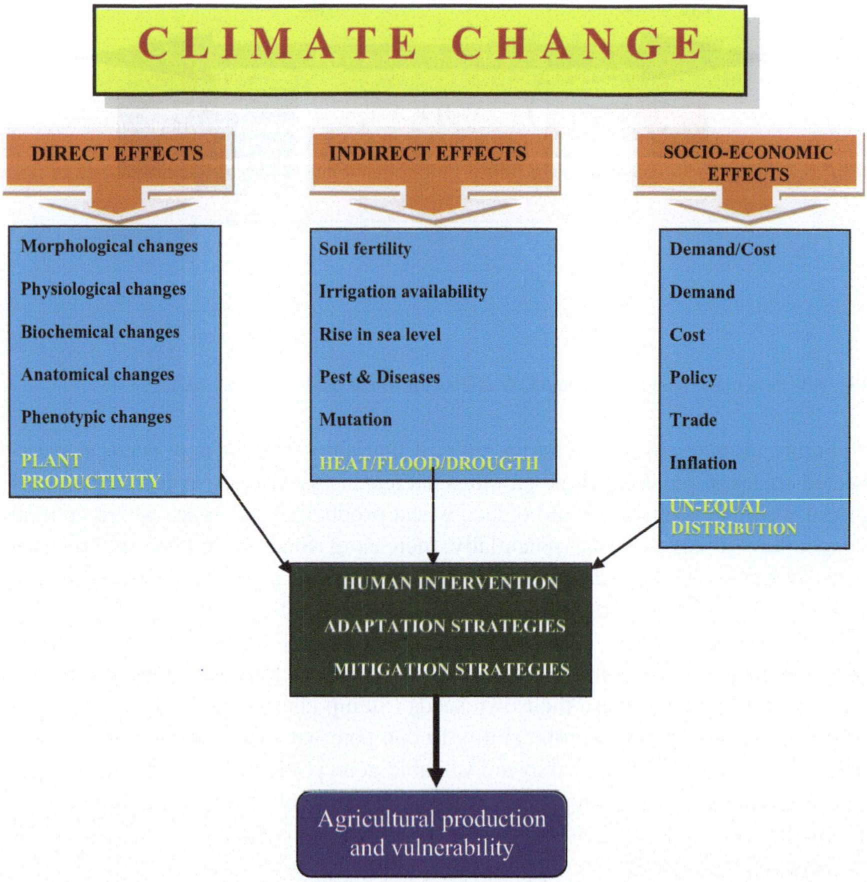

Fig. 9.4 Direct, indirect, and socioeconomic effects of climate change on agricultural production (Rosenzweigh and Hillel 1995; Raza et al. 2019)

with a sowing area of 200 million hectares and constitute approximately 21% of the total nutritional requirements in the world (Anonymous 2018a). Such plants are less affected by CC limitations (as compared to C4) due to CO_2 fertilization, but their yields increase as much as 36% (Uzmen 2007; Mercer and Perales 2010), but afterward, they immediately reduce. The situation may even reduce the photosynthetic activity in other C3 plants outside the grain (Zhai and Zhuang 2009). It is concluded that temperature extremes are complementary to the important physiological parameters in wheat landraces. Frost and heat events cause infertility in bread wheat landraces and cut the grain formation, and the excess heat decreases the number of grains and narrows the formation process. CC's photosynthetic activity with increased photosynthesis rate but increased CO_2 concentration and decreased the WUE values (Dhakhwa and Campbell 1998) was expressed. All the C3 plants,

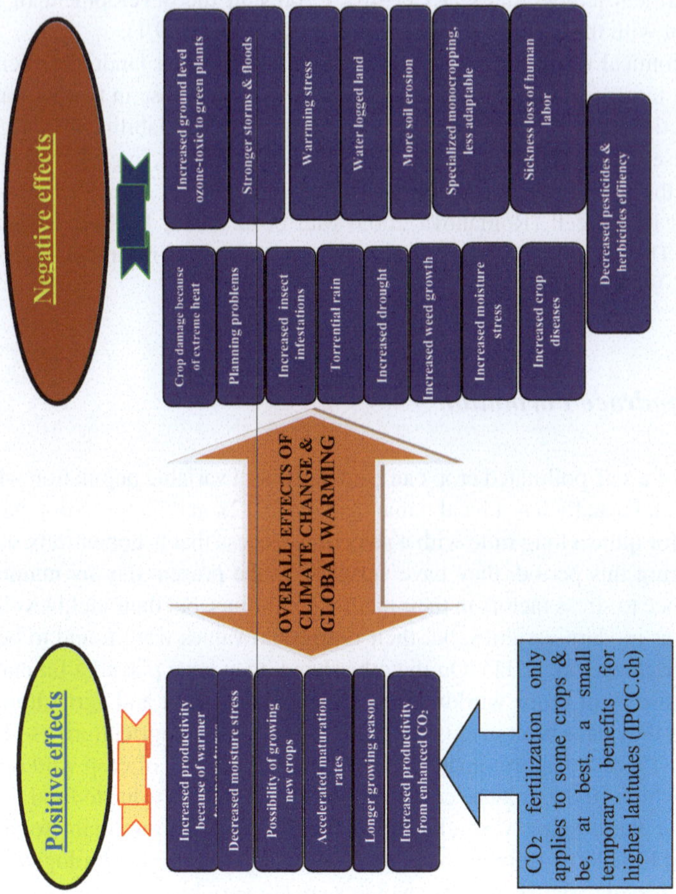

Fig. 9.5 Overall positive and negative effects of climate change and global warming on crops and humans (Raza et al. 2019)

especially local wheat landraces, have different responses to increased CO_2 gas concentration and other climate factors.

Obtained findings showed that elevated CO_2 increases. Findings show that dramatically affect the growth and development of plants against CC and its consequent of GW (Romanova 2005); elevated CO_2 also increases WUE, and is particularly distinct in C3 plants (Cutforth et al. 2007); especially high temperature increasing during the flowering and growing stages decrease grain filling rate, nutrient balance, and all of these adversely affects the fruit and grain formation, especially the critical temperatures at (35–40) °C prevent the development of pollen development with the meiosis division (Fuhrer 2009) (Table 9.1).

The anatomical effects of CC on plants, including wheat landraces, generally result in the increase of the CO_2 concentration and the increase in temperature and their interaction(s). Changes happen in the thickness and viability of leaves as a result of these factors and interaction, decrease in plant height, in growth and development; at the stomata, decrease in the water uptake of the increased amount of chloroplasts in the cell (Romanova 2005; Mei et al. 2007; Ulukan 2008, 2009) (Table 9.2). They rarely fail in the most extremely stressed environments (Ceccarelli 1994) (Figs. 9.6 and 9.7).

9.1.1 Landrace Formation

A landrace of a self-pollinated crop can be defined as a variable population which is identifiable and usually has a local name (Jaradat 2012), and its formation has been carried out for quite a long time with a selection process that is not entirely done by human. During this period, they have survived to the present day by maintaining their resilience to stress factors in their natural conditions, but their yield levels were not as high as modern varieties, but their nutritional values were found to be quite high (Nasserlehaq et al. 2011). On the other hand, they have played a fundamental role in the history of crops worldwide, in crop improvement and agricultural production, and they have been in existence since the origins of agriculture itself (Villa et al. 2006). There are approximately 50,000–60,000 species of crop wild relatives (CWR), of which 10,000 may be considered of high potential value to food security, with 1000 of these being very closely related to the most important food crops (Maxted and Kell 2009; Dempewolf et al. 2014). They have higher biological yields than the cultivated varieties, root dry weights are not very high, but can be increased depending on the situation, transpiration efficiency is higher, soluble carbohydrate concentration is higher than early (early dry matter transfer), early ripening or maturity, grain yields lower (due to earliness) and escape from drought, alternative growing nature, low harvest index, taller and united to low nitrogenous conditions with micronutrients (such as Cu, Fe, Mg, Mn, P, Se, and Zn), especially wheat landraces in the Southeastern Anatolia, the response to fertilizer low, which are not suitable for machine agriculture, sensitive to leaf diseases, adaptation ability is high, grain

Table 9.1 Some agronomic and botanic responses to CC and GW of the wheat landraces' traits. (Modified from Krupa (1997), Dhakhawa and Campbell (1998), Tubiello and Ewert (2002))

Some botanic and agronomic traits	Some CC and GW components			Response
	(Elevated CO_2)	(Elevated UV-A, B)	(Elevated O_3)	
Roots	+	?	?	Root/stem
Photosynthesis	+ in C_3, – in C_4	– in many [C_3 vs C_4]	– in many [C_3 vs C_4]	Yield and respiration
Leaf conductance and leaf development	– in [C_3, C_4]	Majority	– in susceptibles	Leaf area
Water use efficiency (WUE)	+ in [C_3, C_4]	+ in [C_3, C_4]	– in susceptibles	Stomatal conductivity and apertures
Leaf area	More in C_3	– in [C_3, C_4]	– in susceptibles	PAR point
Leaf thickness	+	– in minority	– in susceptibles	?
Maturity and thresh	+	Non-affected	?	Vegetative stage, yield
Flowering (anthesis)	Happens early	Prevents and stimulates	– Flower number and flowering day number	Vegetative stage, yield
Number of days from planting to maturity	?			
Dry matter production	Doubles in C_3, Unknown in C_4	Wide variation	Wide variation	Yield
Susceptibility of species and genus	Varied	Varied	Varied	Yield
Drought resistance	Varied from susceptible to resistance	Varied from susceptible to resistance	Varied from susceptible to resistance	Wiltness, dwarfness, death
Mineral matter	Less response	Some are lots, some are less susceptible	Susceptible to O_3	Dwarfness, death
Vernalization	+	?	?	Vegetative stage
(CO_2/O_2)	?	?	?	Photosynthesis
Respiration rate	?	?	?	Biomass
Seed formation period	?	?	?	Yield
Biomass production	+	?	?	Yield
Internode number	?	?	?	Green part, P. height
Weed distribution	?	?	?	+
Seed germination	+	?	?	Distribution
Rhizomes	+	?	?	Distribution
Seed longevity	+	?	?	?

(continued)

Table 9.1 (continued)

Some botanic and agronomic traits	Some CC and GW components			Response
	(Elevated CO_2)	(Elevated UV-A, B)	(Elevated O_3)	
DNA and sterility	+	?	?	Mutation, death
Ecological factors	+	+	+	Stress

CO_2 carbon dioxide, *O_2* oxygen, *UV-A, B* ultraviolet A, B, *O_3* ozone, + increasing, – decreasing, ? unknown, *WUE* water use efficiency, *PAR* photosynthetic activity radiation

Table 9.2 Some agronomic traits which are based physiologically on wheat landraces (Fischer 2001)

Trait	Related with yield	Heredity	Reflection to genotype
Growth and dry matter distribution			
Growth ratio	No in material	Unknown	Very high
Harvest index	Middle-high	Low-middle	High
Spike index at flowering	Middle	Unknown	Very high
Grain weight at the unit spike	Middle	Unknown	Very high
Leaf effectiveness			
Stoma conductance	Middle	Middle	High
Leaf resistance to air flow	Middle	Middle	Middle
Depression of canopy temp.	Low-middle	Unknown	Low-middle
Distribution of oxygen-18	Middle	Unknown	High
Photosynthetic activity	Low-middle	Low	High
Fluorescence of chlorophyll	Low-middle	Middle	High
Distribution of carbon-13	Low-middle	Middle	High
Leaf greenness	Low	Unknown	Low
Leaf density	Low	Low-middle	Low
Yield components			
Spike number at m^2	No	Middle-high	Low
Spike number	No-low	Low-middle	Middle
Spikelet number in spike	No	Middle-high	Low
Grain number in spike	No-low	Middle	Low-middle
Grain number in spikelet	No-low	Unknown	Low
Grain formation index	No-low	Unknown	Low-middle
Grain weight	No	High	Low
Grain number at m^2	High	Low-middle	High
Morphology			
Mature plant height	Low in 70–100 cm	Very high	Low
Leaf erectness	Unknown-low	Middle-low	Low-middle
Leaf thickness	Unknown	Middle	Low
Awnless	Unknown	Very high	Very low
Yield potential (Yp)			
Yield potential (Yp)	Very high	Low	High

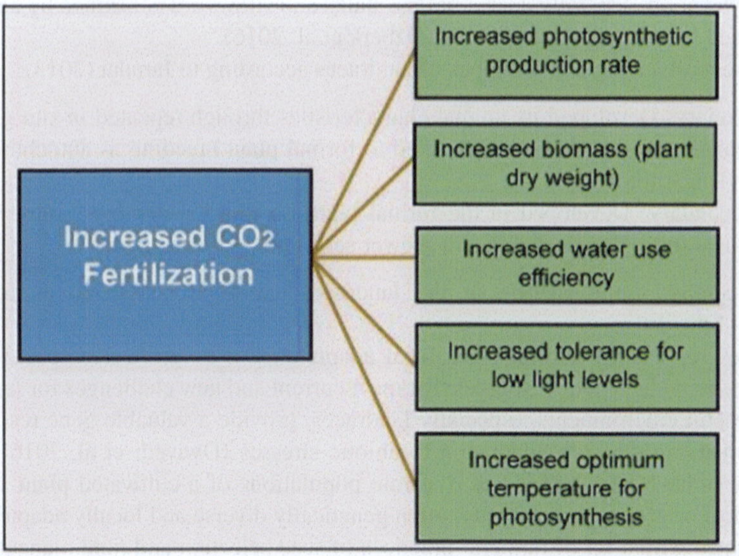

Fig. 9.6 Effect of CO_2 fertilization on wheat landraces' some physiologic components (Lenart et al. 2006)

Fig. 9.7 General impacts of elevated CO_2 on plants (Anonymous 2018b)

quality is good, generally coarse grains, stalks and straws are consumed by animals and liked by animals (Jaradat 2012; Özberk et al. 2016).

Essentially, there are two types of landraces according to Jaradat (2013):

(a) Primary: Developed its unique characteristics through repeated in situ grower selection and never been subjected to formal plant breeding as autochthonous and allochthonous
(b) Secondary: Developed in the formal plant breeding sector but is now maintained through repeated in situ grower selection and seed saving

Generally, the formation of the landraces can be schematized as follows (Fig. 9.8):

They represent heterogeneous, local adaptations of domesticated species, and thereby provide genetic resources that meet current and new challenges for farming in stressful environments, especially landraces, provide a valuable gene resources for enhancing the crop adaptation to abiotic stresses (Dwivedi et al. 2016), and, landraces have been defined as dynamic populations of a cultivated plant with a historical origin, distinct identity, often genetically diverse and locally adapted and associated with a set of farmers' practices of seed selection and field management as well as with a knowledge base (Bellon and Etten 2014). These carry beneficial genes that were not introduced into elite durum cultivars (Kabbaj et al. 2017). Northern landraces evolved a higher tillering capacity, fewer grains per spike and less fertile tillering than those from the south. Our results support the hypothesis that during the Neolithic dispersal of durum wheat from the Fertile Crescent to southern Europe, significant and gradual changes in yield component structure of populations occurred (Akçura 2009). The main threat to the landraces (including wheat landraces) is current minor or major industrial developments such as construction of huge shopping malls, housing, apartments and blocks, and golf fields,

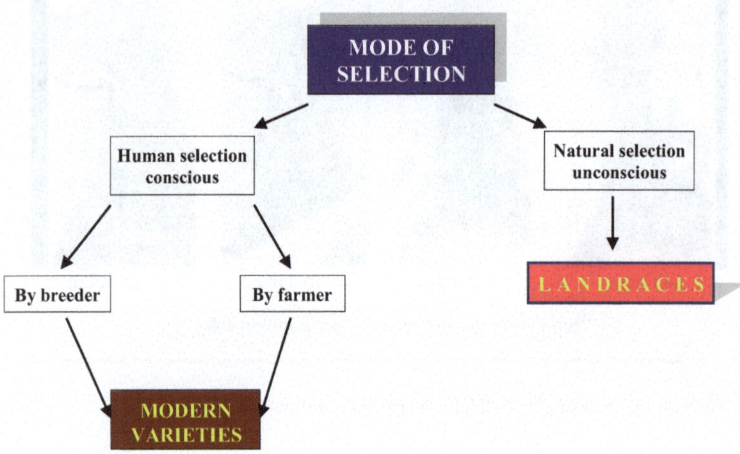

Fig. 9.8 Schematized landrace and modern variety general formation. (Modified Cleveland et al. 1994)

leading to reduction in biodiversity. The landrace gene pool harbors a wide genetic diversity that could be used to enrich the modern wheat genetic repository (Blum et al. 1989; Dotlačil et al. 2010; Ben-David et al. 2014). Wheat landraces generally tolerant to biotic and abiotic stress have been grown under low-input or sustainable farming conditions where they produce reasonable yield (Akçura 2009).

CC and GW effects on wheat landraces, like other cultivated plants, can be itemized as follows (Gray and Brady 2016; Anonymous 2019):

1. Air enriched with CO_2 stimulates growth and development of wheat landraces, thus resulting in the development of more fibrous and voluminous root systems.
2. Leaf size increases and cells expand.
3. More CO_2 (which means elevated CO_2) means less water stress.
4. Helps cope with low levels of essential resources such as light, water, nutrients, etc.
5. In less than optimal conditions, higher CO_2 means more plant growth.
6. Atmospheric CO_2 enrichment increases plant water acquisition.
7. Rising CO_2 enhances plant resource acquisition, such as root system, nitrogen-fixing bacteria, symbiotic soil bacteria, carbon starvation, etc.
8. Promotes the growth of important soil fungi such as *rhizosphere*, *mycorrhizae*, etc.
9. At current CO_2 concentrations (February 2019, which is 411,75 ppm (https://www.co2.earth/), plants are close to starvation.
10. Elevated CO_2 level helps plants to survive environmental stresses such as salinity, pollution, elevated temperatures, etc.
11. Elevated CO_2 level helps in reducing the negative impacts of soil salinity on plant growth.
12. Elevated CO_2 level helps in reducing the negative impacts of high temperatures on plant growth.
13. Elevated CO_2 level helps wheat landraces to survive biological stresses such as weeds, diseases, insects, herbivory, etc.
14. Rising CO_2 does not disappear with time (but it has been observed that in some plant species, foliar N concentrations may decrease; however, in others, it will not).
15. Rising CO_2 enhances carbon sequestration (particularly, this issue is very crucial in terms of the sustainability of agroecosystems, grasslands, and forests).

All the effect(s) of the CC and GW's components were summarized on wheat landraces' traits which were presented generally in Tables 9.1 and 9.2.

Increasing the temperature rises evapotranspiration and drops the soil moisture availability and increases the growth and development of plants, including wheat landraces, due to higher CO_2 concentrations. Wheat landraces, based on their nutritional value, when locally produced can contribute to lower greenhouse gas emissions (0.1 g CO_2 per calorie) as compared to rice (0.43 g CO_2 per calorie) or vegetables (0.57 g CO_2 per calorie) (Jaradat 2013). In addition to this information, Schlenker and Roberts (2009), the yield of low CO_2 concentration, the wheat landrace (36%- (-40))% and (63–70)% depending on the statistical program used.

Hatfield and Prueger (2011) calculated it as 3.8–5.0%, and Fuhrer (2009) stated that the increase of 1 °C in the temperature caused a decrease of 7.0–124.0% in the yield level. At the same time, it has been reported that wheat yield losses in developing countries (producing 66% of the total wheat production) are likely to be around 20–30% due to the increases in temperature caused by climate change (Easterling et al. 2007; Lobell et al. 2008). Under normal conditions and during the development period, the mean temperature increase of 1 °C, causes 6 kg/ha yield losses in *durum* and 12 kg/ha common wheats, and it is expected that the number of wheat yield losses will be between 20 and 30% with a temperature increase of 2–3 °C in developing countries until 2050 (Anonymous 2011; Sayılğan 2016). High temperature (air and soil temperature) and water deficit (drought) are the most important environmental factors that limit plant growth in many huge/mega wheat fields of the world and occur simultaneously (Shah and Paulsen 2003). But its mechanism is still unknown.

The main threat caused by CC and GW is not only increased or elevated CO_2 concentration and temperature but also reduction of the effectiveness of RuBisCo during the production of glucose via PSII stage in photosynthesis. In parallel, the WUE value diminishes. This development leads less water for a less dry matter (that means low yield level), the role of the RuBisCo and indirectly WUE value which is very important. But their mechanisms, etc. are still not fully and clearly known today. In any case, the main aim should be an increase in the WUE values of wheat landraces, especially those grown in arid and semiarid regions.

The enzyme RuBisCo has played a crucial and vital role in photosynthesis and one of the most abundant proteins in leaves of plants. Accelerated development and premature senescence were the primary factors affecting its activity in response to the CO_2 enrichment. This role is very clear during photosynthesis and this enzyme is in close relationship with the WUE of the plant, especially during CC and the GW (Table 9.3).

According to Marin and Nassif (2013), the increase of atmospheric CO_2 concentration increases the gradient that drives the diffusion of CO_2 from the atmosphere to the chloroplasts. And this effect stimulates photosynthesis and reduces stomatal conductance, and a reduction in the transpiration rate happens (Taiz and Zaiger 2013). In Tables 9.1 and 9.2, CC and GW act on C3 plants effects have been with their components. Especially their effects on "vernalization, CO_2/O_2 rate, respiratory rate, seed maturation, sub- and topsoil biomass production, number of internodes, weed distribution, germination of seeds, root/stem or rhizome, seed longevity, DNA molecule and sterility, ecological parameters, etc. " have not yet been fully known. Their effect on the plants is mostly and generally on the biomass, specifically on the leaf and leaf factors. At the selections to be made by taking advantage of the relevant features, which are mentioned or not mentioned in Table 9.2, wheat landraces can be used to complete a gene resource and the missing characters(s) for a valuable genitor or donor.

Water sources and/or soil moisture, which are diminishing due to the decreases in wheat sowing areas and climatic reasons,, have been emphasized that the wheat landraces are an insurance for future agricultural production. Wheat landraces have

Table 9.3 Main agronomic traits of strategically important 18 major field crops (Rötter and Geijn 1999)

	World[a]				
Crops	Production (Mt)	Yield (Hg/Ha)	Origin	Type	WUE
Barley	141,277	30,108	W. Asia	C_3	1.25–2.50
Bean, dry	26,833	9129	S. and Cent. Amer.	C_3	1.40–3.30
Cassava	277,103	118,006	S. and Cent. Amer.	CAM	1.30–3.30
Coconut	59,011	48,493	Africa, Asia	C_3	1.40–3.30
Cotton (seed.)	46,988[a]		S. Amer.	C_3	1.40–3.30
Grape	77,439	109,119	Asia	C_3	1.25–3.30
Maize	1,060,107	56,401	Cent. Amer.	C_4	2.90–6.70
Oats	22,992	24,373	W. Europe	C_3	1.25–2.50
Peanut	–		S. Amer.	C_3	1.40–3.30
Pea, dry	14,363	18,835	W. and N. Asia	C_3	Unknown
Potato	376,827	195,790	S. Amer.	C_3	1.25–2.50
Rice	740,962	46,366	Asia, Africa	C_3	1.40–3.30
Rye	12,944	29,398	W. Asia	C_3	1.25–2.50
Sorghum	63,931	14,279	Africa	C_4	2.90–6.70
Soybean	334,894	27,556	E. Asia?	C_3	1.40–3.30
Sugarcane	1,890,662	706,148	NW Asia, Aust.	C_4	1.25–6.66
Sweet potato	105,191	121,975	S. and Centr. Amer.	C_3	1.40–3.30
Wheat	***794,460***	***34,050***	***Fertile Crescent***	C_3	***1.25–2.50***

W. West, *S.* South, *Cent.* Central, *Amer.* America, *N.* North, *E.* East, *NW* North West, *Aust.* Australia, *WUE* water use efficiency, *CAM* Crassulacean acid metabolism
[a]Anonymous 2018b

agriculturally more undesirable traits such as hulledness in grain, tallness, and low yield level(s) than modern commercial (wheat) varieties. However, cultivation of landraces has been successful for many years without any human intervention under stress conditions. The main contributions of wheat landraces to plant breeding programs have been their desirable traits such as having efficient nutrient uptake and utilization and having useful genes adapted to stressful environments such as water stress, salinity, and higher temperatures (Dwivedi et al. 2016).

In the light of the information that was given, our recommendations are (generally) as follows:

- To make national or international agreements that enable effective use of both CO_2 and water resources.
- To minimize the release of CH_4 from ruminants and nitrous oxide by efficient fertilization (Prasad 2009).
- To avoiding excessive and artificial nitrogenous fertilization.
- To take crop rotation and animal feeding of tuberous plants and legumes (Ulukan 2009).
- To use alternative or clean energy sources.
- To apply minimum soil tillage techniques (Çakır et al. 2009).

- To use organic or environmentally friendly agricultural practices.
- To not burn waste materials at the end of the production.
- Especially in continents, coasts, and oceans, to take the necessary measures without forgetting that the polar regions will warm faster than the equator.
- All activities that cause greenhouse gas emissions should be terminated or minimized.
- Without losing the principle of sustainability, to prevent the destruction of soil, water, and biodiversity.
- When they are used as parent(s) directly in the hybridizations, expands high yielded modern wheat cultivars' and provides the resistance to biotic and abiotic stress factors; in addition, the use of bridge hybridization (Şehirali and Özgen 1987), the production of seeds on a periodically, scientifically and characterization is of great importance (Özberk et al. 2016).
- To develop new (wheat) varieties which are also suitable for the purpose of producing wheat landrace, especially from the elements of genetic variation (Heslop-Harrison 2012). It should be used as rootstock or genitor in breeding studies by utilizing physiological characteristics.
- The development of new varieties using wheat landraces that are more adapted to local biotic and abiotic stresses presents a viable strategy to improve and sustain yields, especially under stresses and future changes in climate (according to Calanca (2017).
- However, landraces with high genetic diversity should be selected and crossed with locally adapted landraces and varieties to achieve breakthroughs in wheat genetic improvement (according to Calanca (2017) and in order to increase tolerance which therefore results in increased yield potential and to respond to climate change (Semenov et al. 2014).

It should be remembered that in all cereals, except wheat landraces, there is an agronomically negative relationship between yield and stress conditions, although higher yield is obtained by cultivation in suitable ecologies with appropriate varieties and cultivation techniques in modern plant breeding programs. However, some agricultural properties that are superior to the various stress conditions can only be achieved by using them as genitors. These genotypes are very important in terms of providing efficiency to the producer and generating income, where stress conditions are common (especially extreme temperature, limited water, etc.) and inadequate. As seen from Table 9.4, *durum*-type local wheat landraces (more than 113 local wheat landraces) are more cultivated than the *aestivum* types. And nearly all the genotypes that are being cultivated are called with their morphological traits such as grain or spike color. Even that, the same local cultivars have different name place to place among the farmers.

Crops of these genotypes are being mostly consumed as regional and healthy dishes (e.g., bulghur, erişte, etc.) due to not only their weak gluten strength but also their nutrition profile (esp. Fe, P, and protein percentage) and poor agronomic traits such as yield level (100–150 kg/da), lateness, short plant height, etc. Wheat landraces are commonly and mainly grown at Black Sea and Central Anatolian regions in

Table 9.4 Turkey's climate regions and total vegetation period length of the wheat cultivars and landraces (Anonymous 2017)

Climate	Vegetation period (days)	Climate	Vegetation period (days)
Overrained Mediterranean	200 (for *common*)	*Erzurum-Kars Pr. Pl.*	322 (for *common*) 115 (for *durum*)
Mediterranean	205 (for *common*)	*Van Pr.*	315 (for *common*) 120 (for *durum*)
P. Mediterranean-1	200 (for *common*)	*Hakkâri Pr.*	310 (for *common*) 110 (for *durum*)
Marmara	252 (for *common*) 130 (for *durum*)	*Southeastern Anatolia-1*	211 (for *common*)
Marmara T.	270 (for *common*)	*Southeastern Anatolia-2*	226 (for *common*)
Cool Black Sea-1	–	*Southeastern Anatolia and T.*	268 (for *common*) 130 (for *durum*)
Cool Black Sea-2	270 (for *common*) 130 (for *durum*)	*Upper Fırat and Murat*	286 (for *common*) 125 (for *durum*)
The warm Black Sea	251 (for *common*) 130 (for *durum*)	*Mediterranean-Southeastern Anatolia T.*	229 (for *common*)
East Black Sea-1	–	*Mountainside and East*	273 (for *common*) 128 (for *durum*)
East Black Sea-2	270 (for *common*) 130 (for *durum*)	*Inner Anatolia and Inner Transition and Cool Black Sea*	270 (for *common*)
Yusufeli Rg.	–	*Mountainside*	270 (for *common*)
East-1	300 (for *common*) 120 (for *durum*)	*Post Mediterrenean-2*	265 (for *common*)
East-2	315 (for *common*) 127 (for *durum*)	*T. Zone*	267 (for *common*)
Iğdır Pr. Microclimate	270 (for *common*) 105 (for *durum*)	*Inner Anatolia*	225 (for *common*)

T. transition, *P.* post, *Pl.* plateau, *Pr.* province, *Rg.* region, minimum days, 105; *Iğdır Pr. microclimate rained climate* (for *durum*); maximum days, 130; (*Marmara and Cool Black Sea-2*) *climates* (for *durum*); minimum days, 200; *P. Mediterranean-1 climate* (for *common*); maximum days, 322; *Erzurum-Kars Pr. Pl. climate* (for *common*)

Turkey. And these regions are mostly marginal and have suffered from stress factors. Particularly, elevated CO_2 is a very critical and crucial factor for them since their water consumption for grain formation is getting reduced during this process. In addition, consuming water for 1 g dry matter is getting lower for C3 plants including wheat landraces. Generally, the CC and GW's effect(s) on wheat landraces are (particularly) at vernalization stage, cellular CO_2/O_2 changeability, respiration ratio, maturity, topsoil and subsoil biomass/root volume(s), internode number, weed distribution, germination, rhizome activity, seed longevity, DNA molecule breaking, sterility and ecological factors where their mechanism, etc. have not been fully known and clear today.

Generally, in the cultivation areas, water is a major determining and limiting factor for agricultural yield. On the other hand, the amount of precipitation and its distribution of the water landrace's vegetation period are ultimately important. Particularly, it is important for the availability of water (directly), nutrient availability, soil fertility, ph value, etc. Water availability has long been known as one of the most important abiotic factors governing crop yield (Boyer 1982; Gray and Brady 2016), and it has played a significant role in plant growth and development processes such as photosynthesis and transpiration. At this point, WUE arises as an important physiological factor which also determines yield. During climate change and global warming, wheat landrace's WUE value is affected and reduced when the CO_2 is elevated and directly linked with the yield. It can be defined as

$$WUE\left(\mathrm{kg\,da^{-1}\,mm^{-1}}\right) = \mathrm{Y / ET_a}$$

where Y is the yield (kg da^{-1}) and ET_a is the actual water consumption (the sum of water consumption for each stage) during the wheat landrace's growth period (mm) as mentioned above.

CC and GW physiologically affect first WUE, which has an important role in the Calvin cycle in the *PSII* stage of photosynthesis. When WUE affects, directly reduces the effectiveness of RuBisCO (ribulose-1,5-bisphosphate carboxylase/oxygenase) enzyme, which links CO_2 and partly O_2 entrance to the chloroplasts, in other words, results in dry matter production and yield reduction. But, this situation (yield reducing) does not happen in C4 grouped plants and increasing yield aggregating CO_2 in C3 grouped plants (incl. wheat landraces): CO_2 fertilization (Uzmen 2007). In a breeding strategy, selection for elevated water use efficiency causes reduced or earlier flowering that results in lower water usage along with lower yield capacity (Blum 2005). Hence, it is vital to produce genotypes having higher WUE as well as higher yields compared to the present varieties (Farooq et al. 2009). But, this point is clear that despite the significant increase in the yield potential of wheat breeding based on yield worldwide, the future success will be determined by the cooperation of plant breeders and plant physiologists and by the support of physiological criteria (Jackson et al. 1996; Sayılğan 2016) (Fig. 9.9).

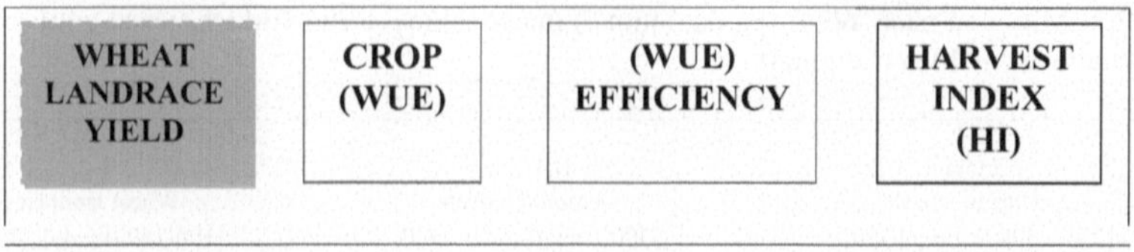

Fig. 9.9 Some major yield components of wheat landraces. (Modified from Richards et al. 2015)

9.2 Status in Turkey

When evaluated in terms of topography and climate, Turkey has a very wide genetic diversity and geographical structure. The most important plant is wheat and wheat landraces are still grown in Turkey. Derived end products such as bread, yufka, noodles, lavash, and bulghur are made from wheat. According to the TUİK-2018 and FAO-2017 statistics, Turkey's wheat sowing area is (7.6–7.7) million ha, its production level is approximately 20 million tons, and its mean yield is (4–4.5) t/ha for common and (2.5–3.0) t/ha durum wheats (Anonymous 2018a, 2019). Wheat landraces in Turkey are usually kept as populations rather than selected as homogenous cultivars. Thus, those populations are characterized by great genetic and phenotypic variations. Landraces even within a single village may show traits such as white, black, or red grain, the presence and absence of awns, tightly or loosely packed spikes, and different abilities to tolerate abiotic conditions (According to Brush 2004 and Karagöz 2014). Wheat landraces are generally grown in small fields and marginal places, in the west and northern transition zones of Central Plateau, and in forest openings of North, Eastern, and Southeastern Anatolia (Akçura 2009), and a full taxonomic list of wheat landraces, which are grown in Turkey, was presented in Table 9.4.

In Turkey, wheat landraces are mostly grown in arid and semiarid regions which are dominated by stress factors such as salinity, drought, and cold (Zencirci et al. 2019). Precipitation, especially in the period of growth of these wheat landraces in arid regions, and high temperatures cause significant decreases in yield level. As mentioned above, their mean yield level is rather lower (100–150 kg/da) compared to modern wheat cultivars' mean yield level (400–450 kg/da). For determining WUE values of wheat cultivars in Turkey, many types of research were carried out in meteorological stations, i.e., more than 259 (Anonymous 2017) at different 28 climate regions. These climate regions and the total vegetation period length of the wheat cultivars and landraces which are grown in Turkey are presented in Table 9.4.

In Turkey, mostly vegetation period length depends on water availability, temperature, and distribution of the precipitation. According to measurements in the meteorological stations (in total 259), a variation of the WUE value in Turkey can be mentioned like this (Anonymous 2017) (Table 9.5):

On the other hand, these findings can be evaluated like the following as well:

As seen from Table 9.6, durum and common wheat WUE values are higher than those in wheat landraces in Turkey. It means that except wheat landraces, wheat

Table 9.5 Min. and max. WUE (kg da^{-1} mm^{-1}) mean values of the Turkish wheat cultivars and wheat landraces in 2019 (Original)

	(A) Durum wheats [250 – 300 (kg/da) / 320]	(B) Common wheats [400 - 450 (kg/da) / 320]	(C) Wheat landraces [100-150 (kg/da) / 488.19]
	(0.78-0.94)	**(1.25-1.41)**	**(0.20-0.31)**
Climate: Type	(*Mediterranean*)	(*P. Mediterranean*)	(*All*)

uses more water and produces more dry matter, resulting in higher yield level. WUE values of landraces are lower than those of durum and common wheat (at least three to four times), and these values should be increased by aggregating the yield. For this purpose, various breeding methods should be used (e.g., mutation breeding) by benefiting from landraces as against stress factors as parents. It was demonstrated that temperature was found to have a positive effect on potential yield as well as earliness within Turkish local wheats, whereas lower drought and heat stress caused varieties from Ethiopia and Syria to have longer spike (Alhajj et al. 2017). But in terms of CC and GW, Turkey is not on dangerous position, but all necessary precautions without delay should be taken into consideration for the sustainability of animal and plant production.

Table 9.6 List of some grown Turkish wheat landraces in 2019 (original)

	Local name	
Region *Province*/district	*T. monoccum* (AA)-*T. durum* (AABB)	*T. aestivum* (AABBDD)
Manisa/Akhisar	Üveyik, Sarıbaşak, Zerun, Akbuğday, Agbuğday, Kırmızı buğday, Sarı buğday, Karakılçık	
Erzurum, Van, Iğdır/Tuzluca	Kırik	
Kastamonu/İhsangazi Seydiler/Merkez/Devrekani	Siyez	
Kütahya/Çavdarhisar	Kocabuğday	
Eskişehir, Balıkesir/Sındırgı	Topbaş, Kırmızı Topbaş, Şahman, Devedişi, Ak 702, Sertak 52, Melez 13, Gernik, Sivas 111/33, Havrani, Köse 220/39, Polatlı/Kobak, Yayla 305, Sürak 1593/51	
Kayseri/Develi	Gacer	
Kütüphane/Çavdarhisar	Koca buğday	
Gümüşhane/Torul	Rus buğdayı	
Karabük/Eflani	Köy buğdayı	
Malatya/Akçadağ	Aşurelik buğday	

(continued)

Table 9.6 (continued)

Region *Province*/district	Local name *T. monoccum* (AA)-*T. durum* (AABB)	*T. aestivum* (AABBDD)
Malatya/Akçadağ, *Elazığ*/ Merkez	Kırmızı Kunduru	
Elazığ/Baskıl	Menceki	
Ağrı/Patnos	Kıraç 70	
Adıyaman/Gölbaşı	Malatya Sarı Bursası	
Tokat/Yeşilyurt	Ormece	
Aksaray/Güzelyurt	Kırmızı Kamçı	
Yozgat/Kadışehri	Çalıbasan	
Van	Tir, Kırmızı, Sarı, Koca, Göderedi	
North East Anatolia	Göle buğdayı, Kelkit buğdayı	
Middle North Anatolia *Ankara, Çankırı, Çorum, Uşak Kırşehir, Yozgat, Bolu, Bilecik, Eskişehir, Kütahya,*	Sarı Buğday Siverek, Çirpuz, Karakılçık, Kunduru, Şahman, Sarı Bursa, Aşurelik Buğday, Ak Başak, Üveyik, Ağ buğdayı	Akbuğday, Sünter, Bindane, Kadiroğlu, Çalıbasan, Köse
Middle East Anatolia *Amasya, Elazığ, Malatya, Sivas, Tokat, Tunceli*	Üveyik, Menceki, Kunduru	Aşure, Akbuğday, Zerun, Gürük, Zerin, Dimenit, Yazlık, Kırik, Köse, Kırmızı, Tercan
Middle South Anatolia *Afyon, Kayseri, Konya, Nevşehir, Niğde*	Bolvadin, Sarı Buğday, Karakılçık	Akbuğday, Akbarnaz, Çomak, Köse, Sivas Buğdayı, Germir, Akevli, Kamçı Wheat, Kızıl Topbaş
North East Anatolia *Ağrı, Artvin, Erzincan, Erzurum, Kars*	Karakılçık, Hazerik,	Kırmızı Buğday, Kırik, Topbaş, Sarıbaş, Kızıl, Köse, Akbuğday
Southeast Anatolia *Bingöl, Bitlis, Hakkâri, Van Mardin, Muş, Siirt, Şanlıurfa,*	Bağacak, Sorgül, Beyaziye, Menceki, İskenderi, Mısri, Havrani, Karakılçık, Sorik Akbaş, Akbaşak, Hamrik	Aşure
Mediterrenean *Antalya, Gaziantep, Hatay, Mersin, Maraş, Adana*	Akbuğday, Karakılçık, Tığrak Buğdayı, Sarı Buğday, Kıbrıs Buğdayı	Yerli Macar, Kırmızı Buğday, Akbuğday, Devedişi, Çavdarlı
Agean *İzmit, Aydın, Muğla, Denizli, Burdur, Isparta, Çanakkale, Manisa, Balıkesir*	Fata, Gökala, Sarı Başak, Kunduru, Menemen, Karakılçık, Sarı Çam, Akbaşak, Akpüsen, Çam Buğdayı, Sarı Buğday, Deve Dişi, Kırmızı Buğday	Kızılca, Akgernaz, Akça Rodos
Marmara *Bursa, Kocaeli, Sakarya, İstanbul, Edirne, Tekirdağ, Kırklareli*	Akbaşak, Karakılçık, Tunus Buğdayı, Sarı Başak, Köse Buğday, Arnavut Buğdayı, Kunduz, Kocabuğday, Kokana	Sünter, Kızılca, Akova, İngiliz Buğdayı, Çapraz Köse Buğday, Çalıbasan,
Black Sea *Rize, Trabzon, Giresun, Ordu, Samsun, Sinop, Gümüşhane, Kastamonu, Zonguldak*	Rumeli/Yunan buğdayı, İlik, Sarı Buğday, Akbuğday, Sarıbaş, Karakılçık, Üveyik, Rumeli, Sarı Hamza, Koçarı, Diş Buğdayı	Mengen, Topbaş, Dimenit, Kırmızı Sünter, Akça

References

Akçura M (2009) Genetic variability and interrelationship among grain yield and some quality traits in Turkish winter durum wheat landraces. Turk J Agric For 33:547–556

Alhajj AS, Tedone L, De Mastro G (2017) Climate variability impact on wheat production in Europe: adaptation and mitigation strategies. In: Ahmed M, Stockle CO (eds) Quantification of climate variability adaptation and migration for agricultural sustainability. Springer, Cham, pp 252–299. https://doi.org/10.1007/978-3-319-32059-5

Anonymous (2011) WHEAT-Global alliance for improving food security and the livelihoods of the resources-poor in the developing world'. Proposal submitted by CIMMYT and ICARDA to the CGIAR consortium board, in collaboration with Bioversity, ICRISAT, IFPRI, ILRI, IRRI, IW MI, 86 NARS Institute, 13 Regional and International Organizations, 71 Universities and Advance Research Institutes, 15 Private Sector Organizations, 14 NGOs and Farmers Cooperatives and 20 Host Countries, p 197

Anonymous (2017) Türkiye'de sulanan bitkilerin bitki su tüketimleri. T.C. Gıda, Tarım ve Hayvancılık Bakanlığı Yayınları, Tarımsal Araştırmalar ve Politikalar Genel Müdürlüğü (TAGEM)-Devlet Su işleri Genel Müdürlüğü (DSİ), Sf. 586, Altındağ, Ankara, Turkey ISBN 978-605-9175-93-7

Anonymous (2018a) FAO statistical database. http://www.fao.org/faostat/en/#data/QC. Access date May 14, 2018

Anonymous (2018b) Plant response to rising carbon dioxide. http://articles.extension.org/pages/58381/plant-response-to-rising-carbon-dioxide. Access date May 20, 2018

Anonymous (2019) Plants need CO_2, carbon dioxide is green. https://plantsneedco2.org/default.aspx/MenuItemID/103/MenuSubID/52/MenuGroup/Home.htme. Access date April 05, 2019

Bellon M, Etten van J (2014) Climate change and on-farm conservation of crop landraces in centers of diversity, In Jackson M, Ford-Lloyd B, Parry M (eds). Plant genetic resources and climate change, pp 1–30.

Ben-David R, Amram A, Nashef K (2014) Genetic improvement of durum wheat establishment under fluctuating environmental conditions. In: Porceddu E, Qualset CO, Damania AB (eds) Proceedings of the International Symposium on Genetics and Breeding of Durum Wheat, vol 110. CIHEAM, Bari, pp 147–149

Blum A (2005) Drought resistance, water-use efficiency, and yield potential: are they compatible, dissonant, or mutually exclusive? Aust J Agr Res 56:1159–1168

Blum A, Golan G, Mayer J, Sinmena B, Shpiler L, Burra J (1989) The drought response of landraces of wheat from the northern Negev desert in Israel. Euphytica 43:87–96

Boyer JS (1982) Plant productivity and the environment. Science, New series 218(4571):443–448

Çakır E, Aykas E, Yalçın H, Dereli İ (2009) The benefit of conservation tillage and its applications in the World and Turkey, 1st Int. Congress on Global Climate Change and Agriculture May 28–30 Tekirdağ, Türkiye, pp 124–134.

Calanca PP (2017) Effects of abiotic stress in crop production. In: Ahmed M, Stockle CO (eds) Quantification of climate variability adaptation and migration for agricultural sustainability. Springer, Cham, pp 166–176. https://doi.org/10.1007/978-3-319-32059-5

Ceccarelli S (1994) Specific adaptation and breeding for marginal conditions. Euphytica 77:205–219

Cleveland DA, Soleri D, Smith SE (1994) Folk crop varieties: do they have a role in sustainable agriculture? BioScience 44:740–751

Cutforth HW, McGinn SM, McPhee E, Miller PR (2007) Adaptation of pulse crops to the changing climate of the Northern Great Plains. Agron J 99:1684–1699

Dempewolf H, Eastwood JR, Guarino L, Khoury KC, Müller VJ, Toll J (2014) Adapting agriculture to climate change: a global initiative to collect, conserve, and use crop wild relatives. Agroecol Sustain Food Syst 38(4):369–377. https://doi.org/10.1080/21683565.2013.870629

Dhakhawa GB, Campbell CL (1998) Potential effects of differential day-night warming in global climate change on crop production. Clim Change 40:647–667

Di Norcia V (2008) Global warming is man-made: key points in the international panel on climate change 2007 Report, Hard like water-ethics in business. Oxford University Press, Oxford, pp 1–4
Dotlačil L, Hermuth J, Stehno Z, Dvořáček V, Bradová J, Leišová L (2010) How can wheat landraces contribute to present breeding? Czech J Genet Plant Breed 46:S70(1)–S74(5)
Dwivedi LS, Ceccarelli S, Blair WM, Upadhyaya DH, Are KA, Ortiz R (2016) Landrace germplasm for improving yield and abiotic stress adaptation. Trends Plant Sci 21(1):31–42
Easterling DR, Horton B, Jones PD, Peterson TC, Karl TR, Parker DE, Salinger MJ, Pazuvayev V, Plummer N, Jamason P, Folland CK (2007) Maximum and minimum temperature trends for the globe. Science 277:364–367
Elissavet GN, Ioannis GM, Athanasios T, Parthenopi R, Christos D, Tokatlidis IS (2014) Wheat landraces are better qualified as potential gene pools at ultra spaced rather than densely grown conditions, Hindawi Publishing Corporation. Scientific World Journal 2014:1–5. https://doi.org/10.1155/2014/957472
Farooq M, Wahid A, Kobayashi N, Fujita D, Basra SMA (2009) Plant drought stress: effects, mechanisms and management. Agron Sustain Dev 29:185–212
Fischer RA (2001) Selection traits for improving yield potential, Chapter 13. In: Reynolds MP, Ortiz-Monasterio JI, McNab A (eds) Application of physiology in wheat breeding. CIMMYT, Mexico, D.F., pp 148–159. p 219
Fuhrer J (2003) Agroecosystem responses to the combination of elevated CO_2, ozone, and global climate change. Agric Ecosyst Environ 97:20
Fuhrer J (2009) Climate change-risks and opportunities for agriculture, 1st Int. Congress on Global Climate Change and Agriculture May 28–30 Tekirdağ, Türkiye, pp 81–87
Gray BS, Brady MS (2016) Plant developmental responses to climate change. Dev Biol 419:64–77
Hatfield JL, Prueger JH (2011) In: Yadav SS, Redden RJ, Hatfield JL, Lotze-Campen H, Hall AE (eds) Agroecology: implications for plant response to climate change. Wiley-Blackwell, West Sussex, pp 27–43
Heslop-Harrison J (2012) Genetics and genomics of crop domestication. University of Leicester, Leicester. 10.10.16/B978-0-12-381466.0001-8
IPCC (2007) New assessment methods and the characterization of future conditions. In: Climate change 2007: impacts, adaptation and vulnerability. The Contribution of Working Group II to the 4th Assessment Report of the Intergovernmental Panel on Climate Change (IPCC). Cambridge University Press, Cambridge, UK, p 976
IPCC (2014) Climate change 2014: synthesis report. The Contribution of Working Groups I, II and III to the 5th Assessment Report of the Intergovernmental Panel on Climate Change Core Writing Team, R.K. Pachauri and L.A. Meyer (Eds.). IPCC, Geneva, pg. 151.
Jackson P, Robertson M, Cooper M, Hammer G (1996) The role of physiological understanding in plant breeding, from a breeding perspective. Field Crop Res 49(1):11–37
Jaradat AA (2012) Wheat landraces: genetic Resources for sustenance and sustainability, USDA-ARS, 803 Iowa, Morris, MN 56267 USA, pp 1–20
Jaradat AA (2013) Wheat landraces: a mini review. Emirates J Food Agric 25(1):20–29
Kabbaj H, Sall AT, Al-Abdallat A, Geleta M, Amri A, Filali-Maltouf A, Belkadi B, Ortiz R, Bassi FM (2017) Genetic diversity within a global panel of durum wheat (*Triticum durum*) landraces and modern germplasm reveals the history of alleles exchange. Front Plant Sci 8(1277):1–13
Karagöz A (2014) Wheat landraces of Turkey. Emirates J Food Agric 26(2):149–156. https://doi.org/10.9755/ejfa.v26i2.16397
Kastner T, Rivas MJI, Koch W, Nonhebel S (2012) Global changes in diets and the consequences for land requirements for food. Proc Natl Acad Sci 109:6868–6872
Krupa SV (1997) Global climate change: processes and products-an overview. Environ Monit Assess 46:73–88
Lenart M, Jones C, Kimball B (2006) Rising carbon dioxide levels and forest management, climate change and variability in the southwest ecosystem series, The University of Arizona, College of Agriculture and Life Sciences Arizona Cooperative Extension, AZ1395, p 4

Lobell DB, Burke MB, Tebaldi C, Mastranddrea MD, Falcon WP, Naylor RL (2008) Supporting online materials for prioritizing climate change adaptation needs for food security in 2030. Science 319:607–610

Lopes MS, El-Basyoni I, Baenziger P, Singh S, Royo C, Ozbek K, Aktas H, Ozer E, Özdemir F, Manickavelu A, Ban T, Vikram P (2015) Exploiting genetic diversity from landraces in wheat breeding for adaptation to climate change. J Exp Bot 66(12):3477–3486. https://doi.org/10.1093/jxb/erv122

Marin F, Nassif DSP (2013) Mudanças climáticas e a cana-de-açúcar no Brasil: fisiologia, conjuntura e cenário futuro. Revista Brasileira Engenharia Agrícola Ambiental 17(2):232–239

Maxted N, Kell S (2009) Establishment of a global network for the *in situ* conservation of crop wild relatives: status and needs, FAO Commission on Genetic Resources for Food and Agriculture, Background Study Paper No 39, pp 1–112, Rome, Italy

Mei H, Chengjun JI, Wenyun Z, Jinsheng HE (2007) Interactive effects of elevated CO_2 and temperature on the anatomical characteristics of leaves in eleven species. Acta Ecol Sin 26:326–333

Mengü GP, Şensoy S, Akkuzu E (2008) Effects of global climate change in agriculture and water resources, Balwois-Ohrid, Republic of Macedonia 27–31 May 2008, p 10

Mercer KL, Perales HR (2010) Evolutionary response of landraces to climate change in centers of crop diversity. Evol Appl 3:481–493

Nasserlehaq N, Amamou A, Taghouti M, Annicchiarico P (2011) Adaptation of Moroccan durum wheat varieties from different breeding eras. J Plant Breed Crop Sci 3(2):34–40

Nema P, Nema S, Roy P (2012) An overview of global climate change in the current scenario and mitigation action. Renew Sustain Energy Rev 16:2329–2336. https://doi.org/10.1016/j.rser.2012.01.044

Özberk F, Karagöz A, Özberk İ, Ayhan AT (2016) Buğday genetik kaynaklarından yerel ve kültür çeşitlerine; Türkiye'de buğday ve ekmek. Tarla Bitkileri Merkez Araştırma Enstitüsü Dergisi 25(2):218–233. [In Turkish]

Prasad PK (2009) Effects of agriculture on climate change: a cross-country study of factors affecting carbon emissions. J Agric Environ 10:72–88

Raza A, Razzaq A, Mehmood SS, Zou X, Zhang X, Lv Y, Xu J (2019) Impact of climate change on crops adaptation and strategies to tackle its outcome: a review. Plants 8(34):1–29

Richards RA, Condon AG, Rebetzke GJ (2015) Chapter 88–100. In: Reynolds MP, Ortiz-Monasterio JI, McNab A (eds) Application of physiology in wheat breeding. CIMMYT, Mexico, D.F, p 219

Rogers HH, Runion GB, Krupa SV (1994) Plant responses to atmospheric CO2 enrichment with emphasis on roots and the rhizosphere. Environ Pollut 83:155–189

Romanova AK (2005) Physiological and biochemical aspects and molecular mechanisms of plant adaptation to the elevated concentration of atmospheric CO_2. Russian Journal of Plant Physiology 52:112–126

Rosenzweigh C, Hillel D (1995) The potential impact of climate change in agriculture and food supply. Consequences 1:24–32

Rötter R, Van de Geijn SC (1999) Climate change effects on plant growth, crop yield and livestock. Clim Change 43:651–681

Sayılğan Ç (2016) Küresel sıcaklık artışının buğdayda beklenen etkileri ve yüksek sıcaklığa toleranslılığın fizyolojik göstergeleri, YYUniv. J Agric Sci 26(3):439–447. [In Turkish]

Schlenker W, Roberts MJ (2009) Nonlinear temperature effects indicate severe damages to U.S. crop yields under climate change. Proc Natl Acad Sci 106:15594–15598

Şehirali S, Özgen M (1987) Bitki genetik kaynakları. Ankara Üniv. Zir. Fak. Yay. No: 1020/, p 294, Ankara [In Turkish]

Semenov MA, Stratonovitch P, Alghabari F, Gooding MJ (2014) Adapting wheat in Europe for climate change. J Cereal Sci 59:245–256

Shah NH, Paulsen GM (2003) Interaction of drought and high temperature on photosynthesis and grain filling of wheat. Plant and Soil 257:219–226

Stamp P, Visser R (2012) The twenty-first century, the century of plant breeding. Euphytica 186:585–591

Taiz L, Zeiger E (2013) Fisiologia vegetal, 5th edn. Porto Alegre, Artmed, p 918

Tricker PJ, Elhabti A, Schmidt J, Fluery D (2018) The physiological and genetic basis of combined drought and heat tolerance in wheat. J Exp Bot 69(13):3195–3210

Tubiello FN, Ewert F (2002) Simulating the effects of elevated CO_2 on crops: approaches and applications for climate change. Eur J Agron 18:57–74

Tubiello FJ, Donatelli M, Rosenzweig C, Stockle CO (2000) Effects of climate change and elevated CO_2 on cropping systems: model predictions at two Italian locations. Eur J Agron 13:179–189

Ulukan H (2008) Agronomic adaptation of some field crops: a general approach. J Agron Crop Sci 194:169–179

Ulukan H (2009) The evolution of cultivated plant species: classical plant breeding versus genetic engineering. Plant Syst Evol 280:133–142

Uzmen R (2007) Global warming and climate changing, is it a catastrophe for humanity ?, Knowledge and Culture Publ., No 221, Ankara, Turkey, p 176. [in Turkish]

Valizadeh J, Ziaei SM, Mazloumzadeh SM (2014) Assessing climate change impacts on wheat production (a case study). J Saudi Soc Agric Sci 13:107–115

Villa CCT, Maxted N, Scholten M, Ford-Lloyd B (2006) Defining and identifying crop landraces. Plant Genet Res 3(3):373–384

Zavarzin AG (2001) The role of biota in global climate change. Russ J Plant Physiol 48(2):265–272

Zencirci N, Ulukan H, Ordu B, Aslan D, Mutlu TH, Örgeç M (2019) Salt, cold, and drought stress on einkorn and bread wheat during germination. Int J SecondMetab 6(2):113–128

Zhai F, Zhuang J (2009) Agricultural impact of climate change: a general equilibrium analysis with special reference to Southeast Asia. ADBI working paper 131. Tokyo, Asian Development Bank Institute, p 21

Chapter 10
Wheat Landraces Versus Resistance to Biotic and Abiotic Stresses

Awatef Ali Shlibak, Mehmet Örgeç, and Nusret Zencirci

10.1 Wheat Landraces

Wheat is a historically significant crop in Turkey, and its landraces have been collected, evaluated, and reported by many researchers from several regions of the world for their diverse traits (Zencirci and Karagoz 2005). Landraces, which have developed under various extreme conditions such as biotic and abiotic stresses, do not only increase yield but also promote stability (Witcombe et al. 1996). With exceptional adaptation capacity in the past, landraces are still out-yielding modern cultivation under low-input production systems (Dwivedi et al. 2016).

The cultivation of resistant crops is based on various biotic and abiotic stress factors with the development of prospective gene pools. Acquiring and selecting new plant types are a dynamic long-term work of scientific teams, offering a high-yielding variety with tolerance levels to specific environmental stresses. A wheat landrace is not necessarily a genetically and phenotypically stable, distinct, and uniform unit. It is related to the diversity of the geographical area and the level of exchange between farmers and short-/long-distance seeds (Jaradat 2006).

A wheat landrace can be defined as a conventional variety of wheat which includes stress and high yield stability with moderate crop yield level under minimal input condition (Zeven 1998; Lodhi et al. 2020). These landraces are the outcome of grain varieties that farmers harvest regularly, and these varieties are resistant to the factors that affect the normal variety of wheat (Zeven 1999). In the same way, these landraces have emerged as a diverse and heterogeneous population because of their resisting capacity against biotic and abiotic stress conditions (Masood et al. 2005). Thus, the genes discovered in landrace varieties can be used for breeding new resistant cultivars (Reynolds et al. 2007). In the present framework of modern

A. A. Shlibak · M. Örgeç · N. Zencirci (✉)
Department of Biology, Faculty of Science and Art, Bolu Abant Izzet Baysal University, Gölköy, Bolu, Turkey

N. Zencirci et al. (eds.), *Wheat Landraces*,
https://doi.org/10.1007/978-3-030-77388-5_10

accelerated farming systems, genetic variation among crop species has acquired a notable significance. Consequently, research on landrace is incredibly important for the discovery of new resistance genes. Since numerous landraces have as well now completely disappeared and have been substituted by new wheat varieties, useful traits related to the yield and tolerance to biotic and abiotic threats need to be differentiated. Moreover, the assembly of landraces has directed the need to characterize the germplasm collected (Lodhi et al. 2018). As a result, wheat landraces generally reflect collections from different geographical regions, which serve as a major source of new genes resistant to rust (Sthapit et al. 2014) and enhance the utilization of plant genetic resources for economically significant traits in wheat breeding programs (Karagöz and Zencirci 2005).

10.1.1 The Distribution of Wheat Landraces

Turkey has given rise to unique plant species represented all over the world. Over 30% of 8800 species found in the country are endemic to Turkey (Kan et al. 2016). Furthermore, the country is the center of origin and a source of genetic diversity for globally important plants which were first domesticated from wild in Turkey (Kan et al. 2016). Moreover, the wild einkorn wheat, *Triticum monococcum* ssp. *monococcum*, originated from *T. boeoticum*, is able to grow in harsh environmental conditions (Ünlü et al. 2018). Emmer (*T. dicoccon*) is another important hulled wheat (Aslan et al. 2016a, b). In addition, the hexaploid spelt (*Triticum spelta* L.) was the predominant cereal food cultivated in Europe from the fifth century and has been substituted by wheat (*Triticum aestivum* L.) since the twentieth century. The spelt yield is much lower, because the husks take up loss of about 30% and the milling procedure requires an extra step for husk separation (Su and Sun 2016). Archeological excavations showed that ancient people living around Şanlıurfa (Karacadag Mountains), a province in the southeastern part of Turkey, planted einkorn wheat (*Triticum boeoticum*) approximately 10,000–12,000 years ago, which is the wild form of today's commercial wheat (Kan et al. 2016).

Moreover, the distribution areas of einkorn types extend in southeast European countries, for instance, Germany, Turkey, Swiss, Spain, and Italy (Wieser et al. 2009). Emmer is a crop that is grown in Ethiopia, Turkey, India, and Italy (Gioia et al. 2015). Spelt is grown in the Czech Republic, Germany, European countries, and Switzerland (Troccoli and Codianni 2005). The emmer and spelt species are grown by farmers not only because they are enriched with nutritional components, but also both are highly resistant to the stress factors present in the environment (Suchowilska et al. 2010). From the past, the domestication area of wheat and barley is distributed also in southeastern Turkey, Syria, Lebanon, Palestine, Jordan, Iraq, and western Iran (Feldman 2001; Russell et al. 2011; Dwivedi et al. 2016).

The cultivation and uses of landraces are dependent on some factors such as market facilities and physical, socioeconomic, and pricing policies as well as climate conditions (Karagöz 2014). Similarly, landraces played an important role in the livelihood of small-scale farmers in Turkey (Karagöz 2014). Turkey's landraces are estimated at 800.000 ha (Karagöz 2014). To sum up, farm breeding facilities, growing landrace mixtures, amendment of seed registration framework and creation of a special registry system for landraces, and the usage of landraces in organic farming systems are recommended for the sustainability of crop landraces which are the optimal solutions for conservation of traditional knowledge linked to landrace sustainable utilization development (Karagöz 2014).

10.1.2 Landraces of Genetic Varieties

Landraces have many beneficial genetic characteristics that many farmers use to cultivate higher yield and/or quality bread. On the other hand, most of the varieties are not genetically stable and do not have some uniform characteristics (Lodhi et al. 2020). Likewise, the combination of homozygous and heterozygous genes may have given rise to this heterogeneous population structure. It is important to characterize the genes that provide stability in wheat. To select genes for more stable expression under the given experimental conditions, two software tools (geNorm and NormFinder) have been formulated. Therefore, several candidate genes have been analyzed for expression stability using these tools (Vandesompele et al. 2002). This data can help to improve unstable and susceptible varieties of wheat. Obsolete cultivars and landraces form an important part of the gene pool because they represent the wide intraspecific genetic diversity of crops from which new cultivars may originate (Newton et al. 2011). Landraces have long been recognized as the source of local adaptation characteristics, stress resistance, the stability of yield, and nutrition of seeds (Dwivedi et al. 2016).

10.1.3 Genetic Diversity of Wheat Landraces Based on the Adaptation to Climate Change

The pivot of variation of bread and durum wheat landraces has been switched by monocultures of natural as well as healthy genotypes. Moreover, traits such as vigor, earliness, growth habit, plant height under drought, and long peduncle, associated with grain yield of landraces under stress, are linked to a short grain filling. Moreover, the association between each of these characteristics and stressed grain yield varied from year to year, showing that all these characteristics are important, but their relative significance depends on the timing, duration, and severity of drought (Ceccarelli et al. 1991). Improving the genetic diversity available by using landraces in breeding programs to counter extreme environmental conditions and end-product quality is a realistic approach. Nevertheless, the present polymorphism needs to be protected in different wheat landraces (Lopes et al. 2015).

Genetic diversity may be assured in the wheat gene pool for continuous improvement of wheat tolerance to abiotic and biotic stresses (Trethowan and Mujeeb-Kazi 2008). An efficient strategy that provides new opportunities to enhance crop adaptability is plant breeding including genetic engineering by biotechnology and other new breeding techniques (NBTs). Nevertheless, the implementation of NBTs is limited by general public interests and complex regulations. It increases the resistance to both biotic and abiotic stresses significantly such as salinity, drought, heat, and cold, enabling data obtained from genetic studies needs to be exploited (Raza et al. 2019). Several experiments have used the activated kinases or phosphatases that can phosphorylate or dephosphorylate specific transcription factors (TFs), thus regulating the expression levels of stress-responsive genes (Lamaoui et al. 2018) (Fig. 10.1).

10.2 Abiotic Stress in Wheat Landraces

Several abiotic factors affect wheat (Rahaie et al. 2013) and decrease its production (Mostek et al. 2015), via stress responses: 31.56% by heat, 26.61% by drought, and 23.38% by salinity (Kamal et al. 2010; Aslan et al. 2016a, b). In addition to salinity, drought, and cold adversely affect the quality of water and water absorption (Pierik and Testerink 2014), reduce soil osmotic potential (Izadi et al. 2014), induce water deficiency, and cause morphological, physiological, and biochemical deterioration, eventually reducing yield (Mehrotra et al. 2014).

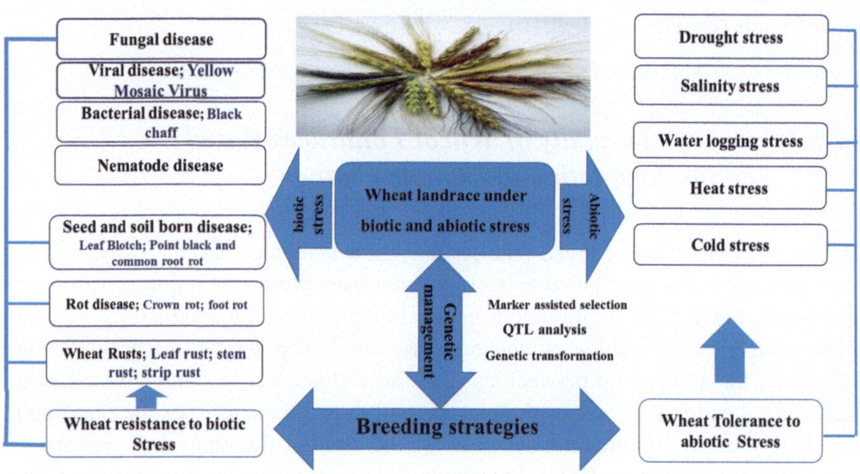

Fig. 10.1 Wheat landrace under abiotic and biotic stresses as well as genetic management; this figure is extended and summarized from the source. Adapted from (Akter and Islam 2017)

10.2.1 Drought Stress

To achieve new improvement strategies for growing stress-tolerant plants, the adaptation of crops to drought stress is a critical concern (Rizhsky et al. 2002). Likewise, water deficiency leads to drought stress which decreases plant vitality (Ashkavand et al. 2018). Drought stress has a detrimental effect on plant growth and development due to global climate change by showing a significant barrier to crop production. The main goal of wheat breeding programs worldwide is to establish the main physio-biochemical factors that restrict agricultural production and the processes associated with crop endurance in a drought period (Naderi et al. 2020). Drought tolerance can be divided into two categories, including avoidance and tolerance of dehydration (Kramer and Boyer 1995). Dehydration tolerance consists of the ability of plants to dehydrate partially and regrow again as precipitates (Salekdeh et al. 2002).

Drought stress can have effects on gene expression, and detection of these genes during water stress remains crucial toward observing their responses (Nezhadahmadi et al. 2013). Several factors can affect plants' responses toward drought stress, for instance, plant genotype, growth stage, severity, as well as exposure time to stress, physiological process of growth (Allahverdiyev et al. 2015), diverse patterns of gene expression (Denby and Gehring 2005), diverse patterns of the activity of respiration (Ribas-Carbo et al. 2005), and activity of photosynthesis machinery which is associated with environmental factors (Flexas et al. 2004). Many physiological bioresearches have been finalized on the causes of drought stress on wheat. In the same way, drought impacts vary from morphological to molecular. Diverse stages of plant development remain influenced by drought. Drought has an impact on three time of duration of plant development: vegetative, pre-anthesis, as well as terminal stage. Physiological responses of plants to drought comprise leaf wilting, reduced leaf area, and leaf abscission, thus reducing water loss through transpiration (Yadav et al. 2020).

Thus, drought stress during flowering and grain filling affects the number of seeds per spike and kernel weight which are the important components of grain yield. Furthermore, as grain yield is a complex trait controlled by many genes, breeders often use indirect selection and use well-correlated traits with the yield for improving grain yield in dry environments (Sallam et al. 2014; Al-Naggar et al. 2020). Moreover, it is common to screen bread wheat genotypes for drought stress, but there is still very limited data for diagnostic physiological parameters associated with improved drought stress yield (Salam et al. 2019). Additionally, many varieties of cultivated wheat are adapted to extreme environmental conditions and are not drought tolerant (Lodhi et al. 2020).

However, yield under drought is related to the collective dry matter at maturity in durum, bread wheat, and barley (Budak et al. 2013). But it was reported that out of 39 wild emmer, 33% showed higher resistance when subjected to osmotic stress in terms of compression with control (Blum 2005). Furthermore, a wide analysis of emmer via allozyme and DNA marker variation has presented greater genetic types linked with environmental factors. In the case of high drought tolerance, wild emmer is found very popular, and some of *T. dicoccoides* have almost thrived fully in arid conditions and arid desert ecology. As compared to durum wheat, emmer has good productivity and stability under limited water conditions. This predicts that wild emmer has allelic traits that can be utilized for a higher yield purpose. Therefore, *T. dicoccoides* can be used for producing drought-tolerant crops. The leaf and root transcript profiling of *T. dicoccoides* TR39477 (tolerant variety) and TTD-22 (sensitive variety) has shown these genes to be responsive in the drought stress mechanism (Lucas et al. 2011; Budak et al. 2013; Baloch et al. 2017).

Zencirci and Kün (1996) reported that stress environments are usually characterized by larger environmental variances than non-stress environments, increasing the difficulties in relating phenotype with genotype. Similarly, Zencirci and Kün (1996), working with durum wheat, reported different main coefficients of variation (C.V.) for 23 provinces in Turkey, using agronomic traits such as number of days to germination, tillering, shooting, and maturity, as well as grain yield. This result is in agreement with that reported by other studies in durum wheat (Zencirci and Kün 1996).

In wheat, using the quantitative trait loci (QTL) tool, a certain tolerance is also identified (Nezhadahmadi et al. 2013). Thus, to develop high-resolution genetic maps and utilize the genetic linkage between markers and essential crop traits, molecular marker systems for crop plants were created (Edwards et al. 1987). Moreover, the usage of molecular markers for marker-assisted selection (MAS) in bread wheat has been encouraged by a significant number of marker X trait associations and gained traction in many countries (Gupta et al. 2008). In addition, wheat genetic engineering can be achieved by modifying gene expression or accumulating certain metabolites that help crop with droughts, such as ABA, mainly hydrophilic proteins, and osmotically active compounds, to assist drought tolerance (Ramachandra-Reddy et al. 2004).

10.2.2 Cold Stress

Cold conditions can delay anthesis before flowering or cause extreme sterility (Sanghera et al. 2011). However, wheat germinates just above 4 °C and shoots shortly thereafter. While germination is inhibited by higher temperatures, it is accelerated by lower temperatures of 2 °C, optimum temperatures of 8 to 10 °C, and maximum temperatures of 20 to 22 °C. Germination and growth are normally controlled by soil temperature until emergence (Vandelook et al. 2008). Thus, the successful germination of seeds is one of the prerequisites for the successful establishment of stands and further growth of crops. Consequently, stronger genetic tools against cold stress are checked, located, and characterized (Zencirci and Karagöz 2005), as also in other characters (Karagöz et al. 2010), cold-tolerant wheat germplasm may help to improve wheat yield as expected (Aslan et al. 2016a, b).

Under the condition of low temperatures, root development greatly decreases in winter wheat with elevated levels of fructose and sugar along with a significant decline in osmotic potential. Winter wheat leaves are smaller and excrete less under low temperatures to survive during cold hardening or acclimatization (Aslan et al. 2016a, b).

During cold hardening, the sum of proteins, for instance, proline, glutathione, TaADF, as well as dehydrins, plays an essential role in decreasing the osmotic potential and acts as increased Cry proteins (Afzal et al. 2015). The heritability of these genes is very high, which is about 60–90%. An absolute standard can be increased for cold tolerance by discovering genetic variation among the wild wheat relatives. Similarly, the transformation of existing genes into commercially acceptable wheat varieties would reasonably increase their cold stress resistance (Kobayashi et al. 2005). Einkorn and emmer wheat have been well suited to high and mountainous regions, hot and cold weather, and infertile soils (Yaman et al. 2019). Frost tolerance is an important trait for breeding in areas with severe winters due to the sensitivity of durum wheat to low temperatures (Longin et al. 2013). As the occurrence of extreme weather events is projected to increase as part of the general phenomenon of climate change, in certain regions of the world, the destruction caused by cold temperatures may increase as well in the Mediterranean area (Marengo and Camargo 2008).

10.2.3 Heat Stress

One of the major environmental factors affecting plant growth, development, and yield is the temperature which may induce heat stress and reduce yield potential (Prasad et al. 2017). In several crop plants, such as wheat, the traditional transcriptomic approach has been used to study gene expression under stress conditions (Qin et al. 2008). The major negative effects of heat stress (HS) during grain production are due to reduced storage compound aggregation, which can harm both seed quality and final yield (Hurkman et al. 2013). In the reaction and acclimation of plants to HS, activation, and development of heat shock factors (HSFs) and heat shock proteins (HSPs) and the rise in reactive oxygen species (ROS), scavenging activity plays a key role (Comastri et al. 2018). For example, in wheat, as a "glue" between the replacement proteins (glutenins and gliadins) and starch (Maestri et al. 2002), HSPs can play a crucial role in gluten creation.

For instance, in 80% of florets, wheat plants, both structurally and functionally, exposed to 30 °C during a three-day cycle around anthesis had abnormal anthers, (Cossani and Reynolds 2012). Yield reductions were due to crop abortion and reduced grain weight during the post-flowering phases (Lobell et al. 2012). After flowering, 5-day heat stress (37 °C) has been recorded by De Leonardis et al. (2015), and the metabolic profiles of durum wheat are impaired. Moreover, grain quality is also highly influenced by heat stress during the grain filling stages (Spiertz et al. 2006).

Heat stress affecting the growth of wheat can be divided into two types: dry-hot wind and high humidity temperatures. Furthermore, metabolomics is an important functional genomics resource for understanding plant's response to heat stress (Guy et al. 2008). This reaction to heat stress was genotype-dependent, with the majority of analyzed metabolites increased in wheat cultivar (cv) Primadur (high in seed carotenoids) and decreased in cv T1303 (high in seed anthocyanin) (Valluru et al. 2017).

In wheat, several chromosomes have been mapped on major QTL clusters linked with drought and heat tolerance (Maulana et al. 2018). In the mapping of heat tolerance genes, Langdon chromosome substitution lines were first used, and associated genes were identified in 1991 on chromosomes 3A, 3B, 4A, 4B, and 6A. Similarly, 3A, 3B, and 3D chromosomes that were correlated with heat tolerance in Hope wheat cultivar were identified. Thus, chromosomes 2A, 3A, 2B, 3B, and 4B of Hope greatly increased heat resistance using chromosome replacement lines between Chinese Spring and Hope (Sun and Quick 1991). Likewise, chromosomes 3A and 3B seemed to harbor essential heat tolerance regulation genes in wheat.

A B-tolerant landrace "G61450" was documented to have contributed to the B toxicity gene Bo4, which was mapped in bread wheat on chromosome 4AL (Paull et al. 1992). To improve thermo-tolerance, crops like wheat, corn, tomato, and rice were genetically engineered, targeting primarily HSPs and HSFs (Trapero-Mozos et al. 2018).

10.2.4 Salinity Stress

Salinity affects almost 7% (950 million hectares) of overall land (Shavrukov et al. 2011), 23% of the cultivated land, and 20% of the world's irrigated land (Vardar et al. 2014). Soil salinity, however, is a significant constraint on the production of wheat in many parts of the world that affects yield losses of up to 60% and causes food insecurity (El-Hendawy et al. 2017). New genetic tools (Karagöz et al. 2010) against abiotic stresses including salt and successful research and screening techniques are desperately needed worldwide to tackle this issue (Munns and James 2003; Mostek et al. 2015). And emmer and einkorn are among the latest genetic opportunities (Karagöz and Zencirci 2005; Zencirci and Karagöz 2005; Munns et al. 2012). Salt stress during germinatio water loss durable embryo development in the final maturation (Masmoudi et al. 2009), and genotype testing against salt tolerance at different growth stages (El-Hendawy et al. 2005).

Therefore, testing wheat at various growth stages or against various salt-sensitive characters results in separate indices of salt resistance and effectively separates resistant and susceptible genotypes (Zencirci et al. 1990; Oyiga et al. 2016). Based on new genetic resources, emmer and einkorn are expected to play an important role (Karagöz and Zencirci 2005; Zencirci and Karagöz 2005; Munns et al. 2012). Experiments on wheat at diverse growth stages or against diverse salt-responsive characters result in diverse salt tolerance indices and differentiate tolerant and susceptible diverse genotypes effectively (Zencirci et al. 1990; Oyiga et al. 2016). However, salinity tolerance is a highly complex quantitative feature requiring plant-specific morphological, physiological, and metabolic mechanisms to withstand salinity stress. Osmotic tolerance, exclusion of harmful ions, and tissue tolerance are usually categorized as these pathways (Gupta and Huang 2014).

Osmotic tolerance includes all plant physiological changes by generating and/or allocating osmoprotectants such as amino acids, for example, proline and sugars (Rhodes et al. 2002), differential preferential K^+ uptake, and K^+ translocation pathways in shoots by different K + -specific channels and transporters. Moreover, the exclusion mechanism primarily aims to reduce the concentration of toxic Na^+ in the cytoplasm of roots and shoots and to retain a high salt K^+/Na^+ ratio (Almeida et al. 2017). In addition, if the salt content in the leaves is high, tissue-tolerant plants reduce the cytoplasm content of Na^+ and thus prevent adverse effects on cell metabolism by sequestrating significant amounts of salts in vacuoles and other cellular compartments (Roy et al. 2014).

10.2.5 Waterlogging Stress

Waterlogging is a major concern for wheat cultivation around the world, where excess water affects about 12% of the crop soil (Oladosu et al. 2020). As a result, about 39–40% yield loss is recorded (Collaku and Harrison 2002). Also, some researchers have reported that the decreased kernel and tiller numbers have a cumulative impact that is responsible for limiting wheat yield in waterlogging (Herzog et al. 2016). All potential crosses were made between spring wheat cultivars that are tolerant to waterlogging and indicated that a few genes regulated waterlogging (Xu et al. 2013). In waterlogged environments, genotypes with well-developed parenchymatous tissues for transportation are known to be tolerant. There are significant variations in genetic diversity between varieties for tolerance, although its frequency is comparatively smaller. The oxygen spills out of the aerenchyma into the roots and underlying soil. Therefore, a limited oxygenated soil system was created. In the same way, this system could produce microorganisms in an aerobic environment and avoid the production of highly toxic soil components such as Fe, Cu, and Mn oxides (Armstrong and Armstrong 1988).

The expression of the Adh gene in wheat, also present in barley and rice, is correlated with the tolerance of waterlogging to ensure the presence of the process of tolerance in wheat. It is recommended that a gradual process is adopted to gain waterlogging tolerance by first adding adaptive traits with established tolerance from local, national, or international germplasm, and then combining other adaptive traits specific to the target setting. Also, the "key synthetic hexaploid wheat" screening was conducted to assess tolerance against waterlogging (Afzal et al. 2015).

10.3 Biotic Stress in Wheat Landraces

Biotic stress in crops is caused by living organisms, primarily viruses, bacteria, fungi, nematodes, insects, arachnids, and weeds. Biotic stress agents effectively deprive their host of its nutrients leading to decreased plant vigor in extreme conditions as well as death of the host plant. Therefore, biotic stress is a major cause of pre- and postharvest losses in the agricultural sector. Moreover, plant loss is an adaptive defense response, otherwise the capacity toward responding to new viruses and memorize previous infections, which include organisms. As a result, several genes for biotic stress tolerance are encoded via plant genomes (Singla and Krattinger 2016).

10.3.1 Wheat Landraces' Fungal and Rust Diseases

Wheat landraces are a great source of resistance alleles for pathogenic fungi (Cavanagh et al. 2013), and resistance to *Septoria tritici* blotch (STB) (Ghaneie et al. 2012; Ferjaoui et al. 2015), which are known as local durum wheat cultivars. Furthermore, landraces with resistance to both stripe rust and stem rust are valuable in the search for diverse resistance genes to achieve durable rust resistance as well as to introduce genetic diversity in resistance to wheat breeding (Sthapit et al. 2014). Further, genetic and phenotypic researches of these landraces will help characterize their resistance to facilitate expansion of genetic diversity for resistance to wheat rusts. Moreover, some landraces with resistance to both stripe and stem rusts have been crossed with the cultivar Avocet S and the line LMPG-6 to develop populations that should help to further elucidate the genetic basis of their resistance.

10.3.1.1 Seed-Borne Diseases

Leaf blight of *Alternaria* spp. – Some cultivars are restricted to this extreme leaf spot and seed-borne disease and are easily managed by resistant cultivars. As a result, most of them are saprophytic with low pathogenicity, but some *Alternaria* spices are nonpathogenic (McIntosh 1998).

Black point – A variety of pathogens, including *Bipolaris sorokiniana* and *A. tenuis*, are responsible for this disease. This disease causes inflammation, dissemination, and discoloration of the end of the embryo wheat kernel, resulting in a decline in industrial quality and productivity (McIntosh 1998; Afzal et al. 2015).

Cephalosporium stripe – Low temperature, low pH, and moist soils tend to the temperature of soil-borne vascular disease. Besides, in the Pacific Northwest states of the USA, outbreaks are the most frequent.

10.3.1.1.1 Soil-Borne Disease

Common root rot is a prevalent disease of wheat and barley in dry temperate areas (Johnson and Lupton 1987). A single recessive gene was found on chromosome 5BS in a study by Larson and Atkinson (1982), and Johnson and Lupton (1987) have documented that there was comparatively little advancement in breeding for resistance, considering the existence of resistance sources.

10.3.1.1.2 Rot Disease

***Fusarium* spp.** causes some diseases such as crown rot, foot rot, and seedling blight. Furthermore, among the most prevalent wheat diseases are crown rot and foot rot which are caused by several species of the *Fusarium* complex (Nelson et al. 1981).

10.3.1.1.3 Wheat Rusts

There are numerous wheat rust diseases which are caused by wheat cultivation losses, for instance, leaf rust caused by *P. triticina*, stem rust caused by *Puccinia graminis* f. sp. *tritici*, and strip rust caused by *P. striiformis* as presented in Table 10.1.

Table 10.1 Wheat landraces' resistance to some diseases

Pathogen	Name patho	Caused by	Wheat landrace	References
Fungal disease	**Wheat rust,** stripe rust, leaf rust, stem rust	*P. striiformis, P. triticina, and Puccinia graminis f.* sp. *triticina*	Sr31 and Sr38 genes; Lr34, Yr 79 spring wheat landrace PI 480035; PI 182103) had resistance to stem rust and stripe rust	Singh et al. (2015), Singla and Krattinger (2016), Sthapit et al. (2014)
Viral disease	Soil-borne wheat mosaic virus **(SBWMV)**	Mosaic virus (SBWMV)	Winter wheat "Karl 92′16," "Pioneer 26R61," "AGS 2020," and "Heyne" 12,17; KS96WGRC40 (*Aegilops tauschii); Anza and wheat-Agropyron crosses*	Liu et al. (2020), Shubing et al. (2020), Hall (2006), Chuang et al. (2017)
Bacterial disease	Bacterial leaf streak **(BLS)**	*Xanthomonas translucens*	*T. turgidum* var. *durum* L; *Triticum aestivum* L	Sapkota et al. (2020), Demir and Üstün (1992)
Nematode disease	Root lesion nematode	*Pratylenchus neglectus, Heterodera filipjevi*	*Triticum dicoccoides*), Tausch's goatgrass (*Aegilops tauschii*)	Holgado et al. (2004), Toktay et al. (2015), Thompson et al. (2016)

Stripe Rust

Wheat landraces have a rich source of genes for rust disease resistance to improve genetic diversity for pre-Green Revolution such as stripe rust, leaf rust, and stem rust (Sthapit et al. 2014; Pasam et al. 2017) (Table 10.1). Sthapit et al. (2017), between 2011 and 2012, studied 165 accessions and found that they are resistant to wheat stripe rust. On the other hand, the total of 652 spring wheat landrace accessions from 54 countries have been previously assessed for resistance to stem rust pathogens (race Ug99) and have been found highly resistant to field stripe rust in Washington (hexaploid spring wheat landrace PI 480035) (Newcomb et al. 2013). In the same study, 30 of the landraces examined had dual resistance to stem rust and stripe rust. A new and different resistant germplasm of both global and regional importance can be created from landraces with resistance to both (stem and stripe) rusts. Genetic study of landraces, using SSR and SNP in which genetically diverse germplasms with rust resistance have been identified (Sthapit et al. 2014).

Consequently, it has been announced that there is a variety of potentially unidentified successful seedlings and adult plant resistance (APR) within wheat landraces, which might indicate new sources of resistance toward rust. Moreover, a novel gene Yr79 and four additional QTLs for all stages and high-temperature APR to stripe rust in wheat landrace PI 182103 were described by Feng et al. (2018).

Stem Rust

Puccinia graminis f. sp. *tritici* is the main fungus that causes stem rust which is considered one of the most dangerous winter wheat fungal diseases (Roelfs et al. 1992) (Table 10.1). Correspondingly, stem rust is widely distributed all over the national domain, even though less common than the other two wheat rust varieties (Singh et al. 2015). Furthermore, symptoms of stem rust infection on the plant parts, for instance, leaf sheaths, stems, glumes, as well as awns of susceptible plants, are masses of dark-red urediniospores (Figueroa et al. 2018). Consequently, stem rust causes wheat yield losses through a reduction of wheat quality. In severe epidemics, agriculturalists will lose harvest returns if susceptible cultivars remain grown in rust hot spot zones. Last century the Green Revolution conducted on breeding for resistance against SR (Figueroa et al. 2016) for the reason that the distribution of stem rust resistance genes, together with the 1BL. Epidemics remained low internationally for the last 30 years except for major epidemics in Ethiopia, in the period from 1993 to 1994 on Enkoy, which carries resistance gene Sr36. The form of the Ug99 race is still resulting economic losses worldwide (Singh et al. 2012). What's concerning remains that (GRRC) experimental examinations have suggested that fungi can infect dozens of lab-developed strains of wheat crops, containing hardy varieties that reportedly remained highly resistant to some diseases (Bhattacharya 2017).

Leaf Rust

Leaf rust occurs globally more than stem or stripe rust, which is linked to infections with *P. triticina* and is attributed to a reduction in kernel number per head and lower kernel weight, and in susceptible cultivars, they can exceed to 40% (Knott 1989) (Table 10.1). Likewise, stripe rust, powdery mildew, and stem rust which is conferred persistent and partial adult plant resistance toward the four biotrophic diseases of leaf rust such as Lr34, Yr18, Pm38, and Sr57 which are linked to as (Lr34). Furthermore, this phenotype can be termed as slow-rusting or slow-mildewing and is correlated with necrosis of the leaf tip, a morphological marker which is associated with processes close to senescence. An ATP-binding cassette (ABC) transporter protein is encoded via Lr34 (Krattinger et al. 2009). Members of this maintained protein family transport various substrates through biological membranes. Therefore, from the time of the domestication of hexaploid bread wheat 8000 years ago, the resistant Lr34 allele, which differs via only two amino acid polymorphisms from the susceptible Lr34 variety, is developed via the acquisition of two gain-of-function mutations (Krattinger et al. 2013).

Four fungal diseases have been subjected to intensive research and breeding efforts. In marker-assisted breeding, for example, gene-specific diagnostic molecular markers engineered for cloned resistance genes can be utilized to detect the existence of multiple resistance genes in breeding programs. It is possible to identify any biotic stress resistance gene according to (1) its visual impact on pathogen development, (2) its race specificity, and (3) its longevity. While complete resistance genes fully disable pathogen growth, partial resistance genes only delay the development of pathogens and can cause the life cycle of the pathogen to be completed. In terms of compression, partial disease resistance genes are widely accepted to exert less genetic influence on the pathogen toward developing virulence against the gene. Therefore, the combination of four to five partial resistance genes will result in disease resistance at close to resistance levels. Race-specific resistance defends against different pathogens, but not all races or strains, whereas non-race-specific resistance genes provide resistance toward all species (Singla and Krattinger 2016).

10.3.2 Wheat Landraces' Viral Diseases

Soil-borne wheat mosaic virus (SBWMV) causes an extreme viral disease in winter wheat worldwide, thus reducing grain yield (Liu et al. 2020; Cao et al. 2020) (Table 10.1). Furthermore, molecular markers are known and used in bread wheat for the identification of barley yellow dwarf virus (Ayala et al. 2001). Moreover, the epidemiology of the wheat yellow mosaic virus (WYMV or WSSMV) is similar to that of SBWMV (Sarwar et al. 2020), but the relative proportions of the viruses differ from place to place. In the same way, the vector of both SBWMV and WYMV

(WSSMV) is *Polymyxa graminis* L., an obligate fungal pathogen of wheat roots (Jiang et al. 2020). For example, *Polymyxa graminis* infects wheat, barley (*Hordeum vulgare* L.), rye (*Secale cereale* L.), *Agropyron repens* (L.) Beauv, *Bromus inermis* Leyss, *B. tectorum* L., *Hordeum jubatum* L., sorghum, and maize released from intracellular sporangia or germinating resting spores, biflagellate zoospores (Webb 2018).

There are no resistance cultivars available for wheat streak mosaic virus (WSMV) or barley yellow dwarf viruses (BYDV), but resistance available in germplasm varies. Furthermore, resistance is very strong in some cultivars, for example, Anza and wheat-*Agropyron* crosses (Chuang et al. 2017). Moreover, using resistant cultivars is the only feasible solution to regain the losses caused by SBWMV. In the same way, using wheat to finely map the resistance gene Sbwm1, 205 wheat accessions were genotyped. Likewise, BeadChips from Infinium with 90 K SNPs are selected as reported by Liu et al. (2020). Similarly, those SNPs were translated into two F6-derived recombinant inbred lines (RIL), Kompetitive Allele-Specific Polymerase (KASP) assays in two resistant cultivars "Wesley" and "Deliver" and a susceptible line "OK03825-5403-6." One more point is that the two Sbwm1 franking SNPs will effectively distinguish the resistant and susceptible lines in a new 159-wheat germplasm diversity panel. To localize SBWMV resistance genes using linkage maps, the QTL mapping technique has been used. Winter wheats "Karl 92′16," "Pioneer 26R61," "AGS 2020," and "Heyne" 12,17 (Shubing et al. 2020) have been registered to the QTL on 5DL (Humbroich 2007). This gene 18 is also borne by the breeding line KS96WGRC40 derived from *Aegilops tauschii* (Hall 2006). In addition, the gene for Tis was designated as Sbwm117. The most frequently distributed DNA sequence variants in a genome are SNPs (Omariba et al. 2020).

10.3.3 Wheat Landraces' Bacterial Diseases

Xanthomonas translucens is a gram-negative bacterium in cereal crops that can cause serious diseases. Because of the transparent lesions on infected leaves, the causal bacterium was called *Bacterium translucens* (Table 10.1). The wheat disease was identified as black chaff, which is referred to the symptoms of the disease on the spikes (Sapkota et al. 2020). Moreover, bacterial leaf streak (BLS) epidemics have been intermittent and occur in warm and humid subtropical regions. In the Near and Middle East, it affects durum (*T. turgidum* var. *durum* L.) and bread wheat in irrigated areas of Turkey (Demir and Üstün 1992). Likewise, most of the cultivars in this area are also highly susceptible; there are no chemical methods required for field BLS control (McMullen and Adhikari 2011). Correspondingly, furthermore, breeding for resistant wheat and barley cultivars is difficult because of the lack of sources of resistance and the cultivar's knowledge linked with host-pathogen interactions (Kandel et al. 2015).

10.3.4 Wheat Landraces' Nematodes

Root lesion nematode (RLN, *Pratylenchus neglectus*) is one of the common and significant plant-parasitic nematodes associated with the roots of wheat and lead to yield losses. Furthermore, *Heterodera filipjevi* is widely spread and has been recorded in European cereal crops (Holgado et al. 2004) and also in the Turkish Central and East Anatolian regions (Toktay et al. 2015; Smiley and Nicol 2009) which is closely related to the climate. Information on CCN pathotypes in Turkey is limited, with populations of *H. filipjevi* from some locations in Central and Southeastern Anatolia close to *H. avenae* pathotype Ha33 (Toktay et al. 2015) and *H. avenae* populations from Eastern Mediterranean and Southeastern Anatolia identified as pathotype Ha21 (Imren et al. 2015).

The use of resistant and tolerant cultivars is the most economical and environmentally friendly way to manage RLNs in wheat (Smiley and Nicol 2009). Moreover, resistant cultivars do not allow RLNs to increase. In contrast, tolerant cultivars allow nematode multiplication as well as perform better than susceptible cultivars in RLN-infested fields (Smiley et al. 2008). Likewise, a cultivar can be intolerant and resistant, tolerant and resistant, tolerant and susceptible, or intolerant and susceptible. Additionally, no wheat cultivar is available which can show both resistance and tolerance to lesion nematodes under the field conditions. One more point is that various cultivars of durum, spring, and winter wheat were evaluated against RLNs along with the breeding lines, landraces, rye, triticale, synthetic hexaploid wheat, and wild wheat accessions (Singh 2020). In addition, several lines of wild emmer wheat (*Triticum dicoccoides*) are presented in Table 10.1. Tausch's goatgrass (*Aegilops tauschii*) and synthetic hexaploid wheat showed resistance against *P. neglectus* (Thompson et al. 2016).

10.4 Conclusion

The production of wheat includes the performance of high-yielding varieties that are allelically enriched to overcome key biotic and abiotic stresses depending on the disrupted area. Generally, drought, salinity, heat, cold, and waterlogging are the abiotic stress constraints that play a key role in wheat manufacturing yield. Biotic stress threats, such as viruses, bacteria, fungi, and nematodes, also play a vital role in improving wheat. Improving the efficacy of the practical applications based on wheat production development of resistance genes/QTLs in the wheat breeding program throughout understanding of the genetics of resistance or tolerance is important. These advances are facilitated by a deep literature review to observe the cloning of large genes and QTLs, such as high-throughput DNA sequencing and microarray analysis, and eventually help to develop resistance/tolerance based on wheat cultivars for biotic and abiotic stress. The rate of discovery of resistance genes should be accelerated by genomic information about both host plants and also pests and pathogens. Therefore, it is important to understand the genetic nature of biotic and abiotic stress resistance/tolerance in aims to discuss problems in the future of maintaining crop productivity in changing environments. Finally, we believe that through using this deep literature review approaches, cultivation can be developed and adapted in less time than when compared to traditional research approaches.

References

Afzal F, Chaudhari SK, Gul A et al (2015) Bread wheat (*Triticum aestivum* L.) under biotic and abiotic stresses: an overview. In: Crop production and global environmental. Springer, Cham, pp 293–317

Akter N, Islam MR (2017) Heat stress effects and management in wheat. A review. Agron Sustain Dev 5:37

Allahverdiyev TI, Talai JM, Huseynova IM et al (2015) Effect of drought stress on some physiological parameters, yield, yield components of durum (Triticum durum desf.) and bread (Triticum aestivum L.) wheat genotypes. Ekin J Crop Breed Gene 1:50–62

Almeida DM, Oliveira MM, Saibo NJ (2017) Regulation of Na+ and K+ homeostasis in plants: towards improved salt stress tolerance in crop plants. Genet Mol Biol 1:326–345

Al-Naggar AM, Abd El-Shafi MA, El-Shal MH et al (2020) Evaluation of Egyptian wheat landraces (Triticum aestivum L.) for drought tolerance, agronomic, grain yield and quality traits. Plant Arch 1:3487–3504

Armstrong J, Armstrong W (1988) Phragmites australis–a preliminary study of soil-oxidizing sites and internal gas transport pathways. New Phytol 4:373–382

Ashkavand P, Zarafshar M, Tabari M et al (2018) Application of SiO2 nanoparticles as pretreatment alleviates the impact of drought on the physiological performance of Prunus mahaleb (Rosaceae). Boletín de la Sociedad Argentina de Botánica 2:207–219

Aslan D, Zencirci N, ETÖZ M et al (2016a) Bread wheat responds salt stress better than einkorn wheat does during germination. Turk J Agric For 5:783–794

Aslan D, Ordu B, Zencirci N (2016b) Einkorn wheat (*Triticum monococcum* spp. *monococcum*) tolerates cold stress better than bread wheat during germination. Tarla Bitkileri Merkez Araştırma Enstitüsü Dergisi 2:182–192. [In Turkish]

Ayala L, Van Ginkel M, Khairallah M et al (2001) Expression of Thinopyrum intermedium-derived Barley yellow dwarf virus resistance in elite bread wheat backgrounds. Phytopathology 1:55–62

Baloch F, Alsaleh A, Shahid M, Çiftçi V, et al (2017) A Whole Genome DArTseq and SNP Analysis for Genetic Diversity Assessment in Durum Wheat from Central Fertile Crescent. Public Library of Science PLoS ONE 12(1):e0167821. https://doi.org/10.1371/journal.pone.0167821

Bhattacharya S (2017) Deadly new wheat disease threatens Europe's crops. Nature 542:7640

Blum A (2005) Drought resistance, water-use efficiency, and yield potential are they compatible, dissonant, or mutually exclusive? Aust J Agric 11:1159–1168

Budak H, Kantar M, Yucebilgili Kurtoglu K (2013) Drought tolerance in modern and wild wheat. Sci World J 2013:1–17

Cao J, Xu Z, Fan X et al (2020) Genetic mapping and utilization analysis of stripe rust resistance genes in a Tibetan wheat (Triticum aestivum L.) landrace Qubaichun. Genet Resour Crop Evol 1–11 , 67(7), 1765–1775

Cavanagh CR, Chao S, Wang S et al (2013) Genome-wide comparative diversity uncovers multiple targets of selection for improvement in hexaploid wheat landraces and cultivars. Proc Natl Acad Sci U S A 110:8057–8062

Ceccarelli S, Acevedo E, Grando S (1991) Breeding for yield stability in unpredictable environments: single traits, interaction between traits, and architecture of genotypes. Euphytica 2:169–185

Chuang WP, Rojas LM, Khalaf LK et al (2017) Wheat genotypes with combined resistance to wheat curl mite, wheat streak mosaic virus, wheat mosaic virus, and triticum mosaic virus. J Econ Entomol 2:711–718

Collaku A, Harrison SA (2002) Losses in wheat due to waterlogging. J Crop Sci 2:444–450

Comastri A, Janni M, Simmonds J et al (2018) Heat in wheat: exploit reverse genetic techniques to discover new alleles within the Triticum durum sHsp26 family. Front Plant Sci 9:1337

Cossani CM, Reynolds MP (2012) Physiological traits for improving heat tolerance in wheat. Plant Physiol 160:1710–1718

De Leonardis AM, Fragasso M, Beleggia R et al (2015) Effects of heat stress on metabolite accumulation and composition, and nutritional properties of durum wheat grain. Int J Mol Sci 16:30382–30404

Demir G, Üstün N (1992) Studies on bacterial streak disease (Xanthomonas campestris pv. translucens (Jones et al.) dye.) of wheat and other gramineae. J Turk Phytopathol 21:33–40

Denby K, Gehring C (2005) Engineering drought and salinity tolerance in plants: lessons from genome-wide expression profiling in Arabidopsis. Trends Biotechnol 11:547–552

Dwivedi SL, Ceccarelli S, Blair MW et al (2016) Landrace germplasm for improving yield and abiotic stress adaptation. Trends Plant Sci 1:31–42

Edwards MD, Stuber CW, Wendell JF (1987) Molecular marker-facilitated investigation of quantitative trait loci in maize. I.Numbers, genomic distribution and types of gene action. Genetics 116:113–125

El-Hendawy SE, Hu Y, Yakout GM et al (2005) Evaluating salt tolerance of wheat genotypes using multiple parameters. Eur J Agron 22:243–253

El-Hendawy SE, Hassan WM, Al-Suhaibani NA et al (2017) Comparative performance of multivariable agro-physiological parameters for detecting salt tolerance of wheat cultivars under simulated saline field growing conditions. Front Plant Sci 8:435

Feldman M (2001) Origin of cultivated wheat. In: Dans Bonjean AP, Angus WJ (eds) The world wheat book: a history of wheat breeding. Intercept Limited, Andover, pp 3–58

Feng JY, Wang MN, See DR et al (2018) Characterization of novel gene Yr79 and four additional quantitative trait loci for all-stage and high-temperature adult-plant resistance to stripe rust in spring wheat PI 182103. J Phytopathol 108:737–747

Ferjaoui SM, Barek SB, Bahri B et al (2015) Identification of resistance sources to Septoria tritici blotch in old tunisian durum wheat germplasm applied for the analysis of the Zymoseptoria tritici-durum wheat interaction. J Plant Pathol 97:471–481

Figueroa M, Upadhyaya NM, Sperschneider J et al (2016) Changing the game: using integrative genomics to probe virulence mechanisms of the stem rust pathogen Puccinia graminis f. sp. tritici. Front Plant Sci 7:205

Figueroa M, Hammond-Kosack KE, Solomon PS (2018) A review of wheat diseases a field perspective. Mol Plant Pathol 6:1523–1536

Flexas J, Bota J, Loreto F et al (2004) Diffusive and metabolic limitations to photosynthesis under drought and salinity in C3 plants. Plant Biol 3:269–279

Ghaneie A, Mehrabi R, Safaie N et al (2012) Genetic variation for resistance to Septoria triticiblotch in Iranian tetraploid wheat landraces. Eur J Plant Pathol 132:191–202

Gioia T, Nagel KA, Beleggia R et al (2015) Impact of domestication on the phenotypic architecture of durum wheat under contrasting nitrogen fertilization. J Exp Bot 18:5519–5530

Gupta B, Huang B (2014) Mechanism of salinity tolerance in plants: physiological, biochemical, and molecular characterization. Int J Genomics 2014:18 p. https://doi.org/10.1155/2014/701596

Gupta PK, Mir RR, Mohan A et al (2008) Wheat genomics: present status and future prospects. J Plant Genomics 2008:36

Guy C, Kaplan F, Kopka J et al (2008) Metabolomics of temperature stress. Physiol Plant 2:220–235

Hall MD (2006) Genetic characterization and utilization of multiple Aegilops tauschii derived pest resistance genes in wheat (Doctoral dissertation, Kansas State University). Search K-REx

Herzog M, Striker GG, Colmer TD et al (2016) Mechanisms of waterlogging tolerance in wheat a review of root and shoot physiology. Plant Cell Environ 5:1068–1086

Holgado R, Andersson S, Rowe JA et al (2004) First record of Heterodera filipjevi in Norway. Nematol Mediterr 32:205–211

Humbroich K (2007) Identification and mapping of resistance genes against soil borne viruses in barley (Hordeum vulgare L.) and wheat (Triticum aestivum L.). Aachen Shaker 2007:1–125

Hurkman WJ, Tanaka CK, Vensel WH et al (2013) Comparative proteomic analysis of the effect of temperature and fertilizer on gliadin and glutenin accumulation in the developing endosperm and flour from Triticum aestivum L. cv. Proteome Sci 11:8

Imren M, Waeyenberge L, Viaene N et al (2015) Morphological and molecular identification of cereal cyst nematode from eastern Mediterranean region of Turkey. Turk J Agric For 39:91–98

Izadi MH, Rabbani J, Emam Y et al (2014) Effects of salinity stress on physiological performance of various wheat and barley cultivars. J Plant Nutr 4:520–531

Jaradat AA (2006) Phenotypic divergence in the meta-population of the Hourani durum wheat landrace. J Food Agric Environ 4:186–191

Jiang C, Kan J, Ordon F, Perovic D, & Yang P (2020) Bymovirus-induced yellow mosaic diseases in barley and wheat: viruses, genetic resistances and functional aspects. Theoretical and Applied Genetics, SpinngerLink, 133:1623–1640

Johnson R, Lupton FGH (1987) Breeding for disease resistance. In: Wheat breeding, its scientific basis. Chapman and Hall, London, pp 369–424

Kamal AHM, Kim KH, Shin KH et al (2010) Abiotic stress responsive proteins of wheat grain determined using proteomics technique. Aust J Crop Sci 3:196

Kan M, Kucukcongar M, Mourgounov A et al (2016) Wheat landraces production on farm level in Turkey; who is growing in where? Pak J Agric Sci 53:159–169

Kandel YR, Glover KD, Osborne LE et al (2015) Mapping quantitative resistance loci for bacterial leaf streak disease in hard red spring wheat using an identity by descent mapping approach. Euphytica 1:53–65

Karagöz A (2014) Wheat landraces of Turkey. Emirates J Food Agric 2:149

Karagöz A, Zencirci N (2005) Variation in wheat (Triticum spp.) landraces from different altitudes of three regions of Turkey. Genet Resour Crop Evol 52:775–785

Karagoz A, Zencirci N, Tan A et al (2010) Conservation and use of plant genetic resources. In: Turkish chambers of agricultural engineers 7th technical congress, vol 1, pp 11–15

Knott DR (1989) The wheat rust: breeding for resistance. Springer-Verlag, Berlin/Heidelberg/New York, 201 p

Kobayashi F, Takumi S, Kume S et al (2005) Regulation by Vrn-1/Fr-1 chromosomal intervals of CBF-mediated Cor/Lea gene expression and freezing tolerance in common wheat. J Exp Bot 413:887–895

Kramer PJ, Boyer JS (1995) Water relations of plants and soils. Academic, pp 1–5

Krattinger SG, Lagudah ES, Spielmeyer W et al (2009) A putative ABC transporter confers durable resistance to multiple fungal pathogens in wheat. Science 5919:1360–1363

Krattinger SG, Jordan DR, Mace ES et al (2013) Recent emergence of the wheat Lr34 multipathogen resistance: insights from haplotype analysis in wheat, rice, sorghum and Aegilops tauschii. Theor Appl Genet 3:663–672

Lamaoui M, Jemo M, Datla R et al (2018) Heat and drought stresses in crops and approaches for their mitigation. Front Chem 6:26

Larson RI, Atkinson TG (1982) Reaction of wheat to common root rot: linkage of a major gene, Crr, with the centromere of chromosome 5B. Can J Genet Cytol 1:19–25

Liu S, Bai G, Lin M et al (2020) Identification of candidate chromosome region of Sbwm1 for soil-borne wheat mosaic virus resistance in wheat. Sci Rep 1:1–11

Lobell DB, Sibley A, Ortiz-Monasterio JI (2012) Extreme heat effects on wheat senescence in India. Climate Change 3:186–189

Lodhi SS, John P, Bux H et al (2018) Resistance potential of Pakistani wheat landrace (Triticum aestivum L.) against stripe rust (Puccinia striiformis) and karnal bunt (Tilletia indica). Pak J Bot 2:801–806
Lodhi SS, Maryam S, Rafique K et al (2020) Overview of the prospective strategies for conservation of genomic diversity in wheat landraces. In: Climate change and food security with emphasis on wheat. (Science Direct). Elsevier, pp 293–309. https://doi.org/10.1016/b978-0-12-819527-7.00021-2
Longin CFH, Sieber AN, Reif JC (2013) Combining frost tolerance, high grain yield and good pasta quality in durum wheat. Plant Breed 132:353–358
Lopes MS, Dreisigacker S, Peña RJ et al (2015) Genetic characterization of the wheat association mapping initiative (WAMI) panel for dissection of complex traits in spring wheat. Theor Appl Genet 3:453–464
Lucas S, Dogan E, Budak H (2011) TMPIT1 from wild emmer wheat: first characterisation of a stress-inducible integral membrane protein. Gene 1-2:22–28
Maestri E, Klueva N, Perrotta C et al (2002) Molecular genetics of heat tolerance and heat shock proteins in cereals. Plant Mol Biol 48:667–681
Marengo JA, Camargo CC (2008) Surface air temperature trends in Southern Brazil for 1960–2002. Int J Climatol 28:893–904
Masmoudi K, Brini F, Feki K et al (2009) Enhancing drought and salinity tolerance in wheat crop grown in the Mediterranean region. J Plant Cell 3:446–449
Masood MS, Javaid A, Rabbani MA et al (2005) Phenotypic diversity and trait association in bread wheat (Triticum aestivum L.) landraces from Baluchistan, Pakistan. Pak J Bot 4:949
Maulana F, Ayalew H, Anderson JD et al (2018) Genome- wide association mapping of seedling heat tolerance in winter wheat. Front Plant Sci 9:1272
McIntosh RA (1998) Euphytica. 1(3):19–34
McMullen MP, Adhikari TB (2011) Bacterial leaf streak and black chaff of wheat. NDSU Extension Service, North Dakota State University
Mehrotra R, Bhalothia P, Bansal P et al (2014) Abscisic acid and abiotic stress tolerance–different tiers of regulation. J Plant Physiol 7:486–496
Mostek A, Börner A, Badowiec A et al (2015) Alterations in root proteome of salt- sensitive and tolerant barley lines under salt stress conditions. J Plant Physiol 174:166–176
Munns R, James R (2003) Screening methods for salinity tolerance: a case study with tetraploid wheat. Plant Soil 253:201–218
Munns R, James RA, Xu B et al (2012) Wheat grain yield on saline soils is improved by an ancestral Na transporter gene. Nat Biotechnol 30:360–364
Naderi S, Fakheri BA, Maali-Amiri R et al (2020) Tolerance responses in wheat landrace Bolani are related to enhanced metabolic adjustments under drought stress. Plant Physiol Biochem 150:244–253
Nelson PE, Toussoun TA, Cook RJ (1981) Fusarium: diseases, biology, and taxonomy. State University Press
Newcomb M, Acevedo M, Bockelman HE et al (2013) Field resistance to the Ug99 race group of the stem rust pathogen in spring wheat landraces. Plant Dis 97:882–890
Newton AC, Johnson SN, Gregory PJ (2011) Implications of climate change for diseases, crop yields and food security. Euphytica 197:3–18
Nezhadahmadi A, Prodhan ZH, Faruq G (2013) Drought tolerance in wheat. Sci World J 2013:610721
Oladosu Y, Rafii MY, Arolu F et al (2020) Submergence tolerance in Rice: review of mechanism, breeding and, future prospects. J MDPI Sustain 4:1632
Omariba G, Xu F, Wang M et al (2020) Genome-wide analysis of MicroRnA-related single nucleotide polymorphisms (Snps) in mouse genome. Sci Rep 1:1–9
Oyiga BC, Sharma RC, Shen J et al (2016) Identification and characterization of salt tolerance of wheat germplasm using a multivariable screening approach. J Agron Crop Sci. https://doi.org/10.1111/jac.12178

Pasam RK, Bansal U, Daetwyler HD et al (2017) Detection and validation of genomic regions associated with resistance to rust diseases in a worldwide hexaploid wheat landrace collection using BayesR and mixed linear model approaches. Theor Appl Genet 4:777–793

Paull JG et al (1992) Physiological and genetic control of the tolerance of wheat to high concentrations of boron and implications for plant breeding. Plant Soil 146:45–52

Pierik R, Testerink C (2014) The art of being flexible: how to escape from shade, salt, and drought. J Plant Physiol 166:5–22

Prasad PV, Bheemanahalli R, Jagadish SK (2017) Field crops and the fear of heat stress pportunities, challenges and future directions. Field Crop Res 200:114–121

Qin D, Wu H, Peng H et al (2008) Heat stress-responsive transcriptome analysis in heat susceptible and tolerant wheat (Triticum aestivum L.) by using Wheat Genome Array. BMC Genomics 1:432

Rahaie M, Xue GP, Schenk PM (2013) The role of transcription factors in wheat under different abiotic stresses. In: Abiotic stress – plant responses and applications in agriculture. InTech, Rijeka, pp 367–385

Ramachandra-Reddy A, Chaitanya KV, Vivekanandan M (2004) Drought-induced responses of photosynthesis and antioxidant metabolism in higher plants. J Plant Physiol 161:1189–1202

Raza A, Razzaq A, Mehmood SS et al (2019) Impact of climate change on crops adaptation and strategies to tackle its outcome: a review. Plants 2:34

Reynolds M, Dreccer F, Trethowan R (2007) Drought-adaptive traits derived from wheat wild relatives and landraces. J Exp Bot 2:177–186

Rhodes D, Nadolska-Orczyk A, Rich PJ (2002) Salinity, osmolytes and compatible solutes. In: Salinity: environment-plants-molecules. Springer, Dordrecht, pp 181–204

Ribas-Carbo M, Taylor NL, Giles L et al (2005) Effects of water stress on respiration in soybean leaves. Plant Physiol 1:466–473

Rizhsky L, Liang H, Mittler R (2002) The combined effect of drought stress and heat shock on gene expression in tobacco. Plant Physiol 3:1143–1151

Roelfs AP, Singh RP, Saari EE (1992) Rust diseases of wheat: concepts and methods of disease management. CIMMYT, Mexico

Roy SJ, Negrão S, Tester M (2014) Salt resistant crop plants. J Biotechnol 26:115–124

Russell IK et al (2011) Analysis of >1000 single nucleotide polymorphisms in geographically matched samples of landrace and wild barley indicates secondary contact and chromosome level differences in diversity around domestication genes. New Phytol 191:564–578

Salam A, Khan MS, Ali A et al (2019) Impact of osmotic stress on seed germination and seedling growth of wheat landraces. In: Conference: 9th national conference on "Innovative trends in wheat: way forward to sustainable wheat production" At: at the College of Agriculture University of Sargodha, Pakistan. https://su.edu.pk/events/post/52/CALL-FOR-PAPER

Salekdeh GH, Siopongco J, Wade LJ et al (2002) Proteomic analysis of rice leaves during drought stress and recovery. Proteomics 9:1131–1145

Sallam A, Hamed ES, Hashad M et al (2014) Inheritance of stem diameter and its relationship to heat and drought tolerance in wheat (Triticum aestivum L.). J Plant Breed Crop Sci 1:11–23. https://doi.org/10.5897/JPBCS11.017

Sanghera GS, Wani SH, Hussain W et al (2011) Engineering cold stress tolerance in crop plants. Curr Genomics 12:30–43

Sapkota S, Mergoum M, Liu Z (2020) The translucens group of Xanthomonas translucens: complicated and important pathogens causing bacterial leaf streak on cereals. Mol Plant Pathol 3:291–302

Sarwar M, Aslam M, Sarwar S et al (2020) Different nematodes and plasmodiophorids as vectors of plant viruses. In: Applied plant virology. Academic, pp 275–290

Shavrukov Y, Shamaya N, Baho M et al (2011) Salinity tolerance and Na+ exclusion in wheat: variability, genetics, mapping populations and QTL analysis. Czech J Genet Plant Breed 47:S85–S93

Shubing L, Guihua B, Lin M et al (2020) Identification of candidate chromosome region of Sbwm1 for soil-borne wheat mosaic virus resistance in wheat. Sci Rep 10:1
Singh G (2020) Resistance Screening and QTL Mapping in Wheat and Triticale Against Root-lesion Nematode (Doctoral dissertation, North Dakota State University)
Singh S, Singh RP, Huerta-Espino J (2012) Stem rust. Disease resistance in wheat See Ref 85:18–32
Singh RP, Hodson DP, Jin Y et al (2015) Emergence and spread of new races of wheat stem rust fungus: continued threat to food security and prospects of genetic control. Phytopathology 7:872–884
Singla J, Krattinger SG (2016) Biotic stress resistance genes in wheat. In: Wrigley CW, Faubion J, Corke H, Seetharaman K (eds) Encyclopedia of food grains. Elsevier, pp 388–392
Smiley RW, Nicol JM (2009) Nematodes which challenge global wheat production. In: Wheat: science and trade. Wiley-Blackwell, Ames, pp 171–187
Smiley RW, Yan GP, Handoo ZA (2008) First record of the cereal cyst nematode Heterodera filipjevi in Oregon. Plant Dis 92:1136
Spiertz JH, Hamer RJ, Xu H et al (2006) Heat stress in wheat (Triticum aestivum L.): effects on grain growth and quality traits. Eur J Agron 2:89–95
Sthapit JK, Krishnan V, Jiwan D et al (2017) Mapping genes for resistance to stripe rust in spring wheat landrace PI 480035. PLoS One 5:e0177898
Sthapit J, Newcomb M, Bonman JM et al (2014) Genetic diversity for stripe rust resistance in wheat landraces and identification of accessions with resistance to stem rust and stripe rust. Crop Sci 5:2131–2139
Su WH, Sun DW (2016) Facilitated wavelength selection and model development for rapid determination of the purity of organic spelt (Triticum spelta L.) flour using spectral imaging. Talanta 155:347–357
Suchowilska E, Kandler W, Sulyok M et al (2010) Mycotoxin profiles in the grain of Triticum monococcum, Triticum dicoccum and Triticum spelta after head infection with Fusarium culmorum. J Sci Food Agric 4:556–565
Sun QX, Quick JS (1991) Chromosomal locations of genes for heat tolerance in tetraploid wheat. Cereal Res Commun 4:431–437
Thompson AL, Smiley RW, Paulitz TC et al (2016) Identification of resistance to Pratylenchus neglectus and Pratylenchus thornei in Iranian landrace accessions of wheat. Crop Sci 2:654–672
Toktay T, Imren M, Ocal A et al (2015) Incidence of cereal cyst nematodes in the East Anatolia region in Turkey. Russ J Nematol 23:29–40
Trapero Mozos A, Morris WL et al (2018) Engineering heat tolerance in potato by temperature-dependent expression of a specific allele of heat- shock cognate 70. Plant Biotechnol J 1:197–207
Trethowan RM, Mujeeb-Kazi A (2008) Novel germplasm resources for improving environmental stress tolerance of Hexaploid wheat. Crop Sci 4:1255
Troccoli A, Codianni P (2005) Appropriate seeding rate for einkorn, emmer, and spelt grown under rainfed condition in southern Italy. Eur J Agron 3:293e300. https://doi.org/10.1016/j.eja.2004.04.003
Ünlü ES, Bataw S, Aslan Şen D et al (2018) Identification of conserved miRNA molecules in einkorn wheat (Triticum monococcum subsp. monococcum) by using small RNA sequencing analysis. Turk J Biol 6:527
Valluru R, Reynolds MP, Davies WJ et al (2017) Phenotypic and genome wide association analysis of spike ethylene in diverse wheat genotypes under heat stress. New Phytol 1:271–283
Vandelook F, Bolle N, Van Assche JA (2008) Seasonal dormancy cycles in the biennial Torilis japonica (Apiaceae), a species with morphophysiological dormancy. Seed Sci Res 18:161–171
Vandesompele J, De Preter K, Pattyn F et al (2002) Accurate normalization of real-time quantitative RT-PCR data by geometric averaging of multiple internal control genes. Genome Biol 7:4–1
Vardar Y, Çifcifci EA, Yagdi K (2014) Salinity effects on germination stage of bread and durum wheat cultivars. Yüzüncü Yıl Üniversitesi Tarım Bilimleri Dergisi 2:127–139

Webb CA (2018) Epidemiological factors impacting the development of Wheat streak mosaic virus outbreaks (Doctoral dissertation)
Wieser H, Mueller KJ, Koehler P (2009) Studies on the protein composition and baking quality of einkorn lines. Eur Food Res Technol 3:523–532
Witcombe JR, Joshi A, Joshi KD et al (1996) Farmer participatory crop improvement. I. Varietal selection and breeding methods and their impact on biodiversity. Exp Agric 4:445e460. https://doi.org/10.1017/S0014479700001526
Xu QT, Yang L, Zhou ZQ et al (2013) Process of aerenchyma formation and reactive oxygen species induced by waterlogging in wheat seminal roots. Planta 5:969–982
Yadav S, Modi P, Dave A et al (2020) Effect of abiotic stress on crops. In: Sustainable crop production. IntechOpen, London
Yaman HM, Ordu B, Zencirci N et al (2019) Coupling socioeconomic factors and cultural practices in production of einkorn and emmer wheat species in Turkey. Environ Dev Sustain 22, 8079–8096 (2020). https://doi.org/10.1007/s10668-019-00562-7
Zencirci N, Karagoz A (2005) Effect of developmental stages length on yield and some quality traits of Turkish durum wheat (Triticum turgidum L. convar. durum (Desf.) Mackey) landraces: influence of developmental stages length on yield and quality of durum wheat. Genet Crop Evol 6:765–774
Zencirci N, Eser V, Baran I (1990) An approach to compare some stability statistics. CRIFC, Ankara. (in Turkish)
Zencirci N, Kün E (1996) Variation in landraces of durum wheat (T. turgidum L. conv. durum (Desf.) M.K.) from Turkey. Euphytica. (92):333–339
Zeven AC (1998) Landraces: a review of definitions and classifications. Euphytica 104:127–139
Zeven AC (1999) The traditional inexplicable replacement of seed and seed ware of landraces and cultivars: a review. Euphytica 3:181–191

Chapter 11
Contribution of Landraces in Wheat Breeding

Muhammad Azhar Nadeem, Mehmet Zahit Yeken, Mehmet Tekin, Zemran Mustafa, Rüştü Hatipoğlu, Husnu Aktaş, Ahmad Alsaleh, Evren Cabi, Ephrem Habyarimana, Nusret Zencirci, Tolga Karaköy, Hakan Özkan, and Faheem Shehzad Baloch

11.1 Introduction of Landraces

Agriculture is one of the oldest livelihood sources of mankind. Humans remained actively involved in the selection of favorable traits which resulted in significant changes in the phenotype and genotype of wild plants. In addition to man's

---

M. A. Nadeem · Z. Mustafa · T. Karaköy · F. S. Baloch (✉)
Faculty of Agricultural Sciences and Technologies, Sivas University of Science and Technology, Sivas, Turkey

M. Z. Yeken
Department of Field Crops, Faculty of Agriculture, Bolu Abant Izzet Baysal University, Bolu, Turkey

M. Tekin
Department of Field Crops, Faculty of Agriculture, Akdeniz University, Antalya, Turkey

R. Hatipoğlu · H. Özkan
Department of Field Crops, Faculty of Agriculture, Çukurova University, Adana, Turkey

H. Aktaş
Artuklu University Vocational Higher School of Kızıltepe, Mardin, Turkey

A. Alsaleh
Science and Technology Application and Research Center (BİLTEM), Yozgat Bozok University, Yozgat, Turkey

E. Cabi
Department of Biology, Faculty of Arts and Science, Tekirdağ Namik Kemal University, Tekirdağ, Turkey

E. Habyarimana
Accelerated Crop Improvement Program, International Crops Research Institute for the Semi-Arid Tropics, Patancheru, Telangana, India
e-mail: E.Habyarimana@cgiar.org

N. Zencirci
Department of Biology, Faculty of Arts and Sciences, Bolu Abant Izzet Baysal University, Bolu, Turkey

N. Zencirci et al. (eds.), *Wheat Landraces*,
https://doi.org/10.1007/978-3-030-77388-5_11

selection activities, environmental factors also played a significant contribution in the selection of various favorable traits suitable for man-made land and gardens. Combination of these activities resulted in the development of distinctive populations called "landraces" (Zeven 1998). Landraces are dynamic populations of cultivated plants having a historical background, genetically diverse and distinct identity, and good adaptation to local environment and that are associated with traditional farming systems (Villa et al. 2005). Dwivedi et al. (2016) stated that landraces are heterogeneous populations of domesticated species having great adaptation to local environment and can serve as a source of genetic variations that can be very helpful to combat the current and new challenges for farming in changing environments. Landraces are found phenotypically diverse and less productive compared to their cultivated types (Mir et al. 2020). However, regarding their quality attributes, landraces have been found highly nutritious compared to their cultivated ones (Azeez et al. 2018). Landraces played a major role in plant breeding by providing novel genes for various agronomic, quality, mineral, biotic, and abiotic traits (Azeez et al. 2018; Lopes et al. 2015). An impressive increase in yields per hectare was the result of the "Green Revolution" due the inclusion of high-yielding varieties (HYVs) having better response to inputs (Mir et al. 2020). After the inclusion of these high-yielding varieties, it was supposed that landraces will inevitably disappear with time (Frankel and Bennett 1970; Zeven 1998). However, these breeding activities led to genetic erosion and emergence of various modern cultivars that are prone to various biotic and abiotic stresses. It is estimated that approximately 75% loss of genetic diversity is observed in the last 100 years (Hammer et al. 1996). Globally, loss of genetic diversity is very alarming because it can be used to combat food scarcity problems in the long term. Therefore, it is very important to pay attention to collect, preserve, and grow these landraces as they guarantee the existence of variations that can be used for breeding of crops for the production of more quantity of food with high quality. Besides the inclusion of HYVs, landraces maintained their position by playing a key role in agricultural production, specifically in those environments where commercial cultivars failed their competitive advantage (Casañas et al. 2017).

11.2 Origin of Wheat Landraces

Wheat is one of the domesticated food crops cultivated in mild temperature and consumed as a staple crop by millions of people (Lodhi et al. 2020). Domestication of wheat is considered a key reason behind increased human population, thereby participating in the emergence of the human civilization (Jaradat 2011). Domestication of wild emmer (*Triticum dicoccoides*), which is the progenitor of all polyploid cultivated wheats, is considered an important event in the emergence of agriculture in Southwest Asia. Domestication of wild emmer occurs in the Fertile Crescent, and it acted as a prerequisite for the evolution of tetraploid durum and hexaploid bread wheat (Jaradat 2011). However, the domestication and the subsequent breeding activities drastically reduced wheat genetic diversity (Dvorak et al. 1998).

Ancient farmers planted diverse assemblages of wheat genotypes (i.e., landraces) aiming to decrease the risk of crop failure and to improve food security because they had limited capacity to control the spatially heterogeneous and temporally unpredictable environment (Jaradat 2006; Peng et al., 2011a, b). This exercise resulted in the development of wheat landrace meta-populations and the evolution of farmers' seed systems through which they accessed and exchanged diverse genetic material. A meta-population structure can be defined as a group of subpopulations that is interconnected through gene flow and seed exchange and favors the evolution of diversity (Jaradat 2011). It is believed that natural interference, human skills, and years of continuous cultivation resulted in great diversity in wheat genotypes (Lodhi et al. 2020). Zeven (2000) stated that previously many farmers used wheat crops to develop new cultivars. Archaeological evidence are present revealing the cultivation of wheat in Iberian Peninsula, since the fifth millennium BC, and the development of wild wheats, traditional wheat varieties, and other crops happened in the Fertile Crescent (Diamond 2002).

Wheat landraces were developed from their older ones having the ability to grow in such conditions which are not feasible for the growth of the regular wheat (Witcombe et al. 1996). Zeven (1999) stated that wheat landraces are crop varieties developed by farmers through human and natural selection and reflect adaptation to local management practices and environmental conditions. Combination of both human and natural selection resulted in changes in the architecture of genotypes having better attributes like drought, salt, cold, or heat tolerance, quality traits, time to heading and maturity, and seed filling duration (Masood et al. 2005). Due to genetically distinct plant populations, wheat landraces are conserved, and some specific names were given by the traditional farmers in order to meet their environmental, cultural, social, and economic needs. Therefore, landraces are also known as farmer's varieties or folk varieties (Belay et al. 1995).

11.3 How Landraces Contributed in Wheat Breeding

Landraces played a significant role in wheat breeding by gaining focus from breeding community. Wheat landraces served as genetic resource for the development of climate-resilient cultivars with high yield (Abu-Zaitoun et al. 2018). An increasing interest has been observed for the usage of landraces as source of nutritional traits and flavor repertoire and landrace cultivation for niche markets (Roselló et al. 2018). Wheat landraces contain higher genetic diversity compared to most modern wheat landraces, and this diversity includes their adaptation to environmental conditions according to the place of origin. Some countries used this characteristic in the development of first improved cultivars through the selection of local landraces. For example, "Aragon 03" was the leading variety in Spain during the period 1960–1976. It was developed from indigenous landrace population "Catalan de Monte" (Gadea

1958) and showed high ability to drought resistance (Royo and Briceño-Félix 2011a; b). Similarly, "Turkey" (syn. "Turkey Red"), a hard red winter wheat having better adaption for cold regions, showed marvelous impact on wheat cultivation in the United States at the turn of the last century due to decreased winterkill, among other traits (Olmstead and Rhode 2002). A Japanese landrace "Akakomugi," containing Rht8c and Ppd-D1, was used by Italian breeder Nazareno Strampelli to improve Italian wheat gene pool (Salvi et al. 2013). The sensational varieties "Ardito and Mentana" developed from the crosses of Strampelli, including Akakomugi, became the basis of most of the new varieties developed in Mediterranean countries, South American countries, and several distant countries such as Russia and China. In Argentina, "Ardito" was used as parent to develop the variety Klein-33, which became the backbone of the former USSR breeding program, generating the variety Bezostaja-1 (Borojevic and Borojevic 2005). Contribution of landraces in wheat breeding for various traits is discussed comprehensively.

11.3.1 Role of Landraces in Adaptive Traits

Adaptive traits suited to target the environment have acted a decisive role during domestication and the spread out of domesticated wheat. Fitting flowering time to the current conditions in the target environments is presumably one of the main important factors during dispersal (Peng et al. 2011a; b; Royo et al. 2020). The first domesticated cereals/old landraces had most probably response to day length and cold temperatures like their wild relatives/progenitors. Motzo and Giunta (2007) hypothesized that old cultivars/landraces had the greatest day-length sensitivity and vernalization in comparison to intermediate and modern ones. However, novel adaptive traits for each target environment were naturally or artificially selected during the domestication and spreading process from the Fertile Crescent to new agricultural areas (Kilian et al. 2009). Especially other yield-related traits such as plant height, waxiness, number of spikes, and weight of spikes and grains were also co-selected by ancient farmers, and many botanical variants have been developed in this process (Peng et al. 2011a; b). Wheat landraces arising from the migration from the Fertile Crescent to the other regions of the world had been grown extensively until the Green Revolution in the early 1970s (Harlan 1975). As a result of the Green Revolution, more productive semidwarf wheat cultivar shaving better response to inputs replaced the landraces/local populations which are generally identified as tall, tended to lodge, sensitive to the foliar diseases, and low yielded (Reynolds and Borlaug 2006a; b; Lopes et al. 2015). Nevertheless, their cultivation has continued in marginal environments and they currently support subsistence farming in many regions of the world (Newton et al. 2010).

The wide range adaptability of wheat is mainly based on three genetic groups such as vernalization (*Vrn*) genes, photoperiod (*Ppd*) genes, and genes controlling "narrow-sense earliness" or "earliness per se" (*Eps*). Vernalization, which is inducting flowering by exposure to cold, basically determines plant growth habit types as winter (strong vernalization requirement) and spring (no vernalization requirement). Vernalization in wheat has very allelic complex and previous studies have presented that *Vrn* allele combinations or frequencies with an adaptive value in target growing areas are varied geographically (Stelmakh 1990; Damania et al. 1996; Iwaki et al. 2001; Zhang et al. 2008; Sun et al. 2009). Kato and Yokoyama (1992) observed the main adaptive traits in 158 bread wheat landraces collected from various climatic regions including Asian and European countries, and they claimed that nearly half of the variation for observed traits was accounted by geographical differences of their origin centers. Kato et al. (1997) also studied geographical variation of wild emmer (*Triticum dicoccoides*) accessions for vernalization response and earliness in comparison to other tetraploid relatives such as cultivated emmer (*T. dicoccum*), durum wheat (*T. durum*), and *T. turgidum*. They concluded that spring growth habit in *T. dicoccoides* could have evolved from a winter type especially in temperate conditions.

Many studies presented that the vernalization requirement in wheat is considered to be genetically controlled by at least three loci, *Vrn-A1* (*Vrn-1*), *Vrn-B1* (*Vrn-2*), and *Vrn-D1* (*Vrn-3*), located in chromosomes 5A, 5B, and 5D, respectively (Pugsley 1971, 1972; Law et al. 1976; Galiba et al. 1995; Dubcovsky et al. 1998; Yan et al. 2003). While *Vrn-A1* has the major impact on transition from vegetative to generative phase, recessive mutants of *Vrn-B1* trigger flowering. While a dominant allele of any *Vrn* genes causes spring growth habit, wheats classified as winter type must have recessive alleles at all *Vrn* loci (Turner et al. 2013). On the other hand, photoperiodic response in wheat is primarily controlled by three major genes, *Ppd-D1* (*Ppd1*), *Ppd-B1* (*Ppd2*), and *Ppd-A1* (*Ppd3*), located in 2DS, 2BS, and 2AS chromosomes, respectively. It is known that *Ppd-D1* plays an important role in regulation of photoperiodic response. In addition, "earliness per se" or "narrow-sense earliness" is the difference in flowering times of genotypes whose vernalization and day-length requirements have been completed (Kato et al. 2001). Earliness per se genes can also affect flowering time independently, but these genes have not been studied in detail because of major effects of vernalization and photoperiod genes on flowering time. Moreover, this trait is highly heritable and can be effectively used in breeding programs (Kato and Wada 1999). Many QTLs have been identified for earliness per se in all three genomes with previous studies (Bullrich et al. 2002; Hanocq et al. 2004; Kamran et al. 2013).

Previous studies with a marker-assisted selection approach have clarified that landraces/accessions have a huge genetic diversity and very allelic complex for vernalization and photoperiod genes. Jiang et al. (2010) found that the frequencies of the dominant *Vrn* genes in 153 Chinese wheat landraces were 60.78% (*Vrn-D1*), 5.88% (*Vrn-A1a*), 5.23% (*Vrn-B1*), and 0 (*Vrn-B3*), respectively. Andeden et al. (2011) determined that Turkish wheat germplasm has mostly the dominant *Vrn-B1* allele followed by *Vrn-D1* and *Vrn-A1*. Derakhshan et al. (2013) reported that the frequencies of dominant *Vrn-D1* and *Vrn-B1* alleles in 395 Iranian wheat landraces were 67.35% and 38.48%, respectively. Manickavelu et al. (2014) characterized 400 wheat landraces genetically collected from different agroecological zones of Afghanistan for adaptive and other yield-related traits, and they reported that 53%

of all landraces were winter types, 43% had one or more dominant *Vrn* alleles, and 4% were either unknown or had *Vrn-A1c* – a rare spring allele. Guo et al. (2015) also studied distribution of the *Vrn-D1b* allele in Chinese wheat accessions and determined that the frequencies of *Vrn-D1a*, *Vrn-D1b*, and *Vrn-D1* alleles were 27.3, 20.6, and 52.1%, respectively, of 689 accessions. They also claimed that *Vrn-D1b* allele originated from Chinese landraces as a result of pedigree analysis. Goncharov (1998) claimed that there is a high rate of the *Vrn-D1* allele in countries near the equator in addition to Pakistan, Afghanistan, and China.

Other important genetic factors like dwarfing genes (*Rht*) are critical against environmental stresses to guarantee both adaptability and grain yield in addition to vernalization, photoperiod, and earliness. It is known that over 30 height-reducing genes have been identified so far (McIntosh et al. 2013). The major dwarfing genes "*Rht-B1*" and "*Rht-D1*" known as the *Reduced height* (*Rht*) loci were introduced during the "Green Revolution" that achieved to improve harvest index by reducing plant height. These genes are known as gibberellic acid (GA)-insensitive dwarfing genes and located on chromosomes 4BS and 4DS, respectively. Another important height-reducing gene is *Rht8* classified as GA-sensitive. *Rht8* is located in chromosome 2D close to *Ppd-D1* and previous studies clarified that *Rht8* and *Ppd-D1a* alleles are often derived together (Worland et al. 1998), but *Ppd-D1a* has pleiotropic effects independently on plant height, grain yield, and yield-related traits (Börner et al. 2002; Chebotar et al. 2013; Zhang et al. 2019a, b). Zhang et al. (2006) determined that *Rht-B1b*, *Rht-D1b*, and *Rht8* are common in autumn-sown Chinese wheat germplasm, while the frequencies of alleles vary from between regions. Kolev et al. (2011) also reported the most frequent alleles as *Ppd-D1b*, *vrn-A1*, *vrn-B1*, *vrn*-D1, and *Rht-B1a* in Bulgarian germplasm including old cultivars and landraces. Rasheed et al. (2016) studied the allelic variation of economically important traits such as *Vrn*, *Ppd* and *Rht*, in 107 wheat landraces collected from different geographic zones of Pakistan. They determined that less than half of the landraces has *Ppd-D1a*, *Rht-B1b*, *Rht-D1b*, and spring-type alleles of *Vrn-A1* and *Vrn-D1*. The studies explained above highlight how these genes from landraces have geographically evolved in the target areas.

11.3.1.1 Success Stories of Wheat Landraces for Adaptive and Yield-Related Traits

There are many successful reports in the development of new wheat varieties with the use of landraces containing different dwarfing genes. A Japanese landrace "Akakomugi," containing *Rht8c* and *Ppd-D1*, was used by Italian breeder Nazareno Strampelli to improve Italian wheat gene pool (Salvi et al. 2013). The crosses between Italian genotypes and Akakomugi resulted into the introgression of new alleles such as *Ppd-D1* and *Rht8c*. The sensational varieties "Ardito and Mentana" developed from the crosses of Strampelli, including Akakomugi, became the basis of most of the new varieties developed in Mediterranean countries, South American countries, and several distant countries such as Russia and China. In Argentina,

"Ardito" was used as parent to develop the variety Klein-33, which became the backbone of the former USSR breeding program, generating the variety Bezostaja-1 (Borojevic and Borojevic 2005). Another variety, Frontana, derived from a cross with Mentana, was part of the pedigree of the varieties Penjamo 62, Yaqui 48, Lerma 50, Escobar, and Supremo. Similarly, many genotypes derived from Mentana were developed in breeding programs of Canada and Australia (Salvi et al. 2013; Tadesse et al. 2016).

A similar success story from the Nobel laureate Norman Borlaug in the mid-20th century was recorded with the Norin 10/Brevor cross containing *Rht-B1* and *Rht-D1*. The lineage of Norin-10, developed by a Japanese breeder G. Inazuka, is tracked back to a Japanese short-straw landrace "Shiro Daruma" containing *Rht-B1* and *Rht-D1* crossed with the American high-yielding varieties Fultz and later Turkey Red (Reitz and Salmon 1968). Norin 10-Brevor 14 cross was sent to N. Borlaug at the International Maize and Wheat Improvement Center (CIMMYT) in Mexico, and this cross and new crosses with Norin/Brevor 14 variants were tested for adaptation in tropical and subtropical climates in the center (Hedden 2003). Wheat varieties developed from semidwarf wheats developed by N. Borlaug and his colleagues in CIMMYT are grown in millions of hectares in many regions of the world.

The story of the Turkey Red brought to America is also very interesting. This bread wheat landrace, which was firstly grown in the USA around Kansas in the 1870s, was introduced to this region by German Mennonites who migrated from Crimea to the USA (Quisenberry and Reitz 1974; Smale 1996). The landrace has thin stem, high plant height, tended to lodging, narrow and dark green leaves, resistance to harsh climate conditions, white grain, high biomass, resistance to rust diseases, and tolerance to other foliar diseases (Quisenberry and Reitz 1974; Lopes et al. 2015). In addition, the landrace "Crimean," introduced at the same time as Turkey Red, was directly included into the Nebraska gene pool. The effects of these two landraces on wheat improvement were indirectly reported with previous studies (Ali et al. 2011; Mengistu et al. 2012). Previous reports reported the investigation of major quantitative trait locus (QTL) related to grain yield on chromosome 3A originated from the cultivar "Wichita" which was obtained from these landraces.

Another important landrace "Chinese Spring" (CS) has also affected the wheat improvement and genetics in depth. This variety is known to be a Sichuan landrace, and Yen et al. (1988) claimed that CS is similar to a Sichuan white landrace "Cheng-du-guang-tou" (CDGT) in terms of morphology, physiology, and cytogenetics-based comparison. The similarity of these two landraces was also presented with RFLP profiling by Ward et al. (1998). CDGT has still been used widely in Sichuan breeding programs because of its high tillering potential, high number of spikelets, and high level of floret fertility (Liu et al. 2018). In addition, the landrace has been widely used to develop wheat-rye translocation lines because of its ready crossability, and therefore, many cultivars and pre-/breeding lines have been developed using CS as parent both in China and many different regions of the world. However, the main important impact of CS is on genetics and molecular breeding of wheat in which CS (IWGSC RefSeq v2.0) was sequenced at genome and single chromosome level and released the genomic data for public access (http://www.wheatgenome.org/News2/IWGSC-RefSeq-v2.0-now-available-at-URGI).

In addition to important wheat landraces mentioned above, several landraces with important adaptive and yield-related traits used by plant breeders in the early twentieth century have intensely been used in pedigrees of modern wheats such as Zeeuwse Witte in the Netherlands, Blount's Lambrigg and Purple Straw in Australia, Marquis and Red Fife in Canada, Kunduru in Turkey, Saragolla in Italy, and Turkey Red in USA, which actually originated from Turkey (Gökgöl 1935; Quisenberry and Reitz 1974; Ozberk et al. 2016; Alsaleh et al. 2016) and became a cornerstone of the early European and indirectly world breeding programs (Smale 1996; Braun et al. 2001). In addition to these examples, wild progenitors/relatives and transition forms of wheat have formed the evolution and distribution of modern wheat landraces and (indirectly) cultivars. Especially, many unique alleles that provide resistance to different diseases and pests, including rust diseases, powdery mildew, *Septoria tritici* blotch, *Septoria nodorum* blotch, tan spot, cyst nematode, root knot nematode, Hessian fly, greenbug, Russian wheat aphid, wheat curl mite, and soil-borne cereal mosaic virus, have been introgressed to modern wheat cultivars (Kishii 2019).

Introgression of new alleles from the locally adapted landraces to modern wheat cultivars should be one of the main breeding targets. Unfortunately, most of landraces have not still been identified both genetically and agronomically. However, the efficient use of landraces in breeding programs requires understanding their genetic diversity and population structure. Baloch et al. (2017) evaluated the genetic diversity of 92 durum wheat landraces from the Central Fertile Crescent including Turkey and Syria with 39,568 DArT-seq and 20,661 SNP markers. As a result of the study, Turkish and Syrian landraces complexly clustered into three groups, and the results illustrated that farmer-mediated selection and lack of the commercial varieties might have concluded in the exchange of genetic materials between two neighboring regions. Soriano et al. (2016) classified 172 durum wheat landraces, using molecular markers, into four genetic populations in relation to their geographic origin: eastern Mediterranean (EM), eastern Balkans and Turkey, western Balkans and Egypt, and western Mediterranean (WM). They determined that the genetic diversity among landraces increased during migration to West Mediterranean basin due to lower genetic diversity in the eastern Mediterranean population. Soriano et al. (2018) also support the theory with an association mapping study that 23 marker alleles in relation to important agronomic traits with different frequencies from east and west regions of Mediterranean basin were identified. With a similar approach, Liu et al. (2017a; b; c) reported a genome-wide association study with 52,303 DArT-seq markers that 723 wheat landraces collected from ten different agroecological zones of China were investigated for 23 agronomic traits in six environments. As a result of the study, all landraces were classified into five clusters based on phenotypic data, and 25 candidate genes associated with significant markers were characterized.

Unveiling the genetic basis of yield-related traits in wheat landraces is vital to ensure global food security because of their higher genetic diversity, large number of alleles, and potency of unique variants of alleles compared to modern wheat varieties. The advent of new technologies about sequencing, mapping, and other related technologies has been facilitating high-quality sequences of wheat and its relatives. The sequences will likely stimulate many new studies on evolution, genetics, and genomics of wheat, and accelerate characterization of novel genes controlling important adaptive and yield-related traits from landraces and wild relatives of wheat.

11.3.2 Role of Landraces in Abiotic Stress

Resistance to abiotic and biotic stresses, productivity, seed quality, seed mineral content, and many other traits will be future breeding aims to meet the world's rapidly increasing food demand. Availability of higher natural genetic diversity to increase selection efficiency is one of the most critical and significant objectives of breeding programs. The abiotic stress factors (salinity, heat, drought, etc.) adversely affect crop production and yield (Jaleel et al. 2009; Thakur et al. 2010; Mantri et al. 2012). Traditional plant breeding is a long-term process that has been used effectively for many years, and molecular tools can be employed to overcome complications and to ensure the improvement of speed breeding strategies (Nadeem et al. 2018; Baloch et al. 2016). In this part, we discussed the role of landraces in different abiotic stress conditions such as salinity, heat, and drought to provide a significant resource for wheat breeders.

11.3.2.1 Wheat Landraces' Role in Salinity Tolerance

Salinity is a major feature that reduces crop production and affects nearly 1 billion hectares of land worldwide (Fageria et al. 2012). Therefore, developing crops providing a satisfactory amount of product in salty soils or different climatic conditions is important to meet the growing food demand. Screening of wheat germplasm for salt tolerance has been conducted by various researchers (Kumar et al. 2017; Arabbeigi et al., 2018). For example, Shahzad et al. (2012) evaluated wheat landrace genotypes using morphological and molecular markers for salinity tolerance at the vegetative stage. The authors proposed that accessions 10793 (Pakistan), 10790 (Pakistan), 10821 (Pakistan), and 11526 (Pakistan) are found salt-tolerant at 200 mM NaCl stress. At 250 mM NaCl stress, accession 11299 (Pakistan) was the most salt-tolerant followed by accessions 11335 (Pakistan), 11370 (Italy), and 11214 (Pakistan). Additionally, accessions 10790 (Pakistan), 10828 (Pakistan), 10823 (Pakistan), and 4098805 (4098805) performed better at both 200 and 250 mM NaCl stresses. In another study, Chaparzadeh et al. (2014) determined the effects of NaCl (control, 75, and 150 mM) on the plant leaves of 18 bread wheat (*Triticum aestivum* L.) landraces from the west area of the Urmia Saline Lake. While accessions 12194 (from Piranshahr), 11199 (from Urmia), and 11488 (from Salmas) were found as the most tolerant with combined salt tolerance indexes for all biochemical and physiological parameters, accessions 11479 (from Mahabad) and 11492 (from Urmia) were determined as the least tolerant. It was suggested that these parameters could be used together as powerful biomarkers to screen for salt-tolerant landraces using the cluster analysis method. Al-maskri et al. (2014) investigated specific stem and leaf structural traits for water conservation. Based on the results of the study, cultivars/landraces were rated according to their degree of drought and salt tolerance as S-24 (from Pakistan) > J-305 (from Oman) > Sarraya (from Northern Asia, Africa, Middle East, Asia Minor) > Senain (from Oman) > Cooley (from Chile and Mongolia) > MH-97 (from Pakistan) > Missani (from the Mediterranean, Middle East Asia, and North Africa) > Hamira (from Oman) > Shwairaa (from Oman). Two of them (S-24 and J-305) are rated as highly tolerant, five moderately tolerant (Sarraya, Senain, Cooley, MH-97, and Missani), and two sensitive (Hamira and Shwairaa). The recent advances in genomic information and technology have opened new horizons and foundations for genetic breeding of salt tolerance. Various QTL mapping studies for salt tolerance in wheat were conducted by Quarrie et al. (2005), Ma et al. (2007), Genc et al. (2010), Hussain et al. (2017), Shamaya et al. (2017), Ren et al. (2018), Devi et al. (2019), and Ilyas et al. (2020). On the other hand, Yu et al. (2020) analyzed in a GWAS using 307 wheat accessions including local landraces and exotic cultivars. Researchers found that some Chinese landraces such as Baihuamai, Youzimai, Beijing 10, Jimai 1, and Zaosui 30 displayed superior salt tolerance. According to kinship analysis, Chinese landraces revealed a source of rare favorable genetic variation. Moreover, many of these landraces have already adapted to the different environments in China (Liu et al. 2017c; Zhou et al. 2018). In addition to these examples, wild relatives of wheat are also potential sources of important genetic materials such as salinity tolerance for wheat breeding. The use of wild relatives of *Triticum* species is one of the main breeding targets and may offer an opportunity to improve salinity tolerance by presenting availability to more variable germplasm (Shavrukov et al. 2009). For this content, researchers investigated the salinity tolerance of various accessions of *Aegilops tauschii*, and determined that the accessions studied are found similar to bread wheat. On the other hand, it was presented that accessions of *Aegilops tauschii* had a much lower Na^+ ratio but higher K^+/Na^+ ratios in their leaves than did durum wheat (Gorham et al. 1987, 1990). Another important wild relative of wheat is jointed goatgrass, *Aegilops cylindrica* Host. (2n = 4x = 28; CCDD) species, which was formed through amphidiploidization of a hybrid or hybrids between *Ae. tauschii Coss*. (2n = 2x = 14; DD) and *Ae. markgrafii* (Greuter) Hammer (2n = 2x = 14; CC). Farooq et al. (1989) screened *Ae. cylindrica* accessions obtained from inland Pakistan and oversea, and determined that some of salinity-tolerant accessions survived at 300 mM NaCl and 400 mM NaCl in treatments using Hoagland solution. Another researcher reviewed the use of wild relatives of wheat for salinity tolerance (Colmer et al. 2006). Arabbeigi et al. (2014) evaluated the physiological response of the highly salinity-tolerant *Ae. cylindrica* genotypes and the SSR and EST-SSR markers linked to the salinity tolerance. As a result of the study, ten most salinity-tolerant genotypes of *Ae. cylindrical* were identified. In addition, Xgwm312, Xwmc170, Xgwm291, and

Xgwm410 microsatellite markers produced a distinguished banding pattern in the ten most salinity-tolerant genotypes in the study. These markers can play important role in wheat breeding programs. Very recently, Ahmadi et al. (2020) investigated the domesticated and ancestral wheat genotypes, including *Ae. triuncialis*, *Ae. neglecta*, *Ae. umbellulata*, *Ae. caudata*, *Ae. speltoides*, *Ae. tauschii*, *T. boeoticum*, *T. durum*, *T. urartu*, and *T. aestivum*, under control and salinity stress to evaluate the mechanisms involved in salinity tolerance. It was found that two neglected (*Ae. triuncialis*) and ancestral (*Ae. tauschii*) wheat genotypes responded better to salinity tolerance than other genotypes. The studies explained above revealed that variation among the wild relatives and landraces of wheat is available for salinity tolerance, and they can be used to develop modern wheat cultivars in breeding studies.

11.3.2.2 Drought and Heat Stress Tolerance in Wheat

Drought and heat stress are important climatic factors that occur in almost all climatic areas of wheat-growing areas and cause a significant crop loss of up to 40% and 60% by drought and heat stresses in fields, respectively (Zampieri et al. 2017; Thirumalaikumar et al. 2018). These factors affect crops at the physiological, morphological, and biochemical levels (Guo et al. 2020); reduce photosynthesis (McKay et al. 2003), cell turgor (Taiz and Zeiger 2006), and chlorophyll fluorescence with a critical reduction of the Fv/Fm ratio (Mohammed and Tarpley, 2009; Izanloo et al. 2008), and impair cell division and elongation (Bal et al. 2010) in sensitive wheat lines compared with tolerant lines. Wheat yield is particularly sensitive to drought and heat stress factors that reduce spikelet productivity, individual grain weight, grain number, and grain filling time during the breeding season (Mahrookashani et al. 2017). The lack of water is not invincible (Ballesta et al. 2019). The adverse effects of drought and heat factors can be overcome by using drought- and heat-resistant cultivars (Van Oosten et al. 2016). The global scenario consists of having a genetic balance of major/minor genes suitable key for these stress factors and developing stress-resistant varieties (Mujeeb-Kazi et al. 2009). Success in plant development commonly depends on the size of genetic variability and the extent to which the beneficial traits are inherited (Kahrizi et al. 2010). Information from the germplasm evaluation will be of great importance for drought- and heat-tolerant genotype selection (Okechukwu et al. 2016). Breeding wheat varieties that tolerate these stressors is currently a major challenge for wheat breeders (Mwadzingeni et al. 2016). Exotic wheat landraces have been shown to be an excellent source of various genes and to function better under stressful conditions (Reynolds et al. 2007). Various studies were conducted to evaluate genetic resources in terms of drought and heat resistance (Hede et al. 1999; Sareen et al. 2014; Pinto et al. 2017; Al Khateeb et al. 2017; Ullah et al. 2018; Korkut et al. 2019). Hede et al. (1999) used a group of 2255 accessions from a Mexican landrace collection in which three landrace accessions (CWI 60155, CWI 59788, and CWI 60391) were determined as having superior and stable leaf chlorophyll content in both environments in 1997. In a study conducted by Sareen et al. (2014), six wheat genotypes (IC 28661, IC 57586,

IC 78856, IC 28938B, IC 36761A, and IC 78869A) were identified as tolerant to drought and heat stresses. Al-maskri et al. (2014) rated cultivars/landraces according to their degree of drought and salt tolerance as S-24 (from Pakistan) > J-305 (from Oman) > Sarraya (from Northern Asia, Africa, Middle East, Asia Minor) > Senain (from Oman) > Cooley (from Chile, Mongolia) > MH-97 (from Pakistan) > Missani (from Mediterranean, Middle East Asia, North Africa) > Hamira (from Oman) > Shwairaa (from Oman). Aktaş (2016) determined the most tolerant genotypes (SEN-DER genotypes G7, G10, landrace group genotype G11 (Sorık)) to be used to improve drought-tolerant varieties. Al Khateeb et al. (2017) used four wheat landraces collected from Jordan and indicated that Karak landrace may be selected as the most tolerant wheat capable of adapting to drought-prone environments. Chaichi et al. (2019) screened 123 Iranian wheat (*Triticum aestivum* L.) landraces (spring and winter genotype) for drought tolerance using morphological and physiological features. They determined L-82 and Marvdasht genotypes as drought-tolerant and sensitive genotypes, respectively. Korkut et al. (2019) determined that some genotypes (Nota, Dropia, CIMMYT-HTN 2014/15-6, CIMMYT-HTN 2014/15-2, CIMMYT HTN 2014/15-10) could be evaluated as genitor(s)/progenitor(s) in the wheat breeding programs for heat tolerance.

Landraces, wild relatives, and traditional varieties are potential reservoirs of novel alleles for improving abiotic stress tolerance (Karan and Subudhi 2012). In this context, a deeper understanding of the genetic mechanisms of drought and heat resistance is important to maintain and further develop the efficiency of wheat breeding programs (Arriagada et al. 2017). The initial genetic investigations of wheat under both drought and heat stress in controlled conditions were conducted in durum wheat and bread wheat by Aprile et al. (2013) and Qaseem et al. (2018), respectively. Merchuk-Ovnat et al. (2016) revealed that introgression of QTLs on chromosomes 1B and 2B of *T. turgidum* into *T. aestivum* can improve drought tolerance in domesticated wheat. B genome has been identified carrying loci controlling water utilization efficiency, associated traits, and grain yield under water stress conditions (Mohammadi et al. 2012; Poersch-Bortolon et al. 2016). In another study conducted by Touzy et al. (2019), a panel of 210 elite European wheat varieties in 35 field trials was evaluated, and GWAS (genome-wide association study) was done with six characters in four different environment types to confirm 590 QTLs, some of which were specific to the different water stress patterns. Schmidt et al. (2020) used 315 spring bread wheat accessions to evaluate in pots with semi-controlled environmental conditions that combined drought and heat stress in 2016 and 2017. Australian and Mexican varieties were rated as having great productivity potential under both stresses, which have been selected for their yield performance and made up about 70% of the spring wheat panels. Nearly one-fifth of the tolerant wheat came from varieties of various origins such as the Middle East, the USA, Central Africa, India, and Canada. In the study, QTLs were determined on all chromosomes, most of which were on chromosomes 3B, 5A, 5B, and 6B. Drought and heat stress factors, which together can lead to significant yield losses, have restricted wheat yields in various wheat-growing areas worldwide, and their combined impact could result in critical yield losses (Toreti et al. 2019). Information about QTLs can help breeders to improve new cultivars tolerant to drought and heat stress in marginal environments in future global margins.

11.3.3 Role of Wheat Landraces in Quality Traits

11.3.3.1 Landraces for Biofortification

"Biofortification" or "biological fortification" is the process of improving the nutritional status of staple crops such as minerals, vitamins, and proteins through traditional breeding, modern biotechnological methods, and agronomic approaches (Garg et al. 2018; Yeken et al. 2018; Saini et al. 2020). It is a long-term and sustainable approach, and a cost-effective way to overcome hidden hunger, which is a progressively severe universal challenge for humanity around the world (De Valença et al. 2017). In low-income countries, micronutrient deficiencies have largely increased in the last decades. Zn and Fe deficiencies in particular are a serious public health problem that negatively affects people's lifespan, health, and productivity (WHO 2009; Khan et al. 2008). People need cereals for their dietary requirements; hence, biofortification of cereals is important worldwide (Saini et al. 2020). Wheat is one of the world's most important crops for global food grain production, which was adversely affected by several biotic and abiotic stresses (Ozer et al. 2020). Annual wheat production is expected to increase in the coming years depending on increases of population (Iizumi et al. 2017). Biofortification can be divided into two categories as agronomic biofortification and genetic biofortification (Saini et al. 2020). The first step of biofortification in food crops for plant breeders is to understand the current genetic diversity in germplasm collections (Baloch et al. 2014). Wheat has a large number of wild relatives that can lead to its genetic development (Dempewolf et al. 2017; Ahmadi et al. 2018; Saini et al. 2020). The most frequently required mineral elements in the human diet can be obtained from genetic variations, which improve the levels of nutrients in crops (White and Broadley 2005; Bouis and Saltzman 2017). Agronomical biofortification techniques include fertilizing crops with different fertilizers containing elements such as zinc, iron, and selenium, while genetic biofortification includes traditional and molecular breeding approaches. These techniques have the potential to increase the levels of these minerals in grains (Saini et al. 2020). Monasterio and Graham (2000) claimed that iron and zinc concentrations especially in some bread wheat genotypes were negatively correlated with *Rht* genes. They also reported that the high-yielding wheat cultivars developed after Green Revolution contained less iron and zinc compared to old cultivars/landraces. Heidari et al. (2016) reported that landraces had higher Fe and Zn concentrations compared with commercial cultivars. Ram and Govindan (2020) clarified that genetic diversity in wheat landraces and wild relatives provides novel alleles for genetic enhancement of Zn and Fe. Lyons et al. (2005) examined 665 wheats (ancestral and wild relatives, landrace accessions, and registered cultivars) in Australia and Mexico for Se concentration in grain. They found that Se concentrations of grains changed between 5 and 720 microgr/kg. Khokhar et al. (2020) studied 245 bread wheat genotypes derived from crosses with landraces and the modern wheat cultivar Paragon to detect grain Zn concentration, and they reached promising results for high level of grain Zn where Zn concentration in wholegrain was positively correlated with Fe concentration and grain protein content. They claimed that landraces have a huge potential to increase the concentration of Zn in whole grain and flour of modern high-yielding bread wheat cultivars.

11.3.3.2 Landraces for Some Important Quality Traits

It is generally known that old landraces or cultivars have a huge diversity for some quality traits such as grain protein content, grain texture (hardness), and gluten strength and quality (glutenin and gliadin subunits) than modern wheat cultivars (Aguiriano et al. 2006; Moragues et al. 2006; Ruiz et al. 2012). The grain protein content (GPC) is a crucial trait in determining the quality of wheat (Veraverbeke and Delcour 2002), and modern wheat grains include inherently low protein levels. Hence, breeding for an increase in the protein levels of grain wheat is required to alleviate hunger and nutrient deficiencies. However, the grain protein content was negatively related to grain yield (Blanco et al. 2006; Iqbal et al. 2007; Klindworth et al. 2009). Avivi (1978) claimed that wild emmer wheat (*T. turgidum* ssp. *dicoccoides*) can be a potential gene source to improve grain protein content in modern wheat. Joppa and Cantrell (1990) also studied this hypothesis that they crossed wild emmer wheat and durum wheat, and obtained substitution lines with high GPC. Joppa et al. (1997) reported that a QTL explained 66% of total variation in these substitution lines for GPC. The QTL was named as *Gpc-B1* (Distelfeld et al. 2004), and Uauy et al. (2006a) also positionally cloned the locus and renamed as *NAM-B1*. Hagenblad et al. (2012) studied 367 bread wheat germplasm with worldwide origin and determined that five accessions had wild-type *NAM-B1* allele where it confers high levels of protein and microelements. They also indicated that several accessions with wild-type *NAM-B1* were traced back to Fennoscandian origin. In addition to landraces, cultivated transitional forms of wheat such as einkorn (*T. monococcum* ssp. *monococcum*), emmer (*T. turgidum* ssp. *dicoccum*), and spelt (*T. aestivum* ssp. *spelta*) and wild relatives have the possibility to contain the wild-type *NAM-B1* allele. Uauy et al. (Uauy et al. 2006a; b) reported that wild emmer accessions and most of cultivated emmer accessions studied had wild-type *NAM-B1* allele. Asplund et al. (2010) also determined that only two spelts had a wild-type *NAM-B1* allele among 62 wheat germplasm displayed at the International Exhibition in London in 1862. It's likely that unique variants for grain protein content can be uncovered due to higher genetic diversity of landraces.

As another important trait, endosperm texture is mainly controlled by the *Hardness* (*Ha*) locus located in 5DS, and it's simply inherited despite the fact that softness is the dominant trait. The lipid binding proteins, puroindoline genes (*Pina-D1* and *Pinb-D1*), which are tightly linked to *Ha* locus, have been used to determine the differences between hard- and soft-textured wheats, and landraces that originated from different geographic regions had different *Puroindoline* allele combinations. As an example of this situation, Ayala et al. (2013) studied 102 lines selected from 15 Mexican landraces and determined that while 16 lines had hard texture, 86 lines were soft-textured. Ten out of 16 lines had presence of both *Pina-D1* and *Pinb-D1* alleles. They concluded that the Mexican old landraces are potential sources for important quality traits to develop new wheat varieties with hard grain texture. Li et al. (2019) also studied 107 Chinese wheat cultivars and landraces in terms of diversity of *Puroindoline* genes and their association with kernel hardness. The most frequent combinations were *Pina-D1a/Pinb-D1a* and *PinaD1a/Pinb-D1b* with 39.3% and 34.6% ratios, respectively. They indicated that Chinese landraces had more allelic than do cultivars and are a valuable source of genetic variability in *Puroindoline* genes. Gluten strength and quality are other important quality traits of wheat. Many studies were conducted to determine the genetic variability of old durum wheat cultivars or landraces for glutenin and gliadin profiles, which affected viscoelastic properties of dough, especially in Mediterranean basin (Melnikova et al. 2010; Xynias et al. 2011; Ribeiro et al. 2011; Ruiz et al. 2012; Janni et al. 2018). Nazco et al. (2012) studied the variability of some quality traits such as protein content, SDS sedimentation, and yellow color index and gluten strength in 154 durum wheat landraces from 20 Mediterranean countries with 18 modern wheat cultivars. They determined that the largest variability for quality traits was observed in landraces from eastern Mediterranean basin followed by landraces from western Mediterranean basin, and identified landraces could be used to improve quality traits especially for gluten strength and grain weight in durum wheat breeding programs. While *Glu-A1c* was the most frequent allele in almost all genetic materials studied for *Glu-A1* locus, but *Glu-A1a* was found at low frequency in Mediterranean basin (Mir Ali et al. 1999; Moragues et al. 2006; Naghavi et al. 2009). In addition to *Glu-A1a*, *Glu-A1b*, and *Glu-A1VI*, encoding the subunits 2* and 2*** were determined at very low frequency. However, Henkrar et al. (2017) reported that in Moroccan genotypes, *Glu-A1a* and *Glu-A1b* were the predominant alleles. On the other hand, at the *Glu-B1* locus, there were more genetic variation between genotypes with *Glu-B1b*, *Glu-B1d*, and *Glu-B1e* alleles encoding the subunits 7+8, 6+8, and 20, respectively. Moreover, the variation varied geographically that while *Glu-B1d* allele was predominant in Algerian, Syrian, and Spanish germplasm (Mir Ali et al. 1999; Moragues et al. 2006; Hamdi et al. 2010), the allele was not present in Iranian landraces that they had more *Glu-B1a*, *Glu-B1e*, and *Glu-B1i* alleles (Naghavi et al. 2009). Similar genetic variation was determined for low molecular weight glutenin subunits (LMW-GS). Li et al. (2009) studied 615 Chinese wheat germplasm including 390 landraces and 225 varieties, for HMW-GS, LMW-GS, Zeleny sedimentation, volume, dough development time, stability time, and strength, and reported that genetic materials with good gluten strength and quality were identified in landraces that did not contain wheat-rye translocation. Wheat-rye (the 1BL/1RS) translocation has been used widely in breeding programs because of its disease resistance genes especially for foliar diseases and increased grain yield in some environments, but it negatively affects bread-making quality of wheat at the same time (Zhao et al. 2012; Oak and Tamhankar 2017).

On the other hand, the new technologies such as sequencing, mapping, and other related technologies have been recently used to reveal genetic diversity and novel variants/alleles among landraces related to quality traits of wheat. For instance, Giraldo et al. (2016) performed an association mapping study with 183 Spanish wheat landraces using 749 DArT markers for 18 agromorphological and grain quality traits including protein content, gluten strength, vitreousness, yellow color index, thousand kernel weight, and test weight. They identified 85 stable MTAs (marker-trait associations) with more than 10% explained phenotypic variation, and claimed that novel MTAs were identified and can provide new information to understand genetic control of complex traits. Roselló et al. (2018) also performed an association mapping study with 165 durum wheat landraces from 21 Mediterranean countries using 1149 DArT markers. Landraces had generally higher GPC than modern ones in this study but lower gluten strength. In addition to this, while eastern landraces showed the highest yellow color index, Balkan landraces had the lowest test weight. They also identified 15 meta-QTL (MQTL) for grain quality traits of wheat.

Various studies about landraces conducted in different countries have been briefly summarized and discussed above. We hope that improving the grain quality via agronomic/genetic biofortification and quality breeding studies and producing wheat genotypes with better quality will be beneficial to prevent hidden hunger and to live healthy. In this regard, collaboration among various specialists from public and private research institutes and universities can accelerate the improvement of wheat varieties with high bread- and pasta-making quality. This section will be helpful for wheat breeders, providing knowledge of the advancement made so far in wheat biofortification and quality.

11.4 Role of Landraces in Biotic Stress

There are many studies conducted to discover resistance properties of wheat landraces for different biotic stresses. Since in the wild the host and the pathogen have co-lived in mutual habitats for long periods of time, they co-evolved together. Thus, the sources of resistance can be found most often at these centers of origin, among the wild relatives and landraces of wheat (McIntosh et al. 1995). Pinpointing the resistance factors and genes in the genome and development of molecular markers to test their presence are of great importance.

11.4.1 Role of Wheat Landraces in Disease Resistance

11.4.1.1 Role of Wheat Landraces in Rust Diseases

11.4.1.1.1 Yellow Rust or Stripe Rust

Rust diseases of wheat are among most important and economically devastating diseases of wheat. Rust diseases of wheat consist of yellow (stripe) rust (YR) caused by *Puccinia striiformis* f. sp. *tritici*, leaf rust (LR) caused by *Puccinia triticina*, and stem rust (SR) caused by *Puccinia graminis* f. sp. *tritici* (Reynolds and Borlaug 2006a; b). Genes that confer resistance to the rust diseases are generally designated as *Yr*, *Lr*, and *Sr* for the effectiveness against yellow rust, leaf rust, and stem rust, respectively. Resistances against rust diseases are the most studied resistance properties in wheat landraces. To date, some genes against rust diseases have been identified from landraces and wild relatives of wheat. Among them, *Sr2* gene, which provides resistance against stem rust, has been incorporated from an emmer wheat landrace (McIntosh et al. 1995). Race-nonspecific resistance genes *Yr52*, *Yr56*, *Yr57*, and *Yr62*, which provide adult plant resistance (APR) against yellow rust, have been also incorporated from landraces (Mondal et al. 2016).

Yellow rust or stripe rust is one of the most prevalent and devastating wheat foliar diseases worldwide (Kumar et al. 2016). It is observed mostly on cool and moist regions and causes lower kernel quality and massive yield losses (Chen et al. 2013). Recently, there are many studies on YR done by genome wide association studies (GWAS) using bread and durum wheat landraces (Tehseen et al. 2020; Long et al. 2019; Liu et al. 2017a; b; c; Manickavelu et al. 2016). Wu et al. (2016) used simple sequence repeats (SSR), sequence-related amplified polymorphism (SRAP), and resistance gene analog polymorphism (RGAP) markers, Ma et al. (2015) used SSR and SRAP markers, while Wang et al. (2010) used SSR markers to find the source of resistance in a known resistant wheat landrace. Kandel et al. (2017) used microsatelite markers to pinpoint the resistance in the genome of known resistant wheat landrace. Wu et al. (2015) used molecular markers to screen wheat landraces to find a suppressor gene of the known resistance gene *Yr18*. Li et al. (2015) used DArT-seq genotyping-by-sequencing (GBS) on 8416 Mexican Creole landrace wheats and found seven accessions from them with less than 20% disease severity after YR inoculation. Gessese et al. (2019) screened resistant landrace Aus27430 with 90K wheat SNP chip array by selective genotyping to locate a new resistance gene "*Yr81*." Yuan et al. (2018), Wang et al. (2019), and Liu et al. (2020) used also wheat SNP chip to locate resistance characteristics of wheat landraces. Bux et al. (2012) evaluated Pakistani wheat landraces phenotypically against the disease; on the other hand Akar et al. (2009) used durum wheat landraces from Turkey to evaluate their resistance against the yellow rust disease. Rola et al. (2019) have found two Lebanese wheat landraces that are resistant to different yellow rust pathogen races, including the devastating Warrior pathotype. Wamalwa et al. (2020) found that Kenyan Kenya Tai landrace shows resistance against many YR races. Mohammadi et al. (2015) screened 380 durum wheat landraces and found 46 accessions to be resistant against YR.

11.4.1.1.2 Leaf Rust

Leaf rust is one of the main wheat diseases seen worldwide, which can affect kernel weight and wheat biomass, causing major yield losses (Herrera-Foessel et al. 2006). There are many studies done on leaf rust resistance. Qureshi et al. (2018) identified a novel resistance gene "*Lr79*," from genotyping analysis of resistant durum wheat by using DArT-seq and 90K chip array, and also developed a Kompetitive Allele Specific Polymerase (KASP) marker to locate the gene. Kolmer et al. (2018) used DArT-seq technology to genotype Uruguayan wheat landrace Americano 44. Qureshi et al. (2017) used DArT-seq markers to locate disease resistance in the genome of two wheat landraces from Portugal. Zhang et al. (2019a), b) screened 46 Chinese wheat landraces for resistance against LR and used molecular markers to find out the presence of known resistance genes in those accessions. Akcura et al. (2017) used Turkish wheat landraces, while Riaz et al. (2017) used 136 wheat landraces from Vavilov Institute of Plant Genetic Resources in Russia to test against YR phenotypically. Andenow et al. (1997) used ten Ethiopian tetraploid wheat (*Triticum turgidum* L.) landraces and found some degree of resistance toward the YR disease.

11.4.1.1.3 Stem Rust

Stem rust is one of the major diseases of wheat which hinders with the nutrient flow to developing ears and result in shriveling of the grain and the breakage of the stem that can cause total yield loss (Roelfs et al. 1992; Leonard and Szabo 2005). Studies on SR have been conducted by Babiker et al. (2015) and Zurn et al. (2014), which used quantitative trait loci (QTL) and linkage map, respectively, to locate the resistance region in the known resistant landrace against stem rust pathogen. Haile et al. (2013) used molecular markers for genotyping the Ethiopian durum wheat landraces. Newcomb et al. (2013) and Toor et al. (2013) have screened the landrace collection phenotypically against the SR disease and genotyped using molecular markers. Denbel and Badebo (2012) screened Ethiopian durum wheat landraces against SR race Ug99. On the other hand, Endresen et al. (2011) used ecogeographic data of landrace accessions to predict the resistance against SR according to climatic factors of their location of origin, while Bonman et al. (2007) studied the geographic origin or the resistant accessions. There are also studies conducted to find multiple rust resistance in wheat landraces. Studies which include resistance against all three rust diseases were conducted by DArT and molecular markers (Rahmatov et al. 2019; Bansal et al. 2013) by GWAS and resistance gene prediction (Kankwatsa et al. 2017; Pasam et al. 2017; Jordan et al. 2015; Daetwyler et al. 2014). Kertho et al. (2015) studied YR and SR resistance traits with GWAS technique, Sthapit et al. (2014) used simple sequence repeat (SSR) markers to study YR and SR resistance, and Aoun et al. (2019) used QTL in durum wheat to locate the resistance region against LR and SR in the known resistant durum wheat landrace.

11.4.1.2 Role of Wheat Landraces in Powdery Mildew (PM)

Powdery mildew (PM) is a foliar fungal disease caused by *Blumeria graminis* f. sp. *tritici*, an obligate biotrophic fungus that causes yield and quality loss in wheat grains (Newton et al. 2011). Chinese wheat landraces known for their PM resistance were screened by microsatelite markers (Xue et al. 2009, Huang et al. 2000), SSR markers (Qie et al. 2019; Sun et al. 2018; Fu et al. 2017; Wang et al. 2015; Xu et al. 2015; Fu et al. 2013; Xue et al. 2012), and RNA-seq SNP markers (Li et al. 2020, Xu et al. 2018) to locate genes in the plants' genome, responsible for the resistance trait. Li et al. (2018a; b) used SSR marker to pinpoint resistance in an Afghan wheat landrace. Tan et al. (2019) and Tan et al. (2018) used single Iranian and Afghan PM-resistant wheat landrace to define new resistance genes "*Pm63*" and "*Pm59*," respectively, using SSR markers. Identification of germplasm strategy (FIGS) was used on wheat landraces in a study conducted by Wang et al. (2015), Bhullar et al. (2010), and Bhullar et al. (2009) to discover new alleles of powdery mildew resistance gene *Pm3*. Huang (1997) also used APR against powdery mildew found in the landrace accession k-15560, and monosomic and hybridological analyses were used to locate the gene (Peusha et al. 2002). Amplified fragment length polymorphism (AFLP) markers and microsatellite markers were used to locate Pm24 resistance gene in a Chinese spring wheat landrace. Li et al. (2012) used SSR markers to test the diversity of the single wheat landrace and its relation to the PM resistance. In their study, Li et al. (2016a; b) used 1,297 landraces from 57 countries to screen for the PM resistance, and molecular markers were used to check the presence of known resistance genes. Hysing et al. (2007) screened 155 Nordic wheat landraces phenotypically and with molecular markers for resistance to PM.

11.4.1.3 Role of Wheat Landraces in Fusarium Head Blight (FHB)

Fusarium head blight (FHB) is caused by the fungal pathogen *Fusarium graminearum* Schwabe and has destructive effects on cereals and especially on wheat production all over the world. Moreover, the diseased plants become contaminated with mycotoxins which are poisonous to mammals (Cetin and Bullerman 2005; Goswami and Kistler 2004). Cai et al. (2019) used meta-analysis of previous QTL studies (MQTL) of five wheat landraces to construct a consensus map, and they also developed 22 KASP markers to ease the MAS in breeding programs. Xiao et al. (2011) located a chromosomal region responsible for FHB resistance by fast-neutron induced chromosome fragment deletion, causing the resistant wheat landrace to lose its resistance and become susceptible. Li et al. (2016a, b) used SSR and sequence-tagged site (STS) markers in 195 wheat accessions to find the presence of known resistance genes, whereas Wei et al. (2005) used microsatellite markers to compare the difference between 20 resistant wheat landraces and 4 susceptible wheat lines. Xiao et al. (2013) used RNA sequencing to determine expression of a resistant wheat landrace during FHB infection. There are also studies where wheat landraces known for their resistance against Fusarium head blight have been screened with SSR markers to pinpoint the resistance source in the genome (Cai et al. 2016; Zhang et al. 2012; Li et al. 2011). Talas et al. (2011) screened 68 Syrian durum wheat landraces and Yu et al. (2008) screened 94 wheat accessions to find new sources of resistance to FHB.

11.4.1.4 Role of Wheat Landraces in Septoria Tritici Blotch (STB)

Septoria tritici blotch (STB) is major foliar wheat disease caused by the fungal pathogen *Zymoseptoria tritici* previously known as *Mycosphaerella graminicola*. It is a major threat to wheat production globally, and it is the most damaging pathogen of wheat in Europe causing loss in chlorophyll, premature death of leaves, and reduction of grain production (O'Driscoll et al. 2014; Ziv and Eyal 1977). Many European and Chinese landraces have been found to contain *Stb6* gene which provides resistance against STB (Chartrain et al. 2005a; b). Kidane et al. (2019) used 318 Ethiopian wheat landraces for GWAS analysis and found four putative loci for STB resistance. Ouaja et al. (2020) screened 304 Tunisian wheat landraces, and Ghaneie et al. (2012) screened 45 tetraploid Iranian wheat landraces to test against STB disease phenotypically and found some promising accessions.

11.4.1.5 Role of Wheat Landraces in Tan Spot

Tan spot is caused by *Pyrenophora tritici-repentis* and is an important foliar wheat disease causing severe loss in the grain yield. The disease causes large-scale chlorosis and tan necrosis on leaves and grain shriveling (Maraite et al., 1997, de Wolf et al. 1998). In their study, Gurung et al. (2011) assessed the resistance of 567 wheat landraces against *P. tritici-repentis* races 1 and 5 using DArT markers and developed association mapping.

11.4.1.6 Role of Wheat Landraces in Eyespot

Eyespot is caused by soilborne necrotrophic funguses *Oculimacula acuformis* and *Oculimacula yallundae*. The disease is seen in temperate areas and affects the stem base of the cereals including wheat, causing premature grain ripening and heavy crop losses (Crous et al. 2003, Fitt et al. 1990, Scott and Hollins 1974). Burt et al. (2014) screened all 1056 hexaploid wheat landraces of Watkins collection against both funguses and found two promising accessions with high level of resistance. They also genotyped the accessions that showed resistance to one or both funguses by SSR, STS, and QTL-linked markers.

11.4.1.7 Role of Wheat Landraces in Stagonospora Nodorum Blotch (SNB)

Stagonospora nodorum blotch (SNB) is caused by *Phaeosphaeria nodorum* and constitutes a serious disease of wheat worldwide (Eyal 1987). SNB disease infects both leaves and glumes, subsequently causing decreased grain quality and yield losses (King et al. 1983). Adhikari et al. (2011a, b) evaluated 567 spring wheat landraces of different origin for resistance to SNB and used DArT markers to genotype and develop association map of the resistance traits.

11.4.1.8 Role of Wheat Landraces in Bacterial Leaf Streak (BLS)

Bacterial leaf streak (BLS) is caused by *Xanthomonas translucens* pv. *undulosa*, the most important wheat bacterial pathogen which can cause major outbreaks in the wheat fields under favorable conditions (Adhikari et al. 2011b, Bragard et al. 1997). Adhikari et al. (2012) screened 566 spring wheat landraces for resistance against BLS and used DArT markers to generate association mapping of the resistance regions. They found five genomic regions which are associated with resistance to the BLS disease.

11.4.1.9 Role of Wheat Landraces in Spot Blotch (SB)

Spot blotch (SB) is caused by *Cochliobolus sativus* which is a fungal disease of wheat and barley, observed globally which results in severe yield losses (Kumar et al. 2002). Adhikari et al. (2012) screened 566 spring wheat landraces also for resistance against SB and used DArT markers to create association mapping of the resistance regions. They found four genomic regions which are associated with resistance to the SB disease.

11.4.1.10 Role of Wheat Landraces in Common Bunt (CB)

Common bunt (CB) is caused by the fungal pathogen *Tilletia tritici* that causes significant yield losses in spring and winter wheat production worldwide (Goates and Peterson 1999). Bonman et al. (2006) investigated 10,759 wheat accessions for resistance against the common bunt disease. Accessions from Bakhtaran province in Iran showed the most resistance.

11.4.1.11 Role of Wheat Landraces in Dwarf Bunt (DB)

Dwarf bunt (DB) is caused by the fungus *Tilletia controversa* in winter wheat in regions where snow is persistent (Goates and Peterson 1999). Bonman et al. (2006) studied 8167 wheat accessions against dwarf bunt resistance. Accessions from Hakkari province in Turkey showed the highest resistance against DB.

11.4.1.12 Role of Wheat Landraces in Wheat Blast (WB)

Wheat blast (WB) is a relatively new emerging disease (mid-1980s) caused by *Triticum* pathotype of *Pyricularia oryzae* fungus. It has immense impacts on wheat production (Inoue et al. 2017). Wang et al. (2018a, b) evaluated 520 landraces of common wheat from different regions of the world for the resistance to Br48 isolate of the fungus and found a unique accession resistant to WB. The resistance was due to combination effect of two genes "*Rmg8*" and newly found "*RmgGR119*" gene.

11.4.2 Role of Wheat Landraces in Pest Resistance

11.4.2.1 Role of Wheat Landraces in Root Lesion Nematodes

Root lesion nematodes *Pratylenchus thornei* and *Pratylenchus neglectus* are the most common root lesion parasites that grow and develop in wheat roots, causing damage and substantial losses in wheat production (Nicol et al. 2002). Thompson and Seymour (2011) analyzed the modes of inheritance of resistance to *P. thornei* in seven wheat accessions that showed resistance against the nematode. Schmidt et al. (2005) studied two resistant Middle Eastern wheat landraces with AFLP and microsatellite markers for QTL analysis of resistance to *P. thornei*. Thompson et al. (2009) screened 207 bread wheat and 102 durum wheat accessions from West Asia and North Africa for resistance against *P. thornei*. Among them, 13 bread wheat and 10 durum wheat showed significant resistance. Thompson et al. (2016) screened 78 Iranian wheat accessions for resistance against *P. thornei* and *P. neglectus*. Among them, 32 showed some degree of resistance to both nematodes.

11.4.2.2 Role of Wheat Landraces in Russian Wheat Aphid (RWA)

Russian wheat aphid (RWA) (*Diuraphis noxia*) is an important wheat pest indigenous to southern Russia and Mediterranean countries which have spread to all continents causing substantial damage to wheat fields (DuToit and Walters 1984; Hewitt et al. 1984). Valdez et al. (2012) have evaluated a resistant Iranian wheat landrace using SSR markers to identify the location of resistance trait. It was found that the trait was due to dominant gene. Similarly, Li et al. (2018a) used an Iranian wheat landrace known for its resistance to RWA to locate the trait in the genome using SSR markers.

11.4.2.3 Role of Wheat Landraces in Wheat Stem Sawfly (WSS)

Wheat stem sawfly (WSS), *Cephus cinctus* Norton, is a major pest insect of wheat observed in North America, with devastating consequences in wheat production (Michael et al. 1992). Mohammadi et al. (2015) evaluated the collection of 380 durum wheat landraces against WSS and found that 33 accessions showed resistance to the pest. Varella et al. (2017) screened 1409 accessions of wheat landraces collected from different regions to WSS. They found 204 accessions that have resistance to the disease. The resistant accessions were screened with KASP markers for QTL analysis. Varella et al. (2019) used four resistant wheat accessions and generated six recombinant inbred lines (RIL) with them and genotyped with 90K iSelect assay to find novel QTL related to WSS resistance.

11.4.2.4 Role of Wheat Landraces in Cereal Cyst Nematodes (CCN)

Cereal cyst nematodes (CCN) *(Heterodera* spp.*)* are a group of 12 known species with *H. avenae*, *H. filipjevi*, and *H. latipons* being the most important ones. The pest is observed in many regions of the world and causes major yield losses in cereals (Nicol et al. 2003). Yavuzaslanoglu et al. (2016) studied the response of 31 Iranian wheat landraces against *H. filipjevi* and found one resistant and five moderately resistant accessions.

11.4.2.5 Role of Wheat Landraces in Cereal Aphids

Cereal aphids cause important yield losses in wheat. There are 14 species of aphids that were observed causing damage to wheat. *Sitobion avenae*, *Rhopalosiphum maidis*, *R. padi*, and *Metopolophium dirhodum* are the most common of these (Popov et al. 1988). Amin et al. (2019) observed 114 wheat landraces for their resistance level against the disease and population dynamics of *R. padi*. They found promising accessions which can be used for breeding of resistant cultivars.

11.5 Landraces and the Future of Wheat Diversity

The world is confronting food scarcity problem due to rapid increase in population and climate change. Previous report showed 6–13% reduction in wheat yield for each °C rise in temperature. Continuously changing climate, extreme weather events, new pathogen strains, and pests further jeopardize linear productivity growth into the future (Mondal et al. 2016). It is believed that the world's population will cross the nine billion mark in 2050. By considering this factor, it is very important to increase wheat production by a rate of 1.6% (Lodhi et al. 2020). To feed the rapidly increasing world's population under changing climatic conditions, more pressure is put on agriculture to produce enough quantity of food. Therefore, it is very important to increase the wheat production to serve enough quantity of food. By considering these factors, it is very important to develop wheat cultivars having higher production and better adaptation to biotic and abiotic stresses (Khan et al. 2013). These targets can be achieved by harnessing wheat genetic diversity. Previous studies explored the existence of higher genetic diversity in wheat landraces compared to its commercial cultivars (Lodhi et al. 2020; Jaradat 2011; Jaradat 2013).

Genetic diversity present in wheat landraces has been successfully utilized for breeding perspectives. Wheat landraces possess a sufficient amount of diversity, including useful genes to adapt to stressful environments such as salinity, heat, and drought (Karagöz and Zencirci 2005; Özkan et al. 2011). The evaluation of genetic diversity in wheat landraces is important for the selection of the suitable landraces as donors of traits in breeding studies (Gurcan et al. 2017; Abbasov et al. 2018). Landraces represent significantly broader genetic diversity than modern varieties (Azeez et al. 2018). For this reason, they can help to increase the genetic source of modern cultivars. However, for their utilization in breeding programs, it is very important that breeders should make crosses among elite lines having the highest likelihood of developing new varieties (Baenziger and DePauw 2009). There is scarcity of information about the successful release of cultivars using wheat landraces. Gerek 79 which is a Turkish variety is developed through crosses with landraces (Smale and McBride 1996). One of the best examples of landraces serving as a source of novel genes is the identification of Rht dwarfing gene that was available through the Japanese variety "Norin 10" originating from a Japanese landrace Shiro Daruma (Reitz and Salmon 1968; Dreisigacker et al. 2005). Dr. Norman E. Borlaug utilized these genes to develop the high-yielding semidwarf wheat varieties that resulted in Green Revolution. Similarly, various wheat landraces served as a foundation in the wheat germplasm pool impotent like: "Cheyenne," a selection from landrace Crimea, founded the Nebraska wheat gene pool. Moreover, "Turkey Red" has been successfully used in winter wheat breeding in the US Great Plains (Lopes et al. 2015). Similarly, previous studies confirmed landrace diversity as a potential source for the breeding of grain yield and climate resilience, for example, the drought-tolerant variety "Aragon 03" was developed from a selection of a landrace population "Catalan de Monte" (Royo and Briceño-Félix 2011a; b). Vikram et al. (2016a; b) stated that a group of Creole wheat landraces (the landraces introduced to Mexico from Europe) has better adaptation to various abiotic stresses including drought because of the presence of rare but beneficial alleles. Further, wheat landraces reflected genetic diversity for various traits like 1000-kernel weight, biomass, and photosynthesis that can be used for cultivar development (Lopes et al. 2015). Various studies have been conducted using wheat landraces as germplasm through molecular markers and explored their potential as a source of novel variations (Sansaloni et al. 2020; Alipour et al. 2017; Lopes et al. 2015; Sofalian et al. 2008;

Alsaleh et al. 2015; Jorgensen et al. 2017; Arystanbekkyzy et al. 2019; Dababat et al. 2020; Ozer et al. 2020). As is obvious from the above-provided information, there is a need to utilize wheat landrace diversity to develop climate-resilient cultivars having high yield. Similarly, some nonbreeding efforts that should be used to promote on-farm dynamic conservation and sustainable utilization of wheat landraces include the following:

1. Awareness should be raised in the farming community about their potential in changing climate.
2. Availability of wheat landrace seeds to the farmers.
3. Development of niche market for landrace products.
4. Involvement of wheat breeders, seed producers, farmers, and end-users, as stakeholders in wheat breeding activities to develop new cultivars (Newton et al. 2011).

References

Abbasov M, Akparov Z, Gross T, Babayeva S, Izzatullayeva V, Hajiyev E, Rustamov K, Gross P, Tekin M, Akar T, Chao S (2018) Genetic relationship of diploid wheat (*Triticum* spp.) species assessed by SSR markers. Genet Resour Crop Evol 65(5):1441–1453

Abu-Zaitoun SY, Chandrasekhar K, Assili S, Shtaya MJ, Jamous RM, Mallah OB, Nashef K, Sela H, Distelfeld A, Alhajaj N, Ali-Shtayeh MS (2018) Unlocking the genetic diversity within a Middle-East panel of durum wheat landraces for adaptation to semi-arid climate. Agronomy 8(10):233

Adhikari TB, Gurung S, Hansen JM, Jackson EW, Bonman JM (2012) Association mapping of quantitative trait loci in spring wheat landraces conferring resistance to bacterial leaf streak and spot blotch. Plant Genome 5(1):1–16

Adhikari TB, Jackson EW, Gurung S, Hansen JM, Bonman JM (2011a) Association mapping of quantitative resistance to Phaeosphaeria nodorum in spring wheat landraces from the USDA National Small Grains Collection. Phytopathology 101(11):1301–1310

Adhikari TB, Hansen JM, Gurung S, Bonman JM (2011b) Identification of new sources of resistance in winter wheat to multiple strains of Xanthomonas translucens pv. undulosa. Plant disease 95(5):582–588

Aguiriano E, Ruiz M, Fite' R, Carrillo JM (2006) Analysis of genetic variability in a sample of the durum wheat (*Triticum durum* Desf.) Spanish collection based on gliadin markers. Genet Resour Crop Evol 53(8):1543–1552

Ahmad M, Shahzad A, Iqbal M, Asif M, Hirani AH (2013) Morphological and molecular genetic variation in wheat for salinity tolerance at germination and early seedling stage. Aust J Crop Sci 7(1):66

Ahmadi J, Pour-Aboughadareh A, Ourang SF, Mehrabi AA, Siddique KHM (2018) Wild relatives of wheat: Aegilops–Triticum accessions disclose differential antioxidative and physiological responses to water stress. Acta Physiol Plant 40:1–14

Ahmadi J, Pour-Aboughadareh A, Ourang SF, Khalili P, Poczai P (2020) Unraveling salinity stress responses in ancestral and neglected wheat species at early growth stage: A baseline for utilization in future wheat improvement programs. Physiol Mol Biol Plants 1–13.

Akar T, Mert Z, Yazar S, Sanal T, Avci M (2009) Sustainable use of winter Durum wheat landraces under Mediterranean conditions. Afr J Biotechnol 8(17)

Akcura M, Kadir A, Hocaoglu O (2017) Biplot analysis of leaf rust resistance in pure lines selected from eastern Anatolian bread wheat landraces of turkey. Turkish J Field Crops 22(2):227–234

Aktaş H (2016) Drought tolerance indices of selected landraces and bread wheat (Triticum aestivum L.) genotypes derived from synthetic wheats. Appl Ecol Environ Res 14(4):177–189

Al Khateeb W, Schroeder D, Musallam I (2017) Phenotypic and molecular variation in drought tolerance of Jordanian durum wheat (Triticum durum Desf.) landraces. Physiol Mol Biol Plants 23(2):311–319

Ali ML, Baenziger PS, Ajlouni ZA, Campbell BT, Gill KS, Eskridge KM, Mujeeb-Kazi A, Dweikat I (2011) Mapping QTLs for yield and agronomic traits on wheat chromosome 3A and a comparison of recombinant inbred chromosome line populations. Crop Sci 51:553–566

Alipour H, Bihamta MR, Mohammadi V, Peyghambari SA, Bai G, Zhang G (2017) Genotyping-by-sequencing (GBS) revealed molecular genetic diversity of Iranian wheat landraces and cultivars. Front Plant Sci 8:1293

Al-maskri A, Hameed M, Ashraf M, Khan MM, Fatima S, Nawaz T, Batool R (2014) Structural features of some wheat (*Triticum spp.*) landraces/cultivars under drought and salt stress. Arid Land Res Manag 28(3):355–370

Al-Naggar AMM, Abd El-Shafi MAE, El-Shal MH, Anany AH (2020) Evaluation of Egyptian wheat landraces (Triticum aestivum L.) for drought tolerance, agronomic, grain yield and quality traits. Plant Archives 20(Supplement 1):3487–3504

Alsaleh A, Baloch FS, Derya M, Azrak M, Kilian B, Özkan H, Nachit M (2015) Genetic linkage map of Anatolian durum wheat derived from a cross of Kunduru-1149× Cham1. Plant Mol Biol Rep 33:209–220

Alsaleh A, Baloch FS, Nachit M, Ozkan H (2016) Phenotypic and genotypic intra-diversity among Anatolian durum wheat "Kunduru" landraces. Biochem Syst Ecol 65:9–16

Amin M, Mahmood K, Nazir N, Kassi AK, Ahmed S (2019) Population dynamics of wheat aphid on different landraces of wheat under field conditions. Plant Protect 3(2):59–66

Andeden EE, Yediay FE, Baloch FS, Shaaf S, Kilian B, Nachit M, Ozkan H (2011) Distribution of vernalization and photoperiod genes (Vrn-A1, Vrn-B1, Vrn-D1, Vrn-B3, Ppd-D1) in Turkish bread wheat cultivars and landraces. Cereal Res Commun 39:352–364

Andenow Y, Hullukal M, Belay G (1997) Resistance and tolerance to leaf rust in Ethiopian tetraploid wheat landraces. Plant breeding 116(6):533–536

Aoun M, Kolmer JA, Rouse MN, Elias EM, Breiland M, Bulbula WD, Chao S, Acevedo M (2019) Mapping of novel leaf rust and stem rust resistance genes in the Portuguese durum wheat landrace PI 192051. G3: Genes, Genomes. Genetics 9(8):2535–2547

Aprile A, Havlickova L, Panna R, Mare C, Borrelli GM, Marone D, Perrotta C, Rampino P, De Bellis L, Curn V, Mastrangelo AM, Rizza F, Cattivelli L (2013) Different stress responsive strategies to drought and heat in two durum wheat cultivars with contrasting water use efficiency. BMC Genomics 14:821–838

Arabbeigi M, Arzani A, Majidi MM, Kiani R, Tabatabaei BES, Habibi F (2014) Salinity tolerance of Aegilops cylindrica genotypes collected from hyper-saline shores of Uremia Salt Lake using physiological traits and SSR markers. Acta Physiol Plant 36(8):2243–2251

Arabbeigi M, Arzani A, Majidi MM, Sayed-Tabatabaei BE, Saha P (2018) Expression pattern of salt tolerance-related genes in Aegilops cylindrica. Physiol Mol Biol Plants 24(1):61–73

Arriagada O, Mora F, Quitral Y, Del Pozo A (2017) Identification of QTL underlying agronomic, morphological and physiological traits in barley under rainfed conditions using SNP markers. Acta Sci Agron 39:321–329

Arystanbekkyzy M, Nadeem MA, Aktas H, Yeken MZ, Zencirci N, Nawaz MA, Ali F, Haider MS, Tunç K, Chung G, Baloch FS (2019) Phylogenetic and taxonomic relationship of Turkish wild and cultivated emmer (Triticum turgidum ssp. dicoccoides) revealed by iPBSretrotransposons markers. Int J Agric Biol 21(1):155–163

Asplund L, Hagenblad J, Leino MW (2010) Re-evaluating the history of the wheat domestication gene *NAM-B1* using historical plant material. J Archaeol Sci 37:2303–2307

Avivi L (1978) High protein content in wild tetraploid *Triticum dicoccoides* Korn. In Ramanujam S (ed). In: Proceedings of the 5th international wheat genetics symposium, New Delhi, 23–28 Feb 1978. Indian Soc Genet Plant Breed, Indian Agric Res Inst, New Delhi, pp 372–380

Ayala M, Guzmán AJB, Peña RJ (2013) Characterization of genetic diversity of puroindoline genes in Mexican wheat landraces. Euphytica 190(1):53–63

Azeez MA, Adubi AO, Durodola FA (2018) Landraces and crop genetic improvement. In Rediscovery of landraces as a resource for the future. IntechOpen.

Babiker EM, Gordon TC, Chao S, Newcomb M, Rouse MN, Jin Y, Wanyera B, Acevedo M, Brown-Guedira G, Williamson S, Bonman JM (2015) Mapping resistance to the Ug99 race group of the stem rust pathogen in a spring wheat landrace. Theor Appl Genet 128(4):605–612

Baenziger PS, Depauw RM (2009) Wheat breeding: Procedures and strategies. In: Wheat science and trade. Wiley-Blackwell, Ames, pp 273–308

Bal W, Kozlowski H, Robbins R, Pettit LD (2010) Improving drought tolerance by exogenous application of glycinebetaine and salicylic acid in sunflower. J Agron Crop Sci 194:193–199

Blanco A, Simeone R, Gadaleta A (2006) Detection of QTLs for grain protein content in durum wheat. Theor Appl Genet 112:1195–1204

Ballesta P, Mora F, Del Pozo A (2019) Association mapping of drought tolerance indices in wheat: QTL-rich regions on chromosome 4A. Sci Agri 77(2)

Baloch FS, Alsaleh A, Andeden EE, Hatipoğlu R, Nachit M, Özkan H (2016) High levels of segregation distortion in the molecular linkage map of bread wheat representing the West Asia and North Africa region. Turk J Agric For 40(3):352–364

Baloch FS, Alsaleh A, Shahid MQ, Çiftçi V, Aasim M, Nadeem MA, Aktaş H, Özkan H, Hatipoğlu R (2017) A whole genome DArTseq and SNP analysis for genetic diversity assessment in durum wheat from central fertile crescent. PLoS One 12(1):e0167821

Baloch FS, Karaköy T, Demirbaş A, Toklu F, Özkan H, Hatipoğlu R (2014) Variation of some seed mineral contents in open pollinated faba bean (Vicia faba L.) landraces from Turkey. Turk J Agric For 38(5):591–602

Bansal UK, Arief VN, DeLacy IH, Bariana HS (2013) Exploring wheat landraces for rust resistance using a single marker scan. Euphytica 194(2):219–233

Belay G, Tesemma T, Bechere E, Mitiku D (1995) Natural and human selection for purple-grain tetraploid wheats in the Ethiopian highlands. Genet Resour Crop Evol 42:387–391

Bhullar NK, Street K, Mackay M, Yahiaoui N, Keller B (2009) Unlocking wheat genetic resources for the molecular identification of previously undescribed functional alleles at the Pm3 resistance locus. Proc Natl Acad Sci U S A 106(23):9519–9524

Bhullar NK, Zhang Z, Wicker T, Keller B (2010) Wheat gene bank accessions as a source of new alleles of the powdery mildew resistance gene Pm3: a large scale allele mining project. BMC Plant Biol 10(1):88

Bonman JM, Bockelman HE, Goates BJ, Obert DE, McGuire PE, Qualset CO, Hijmans RJ (2006) Geographic distribution of common and dwarf bunt resistance in landraces of Triticum aestivum subsp. aestivum. Crop Sci 46(4):1622–1629

Bonman JM, Bockelman HE, Jin Y, Hijmans RJ, Gironella AIN (2007) Geographic distribution of stem rust resistance in wheat landraces. Crop Sci 47(5):1955–1963

Börner A, Schumann E, Fürste A, Cöster H, Leithold B, Röder M, Weber W (2002) Mapping of quantitative trait loci determining agronomic important characters in hexaploid wheat (Triticum aestivum L.). Theor Appl Genet 105(6-7):921–936

Börner A, Worland AJ, Plaschke J, Schumann E, Law CN (1993) Pleiotropic effects of genes for reduced height (Rht) and day-length insensitivity (Ppd) on yield and its components for wheat grown in middle Europe. Plant Breed 111:204–216

Borojevic K, Borojevic K (2005) Historic role of the wheat variety Akakomugi in Southern and Central European wheat breeding programs. Breed Sci 55:253–256

Bouffier B (2014) Genetic and ecophysiological dissection of tolerance to drought and heat stress in bread wheat: from environmental characterization to QTL detection (Doctoral dissertation).

Bouis HE, Saltzman A (2017) Improving nutrition through biofortification: a review of evidence from HarvestPlus, 2003 through 2016. Glob Food Sec 12:49–58

Bragard C, Singer E, Alizadeh A, Vauterin L, Maraite H, Swings J (1997) Xanthomonas translucens from small grains: diversity and phytopathological relevance. Phytopathology 87(11):1111–1117

Braun HJ, Zencirci N, Altay F, Atli A, Avci M, Eser V, Kambertay M, Payne TS (2001) Turkish wheat pool. In: Bonjean AP, Agnus WJ (eds) The world wheat book: A history of wheat breeding. Lavosier, Paris, pp 851–879

Budak H, Shearman RC, Parmaksiz I, Gaussoin RE, Riordan TP, Dweikat I (2004) Molecular characterization of Buffalograss germplasm using sequence-related amplified polymorphism markers. Theor Appl Genet 108:328–334

Bullrich L, Appendino ML, Tranquilli G, Lewis S, Dubcovsky J (2002) Mapping of a thermo-sensitive earliness per se gene on Triticum monococcum chromosome 1Am. Theor Appl Genet 105:585–593

Burt C, Griffe LL, Ridolfini AP, Orford S, Griffiths S, Nicholson P (2014) Mining the Watkins collection of wheat landraces for novel sources of eyespot resistance. Plant Pathol 63(6):1241–1250

Bux H, Ashraf M, Chen X (2012) Expression of high-temperature adult-plant (HTAP) resistance against stripe rust (Puccinia striiformis f. sp. tritici) in Pakistan wheat landraces. Can J Plant Pathol 34(1):68–74

Cai J, Wang S, Li T, Zhang G, Bai G (2016) Multiple minor QTLs are responsible for Fusarium head blight resistance in Chinese wheat landrace Haiyanzhong. PLoS One 11(9):e0163292

Cai J, Wang S, Su Z, Li T, Zhang X, Bai G (2019) Meta-analysis of QTL for Fusarium head blight resistance in Chinese wheat landraces. Crop J 7(6):784–798

Casañas F, Simó J, Casals J, Prohens J (2017) Toward an evolved concept of landrace. Front Plant Sci 8:145

Chaichi M, Sanjarian F, Razavi K, Gonzalez-Hernandez JL (2019) Phenotypic diversity among Iranian bread wheat landraces, as a screening tool for drought tolerance. Acta Physiol Plant 41(6):1–15

Cetin Y, Bullerman LB (2005) Cytotoxicity of Fusarium mycotoxins to mammalian cell cultures as determined by the MTT bioassay. Food Chem Toxicol 43(5):755–764

Chaparzadeh N, Aftabi Y, Dolati M, Mehrnejad F, Pessarakli M (2014) Salinity tolerance ranking of various wheat landraces from the west of the Urmia saline lake in Iran by using physiological parameters. J Plant Nutr 37(7):1025–1039

Chartrain L, Berry ST, Brown JKM (2005a) Resistance of wheat line Kavkaz-K4500 L. 6. A. 4 to Septoria tritici blotch controlled by isolate-specific resistance genes. Phytopathology 95(6):664–671

Chartrain L, Brading PA, Brown JKM (2005b) Presence of the Stb6 gene for resistance to Septoria tritici blotch (Mycosphaerella graminicola) in cultivars used in wheat-breeding programmes worldwide. Plant Pathol 54(2):134–143

Chebotar GO, Chebotar SV, Motsnyy II, Sivolap YM (2013) Clarification of the Rht8–PpdD1 gene linkage on the 2D chromosome of winter bread wheat. Cytol Genet 47:70–74

Chen W, Wellings C, Chen X, Kang Z, Liu T (2014) Wheat stripe (yellow) rust caused by Puccinia striiformis f. sp. tritici. Mol Plant Pathol 15(5):433–446

Chen P, You C, Hu Y, Chen S, Zhou B, Cao A, Wang X (2013) Radiation-induced translocations with reduced Haynaldia villosa chromatin at the Pm21 locus for powdery mildew resistance in wheat. Mol Breed 31(2):477–484

Colmer TD, Flowers TJ, Munns R (2006) Use of wild relatives to improve salt tolerance in wheat. J Exp Bot 57:1059–1078

Crous PW, Groenewald JE, Gams W (2003) Eyespot of cereals revisited: ITS phylogeny reveals new species relationships. Eur J Plant Pathol 109(8):841–850

Dababat A, İmren M, Pridannikov M, Özer G, Zhapayev R, Mokrini F, Otemissova A, Yerimbetova A, Morgounov A (2020. Plant-parasitic nematodes on cereals in northern Kazakhstan. J Plant Dis Prot 1–9

Daetwyler HD, Bansal UK, Bariana HS, Hayden MJ, Hayes BJ (2014) Genomic prediction for rust resistance in diverse wheat landraces. Theor Appl Genet 127(8):1795–1803

Damania AB, Pecetti L, Qualset CO, Humeid BO (1996) Diversity and geographic distribution of adaptive traits in Triticum turgidum L. (durum group) wheat landraces from Turkey. Genet Resour Crop Evol 43:409–422

De Valença AW, Bake A, Brouwer ID, Giller KE (2017) Agronomic biofortification of crops to fight hidden hunger in sub-Saharan Africa. Glob Food Sec 12:8–14

De Wolf ED, Effertz RJ, Ali S, Francl LJ (1998) Vistas of tan spot research. Canadian journal of plant pathology= Revue Canadienne de phytopathologie

Dempewolf H, Baute G, Anderson J, Kilian B, Smith C, Guarino L (2017) Past and future use of wild relatives in crop breeding. Crop Sci 57(3):1070–1082

Denbel W, Badebo A (2012) Valuable sources of resistance in the Ethiopian durum wheat landraces to Ug33 and other stem rust races. Int. J. Agron. Plant Prod 3:191–195

Derakhshan B, Mohammadi SA, Moghaddam M, Jalal Kamali MR (2013) Molecular characterization of vernalization genes in Iranian wheat landraces. Crop Breed J 3:11–14

Devi R, Ram S, Rana V, Malik VK, Pande V, Singh GP (2019) QTL mapping for salt tolerance associated traits in wheat (*Triticum aestivum* L.). Euphytica 215(12):210

Diamond J (2002) Evolution, consequences and future of plant and animal domestication. Nature 418:700–707

Distelfeld A, Uauy C, Olmos S, Schlatter AR, Dubcovsky J, Fahima T (2004) Microcolinearity between a 2-cM region encompassing the grain protein content locus *Gpc-6B1* on wheat chromosome 6B and a 350-kb region on rice chromosome 2. Funct Integr Genomics 4:59–66

Dreisigacker S, Zhang P, Warburton ML, Skovmand B, Hoisington D, Melchinger AE (2005) Genetic diversity among and within CIMMYT wheat landrace accessions investigated with SSRs and implications for plant genetic resources management. Crop Sci 45:653–661

Dubcovsky J, Lijavetzky D, Appendino L, Tranquilli G (1998) Comparative RFLP mapping of Triticum monococcum genes controlling vernalization requirement. Theor Appl Genet 97:968–975

DuToit F, Walters MC (1984) Damage assessment and economic threshold values for the chemical control of the Russian wheat aphid, Diuraphis noxia (Mordvilko) on winter wheat. Technical communication-South Africa, Department of Agriculture

Dvorak J, Luo MC, Yang ZL, Zhang HB (1998) The structure of the Aegilops tauschii genepool and the evolution of hexaploid wheat. Theor Appl Genet 97:657–670

Dvorak J, Noaman MM, Goyal S, Gorham J (1994) Enhancement of the salt tolerance of Triticum turgidum L by the Kna1 locus transferred from Triticum aestivum L. chromosome 4D by homoeologous recombination. Theor Appl Genet 87:872–877

Dwivedi SL, Ceccarelli S, Blair MW, Upadhyaya HD, Are AK, Ortiz R (2016) Landrace germplasm for improving yield and abiotic stress adaptation. Trends Plant Sci 21:31–42

ELshafei AA, Afiah SA, Amer MA, El-enany MAM (2019) Validation of molecular markers linked with salinity tolerance in wheat (Triticum aestivum L.) grown on saline soil. Biosci Res 16(2):963–978

Endresen DTF, Street K, Mackay M, Bari A, DePauw E (2011) Predictive association between biotic stress traits and eco-geographic data for wheat and barley landraces. Crop Sci 51(5):2036–2055

Epstein E, Bloom AJ (2005) Mineral nutrition of plants: principles and perspectives. Sinauer, Sunderland

Eyal Z (1987) The Septoria diseases of wheat: concepts and methods of disease management. CIMMYT

Fageria NK, Stone LF, dos Santos AB (2012) Breeding for salinity tolerance. In plant breeding for abiotic stress tolerance. Springer, Berlin, pp 103–122

Farooq S, Niazi M, Iqbal N, Shah TM (1989) Salt tolerance potential of wild resources of the tribe Triticeae II. Screening of species of the genus Aegilops. Plant and Soil 119:255–260

Fei X, Wen-Wen Z, Xia-Yu D, Yi-Lin Z, Wan-Quan J (2009) Microsatellite mapping of a powdery mildew resistance gene in wheat landrace Xiaobaidong. Acta Agronomica Sinica 35(10):1806–1811

Fitt BD, Goulds A, Hollins TW, Jones DR (1990) Strategies for control of eyespot (Pseudocercosporella herpotrichoides) in UK winter wheat and winter barley. Ann Appl Biol 117(2):473–486

Flowers TJ (2004) Improving crop salt tolerance. J Exp Bot 55:307–319

Frankel OH, Bennett E (1970) Genetic resources in plants-their exploration and conservation. In: Genetic resources in plants-their exploration and conservation. Distributed by Blackwell Scientific, Oxford

Fu B, Chen Y, Li N, Ma H, Kong Z, Zhang L, Jia H, Ma Z (2013) pmX: a recessive powdery mildew resistance gene at the Pm4 locus identified in wheat landrace Xiaohongpi. Theor Appl Genet 126(4):913–921

Fu B, Zhang Z, Zhang Q, Wu X, Wu J, Cai S (2017) Identification and mapping of a new powdery mildew resistance allele in the Chinese wheat landrace Hongyoumai. Molecular Breeding 37(11):133

Gadea M (1958) Trigos cultivados en España y nuevas variedades recomendadas. Ministerio de Agricultura, Madrid

Galiba G, Quarrie SA, Sutka J, Morgounov A, Snape JW (1995) RFLP mapping of the vernalization (Vrnl) and frost resistance (Frl) genes on chromosome 5A of wheat. Theor Appl Genet 90:1174–1179

Garg AK, Kim JK, Owens TG, Ranwala AP, Do Choi Y, Kochian LV, Wu RJ (2002) Trehalose accumulation in rice plants confers high tolerance levels to different abiotic stresses. Proc Natl Acad Sci U S A 99:15898–15903

Garg M, Sharma N, Sharma S, Kapoor P, Kumar A, Chunduri V, Arora P (2018) Biofortified crops generated by breeding, agronomy, and transgenic approaches are improving lives of millions of people around the world. Front Nutr 5:12

Genc Y, Oldach K, Verbyla AP, Lott G, Hassan M, Tester M, Wallwork H, McDonald GK (2010) Sodium exclusion QTL associated with improved seedling growth in bread wheat under salinity stress. Theor Appl Genet 121(5):877–894

Genc Y, Verbyla AP, Torun AA, Cakmak I, Willsmore K, Wallwork H, McDonald GK (2009) Quantitative trait loci analysis of zinc efficiency and grain zinc concentration in wheat using whole genome average interval mapping. Plant and Soil 314:49

Gessese M, Bariana H, Wong D, Hayden M, Bansal U (2019) Molecular mapping of stripe rust resistance gene Yr81 in a common wheat landrace Aus27430. Plant disease 103(6):1166–1171

Ghaneie A, Mehrabi R, Safaie N, Abrinbana M, Saidi A, Aghaee M (2012) Genetic variation for resistance to septoria tritici blotch in Iranian tetraploid wheat landraces. Eur J Plant Pathol 132(2):191–202

Giraldo P, Royo C, Gonzalez M, Carrillo JM, Ruiz M (2016) Genetic diversity and association mapping for agromorphological and grain quality traits of a structured collection of durum wheat landraces including subsp. *durum*, *turgidum* and *diccocon*. PLoS One 11(11):e0166577

Goates BJ, Peterson GL (1999) Relationship between soilborne and seedborne inoculum density and the incidence of dwarf bunt of wheat. Plant Dis 83(9):819–824

Gökgöl M (1935) Turkish wheats, vol I. Ministry of Agriculture, Yesilkoy Seed Breeding Institute Publications. No: 7, Devlet Press, Istanbul. (In Turkish), 436 pp.

Goncharov NP (1998) Genetic resources of wheat related species: The Vrn genes controlling growth habit (spring vs. winter). Euphytica 100(1):371–376

Gonzalez-Hernandez JL, Elias EM, Kianian SF (2004) Mapping genes for grain protein concentration and grain yield on chromosome 5B of Triticum turgidum (L.) var. dicoccoides. Euphytica 139:217–225

Gorham J, Bristol A, Young EM, Wyn Jones RG, Kashour G (1990) Salt tolerance in the Triticeae: K/Na discrimination in barley. J Exp Bot 41:1095–1101

Gorham J (1994) Salt tolerance in the Triticeae: K/Na discrimination in some perennial wheatgrasses and their amphiploids with wheat. J Exp Bot 45:441–447

Gorham J, Bridges J, Dubcovsky J, Dvoák J, Hollington PA, Luo MC, Khan JA (1997) Genetic analysis and physiology of a trait for enhanced K + /Na + discrimination in wheat. New Phytol 137:109–116

Gorham J, Hardy C, WynJones RG, Joppa LR, Law CN (1987) Chromosomal location of a K/Na discrimination character in the D genome of wheat. Theor Appl Genet 74:584–588

Goswami RS, Kistler HC (2004) Heading for disaster: Fusarium graminearum on cereal crops. Mol Plant Pathol 5(6):515–525

Graham RD, Welch RM, Bouis HE (2001) Addressing micronutrient malnutrition through enhancing the nutritional quality of staple foods: principles, perspectives and knowledge gaps. Adv Agron 70:77–142

Guo X, Wang Y, Meng L, Liu H, Yang L, Zhou Y, Zhang H (2015) Distribution of the Vrn-D1b allele associated with facultative growth habit in Chinese wheat accessions. Euphytica 206:1–10

Guo X, Xin Z, Yang T, Ma X, Zhang Y, Wang Z, Ren Y, Lin T (2020) Metabolomics Response for Drought Stress Tolerance in Chinese Wheat Genotypes (Triticum aestivum). Plan Theory 9(4):520

Gurcan K, Demirel F, Tekin M, Demirel S, Akar T (2017) Molecular and agro-morphological characterization of ancient wheat landraces of Turkey. BMC Plant Biol 17(1):171

Gurung S, Mamidi S, Bonman JM, Jackson EW, DelRio LE, Acevedo M, Mergoum M, Adhikari TB (2011) Identification of novel genomic regions associated with resistance to Pyrenophora tritici-repentis races 1 and 5 in spring wheat landraces using association analysis. Theor Appl Genet 123(6):1029

Haile JK, Hammer K, Badebo A, Singh RP, Röder MS (2013) Haplotype analysis of molecular markers linked to stem rust resistance genes in Ethiopian improved durum wheat varieties and tetraploid wheat landraces. Genetic resources and crop evolution 60(3):853–864

Hammer K, Knüpffer H, Xhuveli L, Perrino P (1996) Estimating genetic erosion in landraces—two case studies. Genet Resour Crop Evol 43:329–336

Hanocq E, Niarquin M, Heumez E, Rousset M, Le Gouis J (2004) Detection and mapping of QTL for earliness components in a bread wheat recombinant inbred lines population. Theor Appl Genet 110:106–115

Hao Y, Velu G, Peña RJ, Singh S, Singh RP (2014) Genetic loci associated with high grain zinc concentration and pleiotropic effect on kernel weight in wheat (Triticum aestivum L.). Mol Breed 34:1893–1902

Hagenblad J, Asplund L, Balfourier F, Ravel C, Leino MW (2012) Strong presence of the high grain protein content allele of *NAM-B1* in Fennoscandian wheat. Theor Appl Genet 125:1677–1686

Hamdi O, Bellil I, Branlard G, Khelii D (2010) Genetic variation and geographical diversity for seed storage proteins of seventeen durum wheat populations collected in Algeria. Not Bot Horti Agrobo 38(2):22–32

Harlan JR (1975) Our vanishing genetic resources. Science 188(4188):618–621

Hedden P (2003) The genes of the Green Revolution. Trends Genet 19:5–9

Hede AR, Skovmand B, Reynolds MP, Crossa J, Vilhelmsen AL, Stølen O (1999) Evaluating genetic diversity for heat tolerance traits in Mexican wheat landraces. Genet Resour Crop Evol 46(1):37–45

Heidari B, Padash S, Dadkhodaie A (2016) Variations in micronutrients, bread quality and agronomic traits of wheat landrace varieties and commercial cultivars. Aust J Crop Sci 10:377–384

Henkrar F, El-Haddoury J, Iraqi D, Bendaou N, Udupa SM (2017) Allelic variation at high-molecular weight and low-molecular weight glutenin subunit genes in Moroccan bread wheat and durum wheat cultivars.3. Biotech 7:287

Herrera-Foessel S, Singh R, Huerta-Espino J, Crossa J, Yuen J, Djurle A (2006) Effect of leaf rust on grain yield and yield traits of durum wheats with race-specific and slow-rusting resistance to leaf rust. Plant Dis 90:1065–1072

Hewitt PH, Van Niekerk, GJJ, Walters MC, Kriel CF, Fouche A (1984) Aspects of the ecology of the Russian wheat aphid, Diuraphis noxia, in the Bloemfontein district. I. The colonization and infestation of sown wheat, identification of summer hosts and cause of infestation symptoms. Technical Communication, Department of Agriculture, South Africa, (191), p. 3–13

Huang XQ, Hsam SLK, Zeller FJ (1997) Identification of powdery mildew resistance genes in common wheat (Triticum aestivum L. em Thell.). IX. Cultivars, land races and breeding lines grown in China. Plant Breed 116:233–238

Huang XQ, Hsam SLK, Zeller FJ, Wenzel G, Mohler V (2000) Molecular mapping of the wheat powdery mildew resistance gene Pm24 and marker validation for molecular breeding. Theor Appl Genet 101(3):407–414

Hussain B, Lucas SJ, Ozturk L, Budak H (2017) Mapping QTLs conferring salt tolerance and micronutrient concentrations at seedling stage in wheat. Sci Rep 7(1):1–14

Hysing SC, Merker A, Liljeroth E, Koebner RM, Zeller FJ, Hsam SL (2007) Powdery mildew resistance in 155 Nordic bread wheat cultivars and landraces. Hereditas 144(3):102–119

Iizumi T, Furuya J, Shen Z, Kim W, Okada M, Fujimori S, Hasegawa T, Nishimori M (2017) Responses of crop yield growth to global temperature and socioeconomic changes. Sci Rep 7(1):1–10

Ilyas N, Amjid MW, Saleem MA, Khan W, Wattoo FM, Rana RM, Rana HM, Zahid A, Shah GA, Anwar A, Ahmad MQ, Shaheen M, Riaz H, Ansari MJ (2020) Quantitative trait loci (QTL) mapping for physiological and biochemical attributes in a Pasban90/Frontana recombinant inbred lines (RILs) population of wheat (*Triticum aestivum*) under salt stress condition. Saudi J Biol Sci 27(1):341–351

Inoue Y, Vy TT, Yoshida K, Asano H, Mitsuoka C, Asuke S, Anh VL, Cumagun CJR, Chuma I, Terauchi R, Kato K, Mitchell T, Valent B, Farman M, Yukio TY (2017) Evolution of the wheat blast fungus through functional losses in a host specificity determinant. Science 357(6346):80–83

Iqbal M, Navabi A, Salmon DF, Yang RC, Spaner D (2007) Simultaneous selection for early maturity, increased grain yield and elevated grain protein content in spring wheat. Plant Breed 126:244–250

Iwaki K, Haruna S, Niwa T, Kato K (2001) Adaptation and ecological differentiation in wheat with special reference to geographical variation of growth habit and Vrn genotype. Plant Breed 120:107–114

Izanloo A, Condon AG, Langridge P, Tester M, Schnurbusch T (2008) Different mechanisms of adaptation to cyclic water stress in two South Australian bread wheat cultivars. J Exp Bot 59(12):3327–3346

Jalal A, Shah S, Filho MCMT, Khan A, Shah T, Ilyas M, Rosa PAL (2020) Agro-Biofortification of Zinc and Iron in Wheat Grains. Gesunde Pflanzen 72:227–236

Jaleel CA, Manivannan P, Wahid A, Farooq M, Somasundaram R, Panneerselvam R (2009) Drought stress in plants: a review on morphological characteristics and pigments composition. Int J Agric Biol 11:100–105

Jan SU, Jamil M, Alipour H, Bhatti MF, Gul A (2017) Analysis of salinity tolerance potential in synthetic hexaploid wheat. Pak J Bot 49(4):1269–1278

Janni M, Cadonici S, Bonas U, Grasso A, Dahab AAD, Visioli G, Pignone D, Ceriotti A, Marmiroli N (2018) Gene-ecology of durum wheat HMW glutenin reflects their diffusion from the center of origin. Sci Rep 8:16929

Jaradat A (2006) Phenotypic divergence in the meta-population of the Hourani wheat landrace. J Food Agric Env 4:186–191

Jaradat AA (2011) Wheat landraces: genetic resources for sustenance and sustainability. usda-ars, pp 1–20. http://www.usmarc.usda.gov/SP2UserFiles/Place/36450000/products-wheat/AAJ-wheatlandraces.pdf

Jaradat AA (2013) Wheat landraces: a mini review. Emir J Food Agric 25:20–29

Jiang Y, Huang L, Hu Y (2010) Distribution of vernalization genes in Chinese wheat landraces and their relationship with winter hardness. Sci Agric Sin 43:2619–2632

Joppa LR, Cantrell RG (1990) Chromosomal location of genes for grain protein content of wild tetraploid wheat. Crop Sci 30:1059–1064

Joppa LR, Du C, Hart GE, Hareland GA (1997) Mapping gene(s) for grain protein in tetraploid wheat (*Triticum turgidum* L.) using a population of recombinant inbred chromosome lines. Crop Sci 37:1586–1589

Jordan KW, Wang S, Lun Y, Gardiner LJ, MacLachlan R, Hucl P, Wiebe K, Wong D, Forrest KL, Sharpe AG, Sidebottom CHD, Hall N, Toomajian C, Close T, Dubcovsky J, Akhunova A, Talbert L, Bansal UK, Bariana HS, Hayden MJ, Pozniak C, Jeddeloh JA, Anthony Hall A, Akhunov E (2015) A haplotype map of allohexaploid wheat reveals distinct patterns of selection on homoeologous genomes. Genome Biol 16(1):1–18

Jorgensen C, Luo MC, Ramasamy R, Dawson M, Gill BS, Korol AB, Distelfeld A, Dvorak J (2017) A high-density genetic map of wild emmer wheat from the Karaca Dağ region provides new evidence on the structure and evolution of wheat chromosomes. Front Plant Sci 8:1798

Kahrizi D, Cheghamirza K, Kakaei M, Mohammadi R, Ebadi A (2010) Heritability and genetic gain of some morphophysiological variables of durum wheat (Triticum turgidum var. durum). Afr J Biotechnol 9(30):4687–4691

Kamal NM, Gorafi YSA, Mega R, Tsujimoto H (2018) Physiological response of wheat to chemical desiccants used to simulate post-anthesis drought stress. Agronomy 8(4):44

Kamran A, Iqbal M, Navabi A, Randhawa HS, Pozniak C, Spaner D (2013) Earliness per QTLs and their interaction with photoperiod insensitive allele Ppd-D1a in Cutler x AC Barrie spring wheat population. Theor Appl Genet 126:1965–1976

Kandel JS, Krishnan V, Jiwan D, Chen X, Skinner DZ, See DR (2017) Mapping genes for resistance to stripe rust in spring wheat landrace PI 480035. PloS one 12(5):e0177898

Kankwatsa P, Singh D, Thomson PC, Babiker EM, Bonman JM, Newcomb M, Park RF (2017) Characterization and genome-wide association mapping of resistance to leaf rust, stem rust and stripe rust in a geographically diverse collection of spring wheat landraces. Mol Breed 37(9):113

Karagöz A, Zencirci N (2005) Variation in wheat (Triticum spp.) landraces from different altitudes of three regions of Turkey. Genet Resour Crop Ev 52(6):775–785

Karan R, Subudhi PK (2012) Approaches to increasing salt tolerance in crop plants. In: Abiotic stress responses in plants. Springer, New York, pp 63–88

Kato K, Mori Y, Beiles A, Nevo E (1997) Geographical variation in heading traits in wild emmer wheat, Triticum dicoccoides. I. Variation in vernalization response and ecological differentiation. Theor Appl Genet 95:546–552

Kato K, Taketa S, Ban T, Iriki N, Miura K (2001) The influence of a spring habit gene, Vrn-D1, on heading time in wheat. Plant Breed 120:115–120

Kato K, Wada T (1999) Genetic analysis and selection experiment for narrow-sense earliness in wheat by using segregating hybrid progenies. Breed Sci 49:233–238

Kato K, Yokoyama H (1992) Geographical variation in heading characters among wheat landraces, Triticum aestivum L., and its implication for their adaptability. Theor Appl Genet 84:259–265

Kertho A, Mamidi S, Bonman JM, McClean PE, Acevedo M (2015) Genome-wide association mapping for resistance to leaf and stripe rust in winter-habit hexaploid wheat landraces. PLoS One 10(6):e0129580

Khan MA, Fuller MP, Baloch FS (2008) Effect of soil applied zinc sulphate on wheat *(Triticum aestivum* L.) grown on a calcareous soil in Pakistan. Cereal Res Commun 36:571–582

Khan MH, Bukhari A, Dar ZA, Rizvi SM (2013) Status and strategies in breeding for rust resistance in wheat. Agr Sci 4:292

Khokhar JS, King J, King IP, Young SD, Foulkes MJ, De Silva J et al (2020) Novel sources of variation in grain Zinc (Zn) concentration in bread wheat germplasm derived from Watkins landraces. PLoS One 15(2):e0229107

Kidane YG, Gesesse CA, Hailemariam BN, Desta EA, Mengistu DK, Fadda C, Pe ME, Dell'Acqua M (2019) A large nested association mapping population for breeding and quantitative trait locus mapping in Ethiopian durum wheat. Plant Biotechnol J 17(7):1380–1393

Kidane YG, Hailemariam BN, Mengistu DK, Fadda C, Pè ME, Dell'Acqua M (2017) Genome-wide association study of Septoria tritici blotch resistance in Ethiopian durum wheat landraces. Front Plant Sci 8:1586

Kilian B, Ozkan H, Pozzi C, Salamini F (2009) Domestication of the Triticeae in the Fertile Crescent. In: Feuillet C, Muehlbauer GJ (eds) Genetics and genomics of the Triticeae. USA pp, Springer-Verlag, New York, pp 81–119

King JE, Cook RJ, Melville SC (1983) A review of Septoria diseases of wheat and barley. Ann Appl Biol 103(2):345–373

Kishii M (2019) An update of recent use of Aegilops species in wheat breeding. Front Plant Sci 10:585

Klindworth DL, Hareland GA, Elias EM, Faris JD, Chao S, Xu SS (2009) Agronomic and quality characteristics of two new sets of Langdon durum-wild emmer wheat chromosome substitution lines. J Cereal Sci 50:29–35

Kolev S, Vassilev D, Kostov K, Todorovska E (2011) Allele variation in loci for adaptive response in Bulgarian wheat cultivars and landraces and its effect on heading date. Plant Genet Res 9:251–255

Kolmer JA, Garvin DF, Hayden M, Spielmeyer W (2018) Adult plant leaf rust resistance derived from the wheat landrace cultivar Americano 44d is conditioned by interaction of three QTL. Euphytica 214(3):59

Korkut ZK, Balkan A, Başer İ, Bilgin O (2019) Grain Yield and Some Physiological Traits Associated with Heat Tolerance in Bread Wheat (Triticum aestivum L.) Genotypes. J Agric Sci 25(3):391–400

Kumar J, Jaiswal V, Kumar A, Kumar N, Mir RR, Kumar S, Dhariwala R, Tyagia S, Khandelwale M, Prabhub KV, Prasade R, Balyana HS, Guptaa PK (2011) Introgression of a major gene for high grain protein content in some Indian bread wheat cultivars. Field Crop Res 123:226–233

Kumar J, Schäfer P, Hückelhoven R, Langen G, Baltruschat H, Stein E, Nagarajan S, Kogel KH (2002) Bipolaris sorokiniana, a cereal pathogen of global concern: cytological and molecular approaches towards better control. Mol Plant Pathol 3(4):185–195

Kumar S, Archak S, Tyagi RK, Kumar J, Vikas VK, Jacob SR, Srinivasan K, Radhamani J, Parimalan R, Sivaswamy M, Tyagi S, Yadav M, Kumari J, Deepali SS, Bhagat I, Meeta M, Bains NS, Chowdhury AK, Saha BC, Bhattacharya PM, Kumari J, Singh MC, Gangwar OP, Prasad P, Bharadwaj SC, Gogoi R, Sharma JB, Kumar GMS, Saharan MS, Bag M, Roy A, Prasad TV, Sharma RK, Dutta M, Sharma I, Bansal KC (2016) Evaluation of 19,460 wheat accessions conserved in the Indian national genebank to identify new sources of resistance to rust and spot blotch diseases. PLoS One 12:e0175610

Kumar S, Beena AS, Awana M, Singh A (2017) Physiological, biochemical, epigenetic and molecular analyses of wheat (Triticum aestivum) genotypes with contrasting salt tolerance. Front Plant Sci 8:1151

Law CN, Worland AJ, Giorgi B (1976) The genetic control of ear-emergence time by chromosomes 5A and 5D of wheat. Heredity 36:49–58

Leonard K, Szabo L (2005) Stem rust of small grains and grasses caused by Puccinia graminis. Mol Plant Pathol 6:99–111

Li G, Carver BF, Cowger C, Bai G, Xu X (2018a) Pm223899, a new recessive powdery mildew resistance gene identified in Afghanistan landrace PI 223899. Theor Appl Genet 131(12):2775–2783

Li G, Xu X, Bai G, Carver BF, Hunger R, Bonman JM (2016a) Identification of novel powdery mildew resistance sources in wheat. Crop Sci 56(4):1817–1830

Li G, Xu X, Carver BF, Guo P, Puterka G (2018b) Dn10, a new gene conferring resistance to Russian wheat aphid biotype 2 in Iranian wheat landrace PI 682675. Crop Sci 58(3):1219–1225

Li H, Vikram P, Singh RP, Kilian A, Carling J, Song J, Burgueno-Ferreira JA, Bhavani S, Huerta-Espino J, Payne T, Sehgal D, Wenzl P, Singh S (2015) A high density GBS map of bread wheat and its application for dissecting complex disease resistance traits. BMC Genomics 16(1):1–15.

Li T, Bai G, Wu S, Gu S (2011) Quantitative trait loci for resistance to fusarium head blight in a Chinese wheat landrace Haiyanzhong. Theor Appl Genet 122(8):1497–1502

Li T, Zhang D, Zhou X, Bai G, Li L, Gu S (2016b) Fusarium head blight resistance loci in a stratified population of wheat landraces and varieties. Euphytica 207(3):551–561

Li XJ, Xu X, Yang XM, Li XQ, Liu WH, Gao AN, Li LH (2012) Genetic diversity of the wheat landrace Youzimai from different geographic regions investigated with morphological traits, seedling resistance to powdery mildew, gliadin and microsatellite markers. Cereal Res Commun 40(1):95–106

Li X, Li Y, Zhang M, Yu X, Hu R, Chang J, Yang G, Wang Y, He G (2019) Diversity of *Puroindoline* genes and their association with kernel hardness in Chinese wheat cultivars and landraces. Mol Breed 39(4):1–13

Li Y, Huang C, Sui X, Fan Q, Li G, Chu X (2009) Genetic variation of wheat glutenin subunits between landraces and varieties and their contributions to wheat quality improvement in China. Euphytica 169:159–168

Li Y, Shi X, Hu J, Wu P, Qiu D, Qu Y, Xie J, Wu Q, Zhang H, Yang L, Liu H, Zhou Y, Liu Z, Li H (2020) Identification of a recessive gene PmQ conferring resistance to powdery mildew in wheat landrace Qingxinmai using BSR-Seq analysis. Plant Disease 104(3):743–751

Liu D, Liu Y, Zhang W, Chen X, Zou C (2017a) Agronomic approach of zinc biofortification can increase zinc bioavailability in wheat flour and thereby reduce zinc deficiency in humans. Nutrients 9(5):465

Liu D, Zhang L, Hao M, Ning S, Yuan Z, Dai S, Huang L, Wu B, Yan Z, Lan X, Zheng Y (2018) Wheat breeding in the hometown of Chinese Spring. Crop J 6:82–90

Liu J, Huang L, Wang C, Liu Y, Yan Z, Wang Z, Xiang L, Zhong X, Gong F, Zheng Y, Liu D, Wu B (2019) Genome-Wide Association Study Reveals Novel Genomic Regions Associated With High Grain Protein Content in Wheat Lines Derived From Wild Emmer Wheat. Front Plant Sci 10:464

Liu W, Maccaferri M, Rynearson S, Letta T, Zegeye H, Tuberosa R, Chen X, Pumphrey M (2017b) Novel sources of stripe rust resistance identified by genome-wide association mapping in Ethiopian durum wheat (Triticum turgidum ssp. durum). Front Plant Sci 8:774.1

Liu Y, Lin Y, Gao S, Li Z, Deng M, Chen G, Wei Y, Zheng Y (2017c) A genome-wide association study of 23 agronomic traits in Chinese wheat landraces. Plant J 91:861–873

Liu Y, Qie Y, Li X, Wang M, Chen X (2020) Genome-Wide Mapping of Quantitative Trait Loci Conferring All-Stage and High-Temperature Adult-Plant Resistance to Stripe Rust in Spring Wheat Landrace PI 181410. Int J Mol Sci 21(2):478

Lodhi SS, Maryam S, Rafique K, Shafique A, Yousaf ZA, Talha AM, Gul A, Amir R (2020) Overview of the prospective strategies for conservation of genomic diversity in wheat landraces. In: Climate change and food security with emphasis on wheat. Academic Press, London, pp 293–309

Long L, Yao F, Yu C, Ye X, Cheng Y, Wang Y, Wu Y, Li J, Wang J, Jiang Q, Li W, Ma J, Liu Y, Deng M, Wei Y, Zheng Y, Chen G (2019) Genome-Wide association study for adult-plant resistance to stripe rust in Chinese wheat landraces (Triticum aestivum L.) from the yellow and huai river valleys. Front Plant Sci 10:596

Lopes MS, El-Basyoni I, Baenziger PS, Singh S, Royo C, Ozbek K, Aktas H, Ozer E, Ozdemir F, Manickavelu A, Ban T, Vikram P (2015) Exploiting genetic diversity from landraces in wheat breeding for adaptation to climate change. J Exp Bot 66:3477–3486

Lyons G, Ortiz-Monasterio I, Stangoulis J, Graham R (2005) Selenium concentration in wheat grain: Is there sufficient genotypic variation to use in breeding? Plant and Soil 269:369–380

Ma D, Li Q, Tang M, Chao K, Li J, Wang B, Jing J (2015) Mapping of gene conferring adult-plant resistance to stripe rust in Chinese wheat landrace Baidatou. Mol Breed 35(8):157

Ma L, Zhou E, Huo N, Zhou R, Wang G, Jia J (2007) Genetic analysis of salt tolerance in a recombinant inbred population of wheat (*Triticum aestivum* L.). Euphytica 153(1-2):109–117

Mahrookashani A, Siebert S, Hüging H, Ewert F (2017) Independent and combined effects of high temperature and drought stress around anthesis on wheat. J Agron Crop Sci 203(6):453–463

Manickavelu A, Joukhadar R, Jighly A, Lan C, Huerta-Espino J, Stanikzai AS, Kilian A, Singh RP, Ban T (2016) Genome wide association mapping of stripe rust resistance in Afghan wheat landraces. Plant Sci 252:222–229

Manickavelu A, Niwa S, Ayumi K, Komatsu K, Naruoka Y, Ban T (2014) Molecular evaluation of Afghan wheat landraces. Plant Genet Resour-C 12:S31–S35

Mantri N, Patade V, Penna S, Ford R, Pang E (2012) In: Ahmad P, Prasad MNV (eds) Abiotic stress responses in plants: metabolism, productivity and sustainability. Springer Science+Business Media, LLC, New York

Maraite H, Di Zinno T, Longree H, Daumerie V, Duveiller E (1997) Fungi associated with foliar blight of wheat in warm areas. In Proceedings of the international workshop on helminthosporium diseases of wheat: Spot Blotch and Tan Spot, El Batán (pp. 293–300)

Masood MS, Javaid A, Rabbani MA, Anwar R (2005) Phenotypic diversity and trait association in bread wheat (Triticum aestivum L.) landraces from Baluchistan, Pakistan. Pak J Bot 37:949

McIntosh RA, Wellings CR, Park RF (1995) Wheat rusts: an atlas of resistance genes. CSIRO Publications, East Melbourne (Australia)

McIntosh RA, Yamazaki Y, Dubcovsky J, Rogers J, Morris C, Appels R, Xia XC (2013) Catalogue of gene symbols for wheat. https://shigen.nig.ac.jp/wheat/komugi/genes/download.jsp

McKay JK, Richards JH, Mitchell-Olds T (2003) Genetics of drought adaptation in Arabidopsis thaliana: I. Pleiotropy contributes to genetic correlations among ecological traits. Mol Ecol 12:1137–1151

Melnikova NV, Ganeva GD, Popova ZG, Landjeva SP, Kudryavtsev AM (2010) Gliadins of Bulgarian durum wheat *Triticum durum* Desf. landraces: genetic diversity and geographical distribution. Genet Resour Crop Evol 57:587–595

Mengistu N, Baenziger PS, Eskridge KM, Dweikat I, Wegulo SN, Gill KS, Mujeeb-Kazi A (2012) Validation of QTL for grain yield-related traits on wheat chromosome 3a using recombinant inbred chromosome lines. Crop Sci 52:1622–1632

Merchuk-Ovnat L, BarakV FT, Ordon F, Lidzbarsky GA, Krugman T, Saranga Y (2016) Ancestral QTL alleles from wild emmer wheat improve drought resistance and productivity in modern wheat cultivars. Front Plant Sci 7:452

Mir RA, Sharma A, Mahajan R (2020) Crop landraces: present threats and opportunities for conservation. In: Rediscovery of genetic and genomic resources for future food security 2020. Springer, Singapore, pp 335–349

Mir Ali N, Arabi MIE, Al-Safadi B (1999) Frequencies of high and low molecular weight glutenin subunits in durum wheat grown in Syria. Cereal Res Commun 27:301–305

Mishra VK, Gupta PK, Arun B, Vasistha NK, Vishwakarma MK, SinghYadav P, Kumar H, Joshiac AK (2015) Introgression of a gene for high grain protein content (Gpc-B1) into two leading cultivars of wheat in Eastern Gangetic Plains of India through marker assisted backcross breeding. J Plant Breed Crop Sci 7:292–300

Mitrofanova OP, Khakimova AG (2017) New genetic resources in wheat breeding for increased grain protein content. Russ J Genet Appl Res 7(4):477–487

Mohammadi R, Armion M, Kahrizi D, Amri A (2012) Efficiency of screening techniques for evaluating durum wheat genotypes under mild drought conditions. Int J Plant Prod 4(1):11–24

Mohammadi R, Sadeghzadeh B, Ahmadi H, Bahrami N, Amri A (2015) Field evaluation of durum wheat landraces for prevailing abiotic and biotic stresses in highland rainfed regions of Iran. The Crop Journal 3(5):423–433

Monasterio I, Graham RD (2000) Breeding for trace minerals in wheat. Food Nutr Bull 21:392–396

Mondal S, Rutkoski JE, Velu G, Singh PK, Crespo-Herrera LA, Guzman C, Bhavani S, Lan C, He X, Singh RP (2016) Harnessing diversity in wheat to enhance grain yield, climate resilience, disease and insect pest resistance and nutrition through conventional and modern breeding approaches. Front Plant Sci 7:991

Mora F, Castillo D, Lado B, Matus I, Poland J, Belzile F, Zitzewitz JV, del Pozo A (2015) Genome-wide association mapping of agronomic traits and carbon isotope discrimination in a worldwide germplasm collection of spring wheat using SNP markers. Mol Breed 35(2):69

Moragues M, Zarco-Hernández J, Moralejo MA, Royo C (2006) Genetic diversity of glutenin protein subunits composition in durum wheat landraces [*Triticum turgidum* ssp. turgidum convar. durum (Desf.) MacKey] from the Mediterranean basin. Genet Resour Crop Evol 53(5):993–1002

Motzo R, Giunta F (2007) The effect of breeding on the phenology of Italian durum wheats: From landraces to modern cultivars. Eur J Agron 26:462–470

Mohammed AR, Tarpley L (2009) High nighttime temperatures affect rice productivity through altered pollen germination and spikelet fertility. Agric For Meteorol 149(6–7):999–1008

Mujeeb-Kazi A, Gul A, Ahmad I, Farooq M, Rauf Y, Riaz H (2009) Genetic resources for some wheat abiotic stress tolerances. In: Salinity and water stress. Springer, Dordrecht, pp 149–163

Muqaddasi QH, Reif JC, Li Z, Basnet BR, Dreisigacker S, Röder MS (2017) Genome-wide association mapping and genome-wide prediction of anther extrusion in CIMMYT spring wheat. Euphytica 213(3):73

Mwadzingeni L, Shimelis H, Tesfay S, Tsilo TJ (2016) Screening of bread wheat genotypes for drought tolerance using phenotypic and proline analyses. Front Plant Sci 7:1276

Nadeem MA, Nawaz MA, Shahid MQ, Doğan Y, Comertpay G, Yıldız M, Hatipoğlu R, Ahmad F, Alsaleh A, Labhane N, Özkan H (2018) DNA molecular markers in plant breeding: current status and recent advancements in genomic selection and genome editing. Biotechnol Biotechnol Equip 32:261–285

Naghavi MR, Monfared SR, Ahkami AH, Ombidbakhsh MA (2009) Genetic variation of durum wheat landraces and cultivars using morphological and protein markers. Proceedings of world academy of science. Eng Technol 37:73–75

Nazco R, Villegas D, Ammar K, Pena RJ, Moragues M, Royo C (2012) Can Mediterranean durum wheat landraces contribute to improved grain quality attributes in modern cultivars? Euphytica 185(1):1–17

Newcomb M, Acevedo M, Bockelman HE, Brown-Guedira G, Goates BJ, Jackson EW, Jin Y, Njau P, Rouse MN, Singh RD, Wanyera R, Bonman JM (2013) Field resistance to the Ug99 race group of the stem rust pathogen in spring wheat landraces. Plant disease 97(7):882–890

Newton AC, Akar T, Baresel JP, Bebeli PJ, Bettencourt E, Bladenopoulos KV, Czembor JH, Fasoula DA, Katsiotis A, Koutis K, Koutsika-Sotiriou M, Kovacs G, Larsson H, Pinheiro de Carvalho MAA, Rubiales D, Russell J, Dos Santos TMM, Vaz Patto MC (2010) Cereal landraces for sustainable agriculture. Sustain Agric 2:147–186

Newton AC, Johnson SN, Gregory PJ (2011) Implications of climate change for diseases, crop yields and food security. Euphytica 179:3–18

Nicol J, Rivoal R, Taylor S, Zaharieva M (2003) Global importance of cyst (Heterodera spp.) and lesion nematodes (Pratylenchus spp.) on cereals: distribution, yield loss, use of host resistance and integration of molecular tools. Nematol Monogr Perspect 2:1–19

Nicol JM, Rivoal R, Bolat N, Aktas H, Braun HJ, Mergoum M, Yildrim AF, Bagci A, Eleckcioglu IH, Yahyaoui A (2002) The frequency and diversity of the cyst and lesion nematode on wheat in the Turkish Central Anatolian Plateau. Nematology 4(2):272

Norman A, Taylor J, Tanaka E, Telfer P, Edwards J, Martinant JP, Kuchel H (2017) Increased genomic prediction accuracy in wheat breeding using a large Australian panel. Theor Appl Genet 130(12):2543–2555

O'Driscoll A, Kildea S, Doohan F, Spink J, Mullins E (2014) The wheat–Septoria conflict: a new front opening up? Trends Plant Sci 19(9):602–610

Okechukwu EC, Agbo CU, Uguru MI, Ogbonnaya FC (2016) Germplasm evaluation of heat tolerance in bread wheat in Tel Hadya, Syria. Chil J Agric Res 76(1):9–17

Olmstead AL, Rhode PW (2002) The red queen and the hard reds: productivity grown in American wheat 1800–1940. J Econ Hist 62:929–966

Ouaja M, Aouini L, Bahri B, Ferjaoui S, Medini M, Marcel TC, Hamza S (2020) Identification of valuable sources of resistance to Zymoseptoria tritici in the Tunisian durum wheat landraces. Eur J Plant Pathol 156(2):647–661

Oak MD, Tamhankar SA (2017) 1BL/1RS translocation in durum wheat and its effect on end use quality traits. J Plant Biochem Biot 26(1):91–96

Ozberk I, Atay S, Altay F, Cabi E, Ozkan H, Atli A (2016) The Wheat Atlas of Turkey. (World Wildlife Fund), Istanbul (in Turkish)

Ozer G, Paulitz TC, Imren M, Alkan M, Muminjanov H, Dababat AA (2020) Identity and Pathogenicity of Fungi Associated with Crown and Root Rot of Dryland Winter Wheat in Azerbaijan. Plant Disease 104:2149–2157

Özkan H, Brandolini A, Schäfer-Pregl R, Salamini F (2002) AFLP analysis of a collection of tetraploid wheats indicates the origin of emmer and hard wheat domestication in southeast Turkey. Mol Biol Evol 19(10):1797–1801

Özkan H, Willcox G, Graner A, Salamini F, Kilian B (2011) Geographic distribution and domestication of wild emmer wheat (Triticum dicoccoides). Genet Resour Crop Evol 58(1):11–53

Pasam RK, Bansal U, Daetwyler HD, Forrest KL, Wong D, Petkowski J, Willey N, Randhawa M, Chhetri M, Miah H, Tibbits J, Bariana H, Hayden MJ (2017) Detection and validation of genomic regions associated with resistance to rust diseases in a worldwide hexaploid wheat landrace collection using BayesR and mixed linear model approaches. Theor Appl Genet 130(4):777–793

Peng JH, Sun D, Nevo E (2011a) Domestication evolution, genetics and genomics in wheat. Mol Breed 28:281–301

Peng ZS, Li X, Yang ZJ, Liao ML (2011b) A new reduced height gene found in the tetraploid semi-dwarf wheat landrace Aiganfanmai. Genet Mol Res 10:2349–2357

Peusha H, Lebedeva T, Prilinn O, Enno T (2002) Genetic analysis of durable powdery mildew resistance in a common wheat line. Hereditas 136:201–206

Pinto RS, Molero G, Reynolds MP (2017) Identification of heat tolerant wheat lines showing genetic variation in leaf respiration and other physiological traits. Euphytica 213(3):76

Poersch-Bortolon LB, Pereira JF, Nhani Junior A, Gonzáles HHS, Torres GAM, Consoli L, Arenhart RA, Bodanese-Zanettini MH, Margis-Pinheiro M (2016) Gene expression analysis reveals important pathways for drought response in leaves and roots of a wheat cultivar adapted to rainfed cropping in the Cerrado biome. Genet Mol Biol 39(4):629–645

Popov C, Hondru N, Bărbulescu A, Vonica I, Mărgărit G (1988) Species of aphids attacking wheat and barley crops. Analele Institutului de Cercetări pentru Cereale şi Plante Tehnice, Fundulea 56:379–384

Pu Z, Pei Y, Yang J, Ma J, Li W, Liu D, Wang J, Wei Y, Zheng Y (2018) A QTL located on chromosome 3D enhances the selenium concentration of wheat grain by improving phytoavailability and root structure. Plant and Soil 425(1-2):287–296

Pugsley AT (1971) A genetic analysis of the spring-winter habit of growth in wheat. Aust J Agr Res 22:21–23

Pugsley AT (1972) Additional genes inhibiting winter habit in wheat. Euphytica 21:547–552

Qaseem MF, Qureshi R, Muqaddasi QH, Shaheen H, Kousar R, Röder MS (2018) Genome-wide association mapping in bread wheat subjected to independent and combined high temperature and drought stress. PLoS One 13:6

Qiang LI, Wang ZR, Ding LI, Wei JW, Qiao WC, Meng XH, Sun SI, Li HM, Zhao MH, Chen XM, Zhao FW (2018) Evaluation of a new method for quantification of heat tolerance in different wheat cultivars. J Integr Agric 17(4):786–795

Qie Y, Sheng Y, Xu H, Jin Y, Ma F, Li L, Li X, An D (2019) Identification of a new powdery mildew resistance gene pmDHT at or closely linked to the Pm5 locus in the Chinese wheat landrace Dahongtou. Plant disease 103(10):2645–2651

Quarrie SA, Steed A, Calestani C, Semikhodskii A, Lebreton C, Chinoy C, Steele N, Pljevljakusić D, Waterman E, Weyen J, Schondelmaier J, Habash DZ, Farmer P, Saker L, Clarkson DT, Abugalieva A, Yessimbekova M, Turuspekov Y, Abugalieva S, Tuberosa R, Sanguineti MC, Hollington PA, Aragués R, Royo A, Dodig D (2005) A high-density genetic map of hexaploid wheat (*Triticum aestivum* L.) from the cross Chinese Spring× SQ1 and its use to compare QTLs for grain yield across a range of environments. Theor Appl Genet 110(5):865–880

Quisenberry KS, Reitz LP (1974) Turkey wheat: The cornerstone of an empire. Agric Hist 48:98–110

Qureshi N, Bariana H, Kolmer JA, Miah H, Bansal U (2017) Genetic and molecular characterization of leaf rust resistance in two durum wheat landraces. Phytopathology 107(11):1381–1387

Qureshi N, Bariana H, Kumran VV, Muruga S, Forrest KL, Hayden MJ, Bansal U (2018) A new leaf rust resistance gene Lr79 mapped in chromosome 3BL from the durum wheat landrace Aus26582. Theor Appl Genet 131(5):1091–1098

Rahmatov M, Otambekova M, Muminjanov H, Rouse MN, Hovmøller MS, Nazari K, Steffenson BJ, Johansson E (2019) Characterization of stem, stripe and leaf rust resistance in Tajik bread wheat accessions. Euphytica 215(3):1–22

Ram S, Govindan V (2020) Improving wheat nutritional quality through biofortification. In: Igrejas G, Ikeda TM, Guzman C (eds) Wheat quality for improving processing and human health. Springer, Switzerland, pp 205–224

Rasheed A, Xia X, Mahmood T, Quraishi UM, Bux AAH, Mahmood Z, Mirza JI, Mujeeb-Kazi A, He Z (2016) Comparison of economically important loci in landraces and improved wheat cultivars from Pakistan. Crop Sci 56:287–301

Rawson HM, Richards RA, Munns R (1988) An examination of selection criteria for salt tolerance in wheat, barley and triticale genotypes. Aust J Agr Res 39:759–772

Reitz LP, Salmon SC (1968) Origin, history, and use of Norin 10 wheat. Crop Sci 8:686–689

Ren Y, Xu Y, Teng W, Li B, Lin T (2018) QTLs for seedling traits under salinity stress in hexaploid wheat. Cienc Rural 48(3):e20170446

Reynolds M, Dreccer F, Trethowan R (2007) Drought-adaptive traits derived from wheat wild relatives and landraces. J Exp Bot 58(2):177–186

Reynolds MP, Borlaug NE (2006a) Applying innovations and new technologies for international collaborative wheat improvement. J Agr Sci-Cambridge 144:95

Reynolds MP, Borlaug NE (2006b) Impacts of breeding on international collaborative wheat improvement. J Agr Sci-Cambridge 144:3–17

Riaz A, Athiyannan N, Periyannan S, Afanasenko O, Mitrofanova O, Aitken EA, Lagudah E, Hickey LT (2017) Mining Vavilov's treasure chest of wheat diversity for adult plant resistance to Puccinia triticina. Plant disease 101(2):317–323

Ribaut JM, Hoisington D (1998) Marker-assisted selection: new tools and strategies. Trends Plant Sci 3(6):236–239

Ribeiro M, Carvalho C, Carnide V, Guedes-Pinto H, Igrejas G (2011) Towards allelic diversity in the storage proteins of old and currently growing tetraploid and hexaploid wheats in Portugal. Genet Resour Crop Evol 58:1051–1073

Roelfs A, Huerto-Espino J, Marshall D (1992) Rust diseases of wheat: concepts and methods of disease management. CIMMYT, Mexico

Rola EL, De Vallavieille-Pope C, Leconte M, Nazari K (2019) Diversity of genes for resistance to stripe rust in wheat elite lines, commercial varieties and landraces from Lebanon and Syria. Phytopathologia Mediterranea 58(3):607–627

Roselló M, Royo C, Álvaro F, Villegas D, Nazco R, Soriano JM (2018) Pasta-making quality QTLome from Mediterranean durum wheat landraces. Front Plant Sci 9:1512

Roshanzamir H, Kordenaeej A, Bostani A (2013) Mapping QTLs related to Zn and Fe concentrations in bread wheat (Triticum aestivum) grain using microsatellite markers. Iran J Genet Plant Breed 2:10–17

Royo C, Briceño-Félix GA (2011a) Spanish wheat pool. In: Bojean AP, Angus WJ, van Ginkel M (eds) The world wheat book. A history of wheat breeding, vol 2. Lavoisier, Paris, pp 121–154

Royo C, Dreisigacker S, Soriano JM, Lopes MS, Ammar K, Villegas D (2020) Allelic variation at the vernalization response (Vrn-1) and photoperiod sensitivity (Ppd-1) genes and their association with the development of durum wheat landraces and modern cultivars. Front Plant Sci 11:838

Royo C, Briceño-Félix GA (2011b) Spanish wheat pool. In: Bojean AP, Angus WJ, van Ginkel M (eds) The world wheat book. A history of wheat breeding. Lavoisier Publishing, Paris, pp 121–154

Ruiz M, Giraldo P, Royo C, Villegas D, Jose Aranzana M, Carrillo JM (2012) Diversity and genetic structure of a collection of spanish durum wheat landraces. Crop Sci 52(5):2262–2275

Saini DK, Devi P, Kaushik P (2020) Advances in genomic interventions for wheat biofortification: a review. Agronomy 10(1):62

Salvi S, Porfiri O, Ceccarelli S (2013) Nazareno Strampelli, the 'Prophet' of the green revolution. J Agr Sci-Cambridge 151:1–5

Sansaloni C, Franco J, Santos B, Percival-Alwyn L, Singh S, Petroli C, Campos J, Dreher K, Payne T, Marshall D, Kilian B (2020) Diversity analysis of 80,000 wheat accessions reveals consequences and opportunities of selection footprints. Nat Commun 11:1–2

Sareen S, Tyagi BS, Sarial AK, Tiwari V, Sharma I (2014) Trait analysis, diversity, and genotype x environment interaction in some wheat landraces evaluated under drought and heat stress conditions. Chil J Agric Res 74(2):135–142

Schmidt AL, McIntyre CL, Thompson J, Seymour NP, Liu CJ (2005) Quantitative trait loci for root lesion nematode (Pratylenchus thornei) resistance in Middle-Eastern landraces and their potential for introgression into Australian bread wheat. Aust J Agr Res 56(10):1059–1068

Schmidt J, Tricker PJ, Eckermann P, Kalambettu P, Garcia M, Fleury DL (2020) Novel alleles for combined drought and heat stress tolerance in wheat. Front Plant Sci 10:1800

Scott PR, Hollins TW (1974) Effects of eyespot on the yield of winter wheat. Ann Appl Biol 78(3):269–279

Shahzad A, Ahmad M, Iqbal M, Ahmed I, Ali GM (2012) Evaluation of wheat landrace genotypes for salinity tolerance at vegetative stage by using morphological and molecular markers. Genet Mol Res 11(1):679–692

Shamaya NJ, Shavrukov Y, Langridge P, Roy SJ, Tester M (2017) Genetics of Na+ exclusion and salinity tolerance in Afghani durum wheat landraces. BMC Plant Biol 17(1):209

Sharma DK, Torp AM, Rosenqvist E, Ottosen CO, Andersen SB (2017) QTLs and potential candidate genes for heat stress tolerance identified from the mapping populations specifically segregating for Fv/Fm in wheat. Front Plant Sci 8:1668

Shavrukov Y, Langridge P, Tester M (2009) Salinity tolerance and sodium exclusion in genus Triticum. Breed Sci 59(5):671–678

Shi R, Li H, Tong Y, Jing R, Zhang F, Zou C (2008) Identification of quantitative trait locus of zinc and phosphorus density in wheat (Triticum aestivum L.) grain. Plant and Soil 306:95–104

Singh R, Govindan V, Andersson MS (2017) Zinc-Biofortified Wheat: Harnessing Genetic Diversity for Improved Nutritional Quality. Sci Br Biofortif Ser 1:1–4

Smale M (1996) Understanding global trends in the use of wheat diversity and international flows of wheat genetic resources. CIMMYT, Mexico

Smale, M. and McBride, T., 1996. Understanding global trends in the use of wheat diversity and international flows of wheat genetic resources: part 1. CIMMYT 1995/96 World Wheat Facts and Trends: Understanding Global Trends in the Use of Wheat Diversity and International Flows of Wheat Genetic Resources (No. Look under series title. CIMMYT.). Centro Internacional de Mejoramiento de Maiz y Trigo (CIMMYT), Mexico

Sofalian O, Chaparzadeh N, Javanmard A, Hejazi MS (2008) Study the genetic diversity of wheat landraces from northwest of Iran based on ISSR molecular markers. Int J Agric Biol 10:466–468

Soriano JM, Villegas D, Aranzana MJ, García del Moral LF, Royo C (2016) Genetic structure of modern durum wheat cultivars and Mediterranean landraces matches with their agronomic performance. PLoS One 11:e0160983

Soriano JM, Villegas D, Sorrells MR, Royo C (2018) Durum wheat landraces from east and west regions of the Mediterranean basin are genetically distinct for yield components and phenology. Front Plant Sci 9:80

Stelmakh AF (1990) Geographic distribution of Vrn-genes in landraces and improved varieties of spring bread wheat. Euphytica 45:113–118

Sthapit J, Newcomb M, Bonman JM, Chen X, See DR (2014) Genetic diversity for stripe rust resistance in wheat landraces and identification of accessions with resistance to stem rust and stripe rust. Crop Sci 54(5):2131–2139

Sun H, Hu J, Song W, Qiu D, Cui L, Wu P, Zhang H, Liu H, Li L, Qu Y, Li Y, Li T, Cheng W, Zhou Y, Liu Z, Li J, Li H (2018) Pm61: A recessive gene for resistance to powdery mildew in wheat landrace Xuxusanyuehuang identified by comparative genomics analysis. Theor Appl Genet 131(10):2085–2097

Sun QM, Zhou RH, Gao LF, Zhao GY, Jia JZ (2009) The characterization and geographical distribution of the genes responsible for vernalization requirement in Chinese bread wheat. J Integr Plant Biol 51:423–432

Tadesse W, Amri A, Ogbonnaya FC, Sanchez-Garcia M, Sohail Q, Baum M (2016) Wheat. In: Mohar S, Upadhyaya H (eds) Genetic and genomic resources for grain cereals improvement. Academic Press, Oxford, UK, pp 81–124

Taiz L, Zeiger E (2006) Plant Physiology, 4th edn. Sinauer Associates Inc Publishers, Sunderland

Talas F, Longin F, Miedaner T (2011) Sources of resistance to Fusarium head blight within Syrian durum wheat landraces. Plant breeding 130(3):398–400

Tan C, Li G, Cowger C, Carver BF, Xu X (2018) Characterization of Pm59, a novel powdery mildew resistance gene in Afghanistan wheat landrace PI 181356. Theor Appl Genet 131(5):1145–1152

Tan C, Li G, Cowger C, Carver BF, Xu X (2019) Characterization of Pm63, a powdery mildew resistance gene in Iranian landrace PI 628024. Theor Appl Genet 132(4):1137–1144

Tehseen MM, Tonk FA, Tosun M, Amri A, Sansaloni CP, Kurtulus E, Yazbek M, Al-Sham'aa K, Ozseven I, Safdar LB, Shehadeh A, Nazari K (2020) Genome Wide Association Study of Resistance to PstS2 and Warrior Races of Stripe (Yellow) Rust in Bread Wheat Landraces. bioRxiv

Thakur P, Kumar S, Malik JA, Berger JD, Nayyar H (2010) Cold stress effects on reproductive development in grain crops: an overview. Environ Exp Bot 67(3):429–443

Thirumalaikumar VP, Devkar V, Mehterov N, Ali S, Ozgur R, Turkan I, Mueller-Roeber B, Balazadeh S (2018) NAC transcription factor JUNGBRUNNEN 1 enhances drought tolerance in tomato. Plant Biotechnol J 16(2):354–366

Thompson AL, Smiley RW, Paulitz TC, Garland-Campbell K (2016) Identification of resistance to Pratylenchus neglectus and Pratylenchus thornei in Iranian landrace accessions of wheat. Crop Sci 56(2):654–672

Thompson JP, O'reilly MM, Clewett TG (2009) Resistance to the root-lesion nematode Pratylenchus thornei in wheat landraces and cultivars from the West Asia and North Africa (WANA) region. Crop Pasture Sci 60(12):1209–1217

Thompson JP, Seymour NP (2011) Inheritance of resistance to root-lesion nematode (Pratylenchus thornei) in wheat landraces and cultivars from the West Asia and North Africa (WANA) region. Crop Pasture Sci 62(1):82–93

Toor AK, Bansal UK, Bhardwaj S, Badebo A, Bariana HS (2013) Characterization of stem rust resistance in old tetraploid wheat landraces from the Watkins collection. Genet Resour Crop Evol 60(7):2081–2089

Toreti A, Cronie O, Zampieri M (2019) Concurrent climate extremes in the key wheat producing regions of the world. Sci Rep 9:5493

Touzy G, Rincent R, Bogard M, Lafarge S, Dubreuil P, Mini A, Deswarte JC, Beauchene K, Le Gouis J, Praud S (2019) Using environmental clustering to identify specific drought tolerance QTLs in bread wheat (T. aestivum L.). Theor Appl Genet 132(10):2859–2880

Turner AS, Faure S, Zhang Y, Laurie DA (2013) The effect of day-neutral mutations in barley and wheat on the interaction between photoperiod and vernalization. Theor Appl Genet 126:2267–2277

Uauy C, Brevis JC, Dubcovsky J (2006a) The high grain protein content gene Gpc-B1 accelerates senescence and has pleiotropic effects on protein content in wheat. J Exp Bot 57:2785–2794
Uauy C, Distelfeld A, Fahima T, Blechl A, Dubcovsky J (2006b) A NAC gene regulating senescence improves grain protein, zinc, and iron content in wheat. Science 314:1298–1301
Ullah S, Bramley H, Daetwyler H, He S, Mahmood T, Thistlethwaite R, Trethowan R (2018) Genetic contribution of emmer wheat (Triticum dicoccon Schrank) to heat tolerance of bread wheat. Front Plant Sci 9:1529
Valdez VA, Byrne PF, Lapitan NL, Peairs FB, Bernardo A, Bai G, Haley SD (2012) Inheritance and genetic mapping of Russian wheat aphid resistance in Iranian wheat landrace accession PI 626580. Crop Sci 52(2):676–682
Valluru R, Reynolds MP, Davies WJ, Sukumaran S (2017) Phenotypic and genome-wide association analysis of spike ethylene in diverse wheat genotypes under heat stress. New Phytol 214(1):271–283
Van Oosten MJ, Costa A, Punzo P, Landi S, Ruggiero A, Batelli G, Grillo S (2016) Genetics of drought stress tolerance in crop plants. In Drought Stress Tolerance in Plants, vol 2. Springer, Cham, pp 39–70
Varella AC, Weaver DK, Blake NK, Hofland ML, Heo HY, Cook JP, Lamb PF, Jordan KW, Akhunov E, Chao S, Talbert LE (2019) Analysis of recombinant inbred line populations derived from wheat landraces to identify new genes for wheat stem sawfly resistance. Theor Appl Genet 132(8):2195–2207
Varella AC, Weaver DK, Cook JP, Blake NK, Hofland ML, Lamb PF, Talbert LE (2017) Characterization of resistance to the wheat stem sawfly in spring wheat landrace accessions from targeted geographic regions of the world. Euphytica 213(7):153
Vats P, Banerjee UC (2004) Production studies and catalytic properties of phytases (myo-inositol hexakisphosphate phosphohydrolases): an overview. Enzyme Microb Technol 35(1):3–14
Veraverbeke WS, Delcour JA (2002) Wheat protein composition and properties of wheat glutenin in relation to breadmaking functionality. Crit Rev Food Sci Nutr 42:179–208
Vikram P, Franco J, Burgueño-Ferreira J, Li H, Sehgal D, Saint Pierre C, Ortiz C, Sneller C, Tattaris M, Guzman C, Sansaloni CP (2016a) Unlocking the genetic diversity of Creole wheats. Sci Rep 6:23092
Vikram P, Franco J, Burgueño-Ferreira J, Li H, Sehgal D, Saint Pierre C, Cynthia Ortiz C, Clay Sneller C, Maria Tattaris M, Carlos Guzman C, Carolina Paola Sansaloni CP, Ellis M, Fuentes-Davila G, Reynolds M, Sonder K, Singh P, Payne T, Wenzl P, Sharma A, Bains NS, Singh GP, Crossa J, Singh S (2016b) Unlocking the genetic diversity of Creole wheats. Sci Rep 6:23092
Villa TC, Maxted N, Scholten M, Ford-Lloyd B (2005) Defining and identifying crop landraces. Plant Genet Res 3(3):373–384
Vishwakarma MK, Arun B, Mishra VK, Yadav PS, Kumar H, Joshi AK (2016) Marker-assisted improvement of grain protein content and grain weight in Indian bread wheat. Euphytica 208:313–321
Von Rünker K (1908) Die Systematischeeinteilung und Benen-ung der Getreidesortenfu¨rpr aktische Zwecke. Jahrbuch der Deutschenlandwirtschafts-Gesellschaft 23:137–167
Wamalwa MN, Owuoche J, Ogendo J, Wanyera R (2019) Multi-Pathotype Testing of Selected Kenyan Wheat Germplasm and Watkin Landraces for Resistance to Wheat Stripe Rust (Puccinia striiformis f. sp tritici). Races Agronomy 9(11):770
Wamalwa M, Tadesse Z, Muthui L, Yao N, Zegeye H, Randhawa M, Wanyera R, Uauy C, Shorinola O (2020) Allelic diversity study of functional genes in East Africa bread wheat highlights opportunities for genetic improvement. Mol Breedi 40(11):1–14
Wang S, Asuke S, Vy TTP, Inoue Y, Chuma I, Win J, Kato K, Tosa Y (2018a) A new resistance gene in combination with Rmg8 confers strong resistance against Triticum isolates of Pyricularia oryzae in a common wheat landrace. Phytopathology 108(11):1299–1306
Wang Y, Peng H, Liu G, Xie C, Ni Z, Yang T, Liu Z, Sun Q (2010) Identification and molecular mapping of a leaf rust resistance gene in spelt wheat landrace Altgold. Euphytica 174(3):371–375

Wang Z, Huang L, Wu B, Hu J, Jiang Z, Qi P, Zheng Y, Liu D (2018b) Characterization of an integrated active Glu-1Ay allele in common wheat from wild emmer and its potential role in flour improvement. Int J Mol Sci 19:E923
Wang Z, Li H, Zhang D, Guo L, Chen J, Chen Y, Wu Q, Xie J, Zhang Y, Sun Q, Dvorak J, Luo M, Liu Z (2015) Genetic and physical mapping of powdery mildew resistance gene MlHLT in Chinese wheat landrace Hulutou. Theor Appl Genet 128(2):365–373
Wang Z, Ren J, Du Z, Che M, Zhang Y, Quan W, Jiang X, Ma Y, Zhao Y, Zhang Z (2019) Identification of a major QTL on chromosome arm 2AL for reducing yellow rust severity from a Chinese wheat landrace with evidence for durable resistance. Theor Appl Genet 132(2):457–471
Ward RW, Yang ZL, Kim HS, Yen C (1998) Comparative analyses of RFLP diversity in landraces of Triticum aestivum and collections of T. tauschii from China and Southwest Asia. Theor Appl Genet 96:312–318
Wei YM, Hou YC, Yan ZH, Wu W, Zhang ZQ, Liu DC, Zheng YL (2005) Microsatellite DNA polymorphism divergence in Chinese wheat (Triticum aestivum L.) landraces highly resistant to Fusarium head blight. J Appl Genet 46(1):3–9
White PJ, Broadley MR (2005) Biofortifying crops with essential mineral elements. Trends Plant Sci 10(12):586–593
WHO (2009) Global health risks, mortality and burden of disease attributable to selected major risks. Geneva, Switzerland, WHO
Witcombe JR, Joshi A, Joshi KD, Sthapit BR (1996) Farmer Participatory Crop Improvement. I. Varietal Selection and Breeding Methods and Their Impact on Biodiversity. Exp Agric 32(04):445–460
Worland AJ, Korzun V, Röder MS, Ganal MW, Law CN (1998) Genetic analysis of the dwarfing gene Rht8 in wheat. Part II. The distribution and adaptive significance of allelic variants at the Rht8 locus of wheat as revealed by microsatellite screening. Theor Appl Genet 96:1110–1120
Wu L, Xia X, Rosewarne GM, Zhu H, Li S, Zhang Z, He Z (2015) Stripe rust resistance gene Yr18 and its suppressor gene in Chinese wheat landraces. Plant Breed 134(6):634–640
Wu XL, Wang JW, Cheng YK, Ye XL, Li W, Pu ZE, Jiang QT, Wei YM, Deng M, Zheng YL, Chen GY (2016) Inheritance and molecular mapping of an all-stage stripe rust resistance gene derived from the Chinese common wheat landrace "Yilongtuomai". J Hered 107(5):463–470
Xiao J, Jia X, Wang H, Zhao R, Fang Y, Gao R, Wu Z, Cao A, Wang J, Xue Z, Zhao W, Kang J, Chen Q, Chen P, Wang X (2011) A fast-neutron induced chromosome fragment deletion of 3BS in wheat landrace Wangshuibai increased its susceptibility to Fusarium head blight. Chromosome Res 19(2):225–234
Xiao J, Jin X, Jia X, Wang H, Cao A, Zhao W, Pei H, Xue Z, He L, Chen Q, Wang X (2013) Transcriptome-based discovery of pathways and genes related to resistance against Fusarium head blight in wheat landrace Wangshuibai. BMC Genomics 14(1):1–19
Xu H, Yi Y, Ma P, Qie Y, Fu X, Xu Y, Zhang X, An D (2015) Molecular tagging of a new broad-spectrum powdery mildew resistance allele Pm2c in Chinese wheat landrace Niaomai. Theor Appl Genet 128(10):2077–2084
Xu X, Li Q, Ma Z, Fan J, Zhou Y (2018) Molecular mapping of powdery mildew resistance gene PmSGD in Chinese wheat landrace Shangeda using RNA-seq with bulk segregant analysis. Mol Breed 38(3):23
Xu X, Liu W, Liu Z, Fan JR, Zhou Y (2020) Mapping powdery mildew resistance gene pmYBL on chromosome 7B of Chinese Wheat (Triticum aestivum L.) Landrace Youbailan. Plant Dis 104(9):2411–2417
Xu Y, An D, Liu D, Zhang A, Xu H, Li B (2012) Molecular mapping of QTLs for grain zinc, iron and protein concentration of wheat across two environments. Field Crop Res 138:57–62
Xue F, Wang C, Li C, Duan X, Zhou Y, Zhao N, Wang Y, Ji W (2012) Molecular mapping of a powdery mildew resistance gene in common wheat landrace Baihulu and its allelism with Pm24. Theor Appl Genet 125(7):1425–1432

Xue F, Zhai WW, Duan XY, Zhou YL, Ji WQ (2009) Microsatellite mapping of powdery mildew resistance gene in wheat landrace Xiaobaidong. Acta Agron Sin 34:1193–1198

Xynias IN, Kozub NA, Sozinov IA (2011) Analysis of hellenic durum wheat *Triticum turgidum* L. var. *durum* germplasm using gliadin and high-molecular-weight glutenin subunit loci. Cereal Res Commun 39:415–425

Yan L, Loukoianov A, Tranquilli G, Helguera M, Fahima T, Dubcovsky J (2003) Positional cloning of the wheat vernalization gene VRN1. Proc Natl Acad Sci U S A 100:6263–6268

Yeken MZ, Akpolat H, Karaköy T, Çiftçi V (2018) Assessment of Mineral Content Variations for Biofortification of the Bean Seed. Int J Agri Wild Sci 4(2):261–269

Yen C, Luo MC, Yang JL (1988) The origin of the Tibetan weedrace of hexaploid wheat, Chinese Spring, Chengdu-guang-tou and other landraces of the white wheat complex from China. In: Miller TE, Koebner RMD (eds.) Proceedings of the 7th International Wheat Genetics Symposium, Cambridge, pp 175–179

Yu S, Wu J, Wang M, Shi W, Xia G, Jia J, Kang Z, Han D (2020) Haplotype variations in QTL for salt tolerance in Chinese wheat accessions identified by marker-based and pedigree-based kinship analyses. Crop J. https://doi.org/10.1016/j.cj.2020.03.007

Yuan FP, Zeng QD, Wu JH, Wang QL, Yang ZJ, Liang BP, Kang ZS, Chen XH, Han DJ (2018) QTL mapping and validation of adult plant resistance to stripe rust in Chinese wheat landrace Humai 15. Front Plant Sci 9:968

Zampieri M, Ceglar A, Dentener F, Toreti A (2017) Wheat yield loss attributable to heat waves, drought and water excess at the global, national and subnational scales. Environ Res Lett 12(6):064008

Zeven AC (1998) Landraces: a review of definitions and classifications. Euphytica 104:127–139

Zeven AC (1999) The traditional inexplicable replacement of seed and seed ware of landraces and cultivars: a review. Euphytica 110:181–191

Zeven AC (2000) Traditional maintenance breeding of landraces: 1. Data by crop. Euphytica 116:65–85

Zhang K, Wang J, Qin H, Wei Z, Hang L, Zhang P, Reynolds M, Wang D (2019a) Assessment of the individual and combined effects of Rht8 and Ppd-D1a on plant height, time to heading and yield traits in common wheat. The Crop Journal 7:845–856

Zhang P, Gebrewahid TW, Zhou Y, LI Q, LI Z, LIu D (2019b) Seedling and adult plant resistance to leaf rust in 46 Chinese bread wheat landraces and 39 wheat lines with known Lr genes. J Integr Agric 18(5):1014–1023

Zhang X, Pan H, Bai G (2012) Quantitative trait loci responsible for Fusarium head blight resistance in Chinese landrace Baishanyuehuang. Theor Appl Genet 125(3):495–502

Zhang X, Yang S, Zhou Y, He Z, Xia X (2006) Distribution of the Rht-B1b, Rht-D1b and Rht8 reduced height genes in autumn-sown Chinese wheats detected by molecular markers. Euphytica 152:109–116

Zhang XK, Xiao YG, Zhang Y, Xia XC, Dubcovsky J, He ZH (2008) Allelic variation at the vernalization genes Vrn-A1, Vrn-B1, Vrn-D1, and Vrn-B3 in Chinese wheat cultivars and their association with growth habit. Crop Sci 48:458–470

Zhao C, Cui F, Wang X, Shan S, Li X, Bao Y, Wang H (2012) Effects of 1BL/1RS translocation in wheat on agronomic performance and quality characteristics. Field Crops Res 127:79–84

Zhao FJ, Su YH, Dunham SJ, Rakszegi M, Bedo Z, McGrath SP, Shewry PH (2009) Variation in mineral micronutrient concentrations in grain of wheat lines of diverse origin. J Cereal Sci 49:290–295

Zhou Y, Chen Z, Cheng M, Chen J, Zhu T, Wang R, Liu Y, Qi P, Chen G, Jiang Q, Wei Y, Luo MC, Nevo E, Allaby RG, Liu D, Wang J, Dvorak J, Zheng Y (2018) Uncovering the dispersion history, adaptive evolution and selection of wheat in China. Plant Biotechnol J 16(1):280–291

Ziv O, Eyal Z (1977) Assessment of yield component losses caused in plants of spring wheat cultivars by selected isolates of septoria tritici. Phytopathology 68:791–796

Zurn JD, Newcomb M, Rouse MN, Jin Y, Chao S, Sthapit J, See DR, Wanyera R, Njau P, Bonman JM, Brueggeman R, Acevedo M (2014) High-density mapping of a resistance gene to Ug99 from the Iranian landrace PI 626573. Molecular breeding 34(3):1

Index

A
Abiotic stress
cold stress, 198
collective dry matter, 197
drought stress, 197
heat stress, 199
responses, 196
salinity stress, 200
waterlogging stress, 201
Afghanistan wheat landraces, 70
AFLP molecular markers, 70
Agricultural diversity, 52
Agriculture, 35, 215
Agrobiodiversity, 23
Agroecological zones, 76
Agroecosystems, 26
Allele mining, 27
AMOVA, 74
Amplified fragment length polymorphism (AFLP) markers, 13, 15, 21
Ancestor-descendant relationships, 73
Animal feed, 7
Antioxidant compounds, 156
Arabinoxylans, 159
Ash-rich einkorn wheat, 84

B
Bacterial leaf streak (BLS), 206, 235
Beriberi disease, 149, 150
Biodiversity, 121–123, 129, 134
Biofortification, 102, 227
Biological diversity, 51
Biological yield (BY), 60
Biotic and abiotic stress, 128
Biotic stress
CB, 235
DB, 236
eyespot, 234
FHB, 234
fungal diseases, 205
genetic and phenotypic researches, 202
leaf rust, 205, 232
nematode, 207
pathogen, 230
PM resistance, 233
resistant and tolerant cultivars, 207
rot diseases, 203
SB, 235
SBWMV and WYMV, 205
seed borne diseases, 202
SNB, 235
soil borne disease, 203
STB, 234
stem rust, 204, 232
stripe rust, 204
tan spot, 234
WB, 236
wheat rust diseases, 203
Botanic gardens, 45
Breeding, 25
Breeding studies, wheat landraces
cereals, 85
copper content, cereals, 88
dietary Zn deficiency, 88
durum wheat, mineral content, 86
Einkorn wheat, 86
Einkorn X durum wheat cross, 87

N. Zencirci et al. (eds.), *Wheat Landraces*,
https://doi.org/10.1007/978-3-030-77388-5

Breeding studies, wheat landraces (*cont.*)
genetic material, 85
genitors, 85
genotypes, 87
Kamut wheat, 86
Kamut X bread wheat cross, 86, 87
macro- and micronutrient contents, 88
microelement contents, 89
mineral content, 85
primary gene source, 85
Triticum species, 88
wheat gene pool, 88
winter wheat breeding program, Turkey, 85
Bread wheat, 21
Bread wheat landrace, 18, 70
agronomic characters, 65
genetic composition, 65
genotypes, 65
populations, 65
properties and resistance, 65
qualitative and quantitative characters, 65
relative variable character, 66
surveyed morphological differences, 65
Brittle rachis, 61

C
Calcium, 84
Carbohydrates, 150, 156
Carotenes, 151–153
Carotenoids, 155, 159
Celiac disease (CD), 159
Cereal aphids, 237
Cereal cyst nematodes (CCN), 237
Cereals, 85, 155
CGIAR Research Programs (CRPs), 36
Chinese Spring (CS), 221
Chinese wheat landraces, 72
Chronic diseases, 156
Climate change (CC), 26, 169
abiotic stress, 171
on agricultural production, 172
agricultural sector, 169
anatomical effects, 174
CO_2 gas, 169, 174
GHG accumulation, 169
and GW, 170, 171
greenhouse gases, 170
positive and negative effects, 173
Turkey, 185
Cold stress
frost tolerance, 199
germination, 198
temperatures, 199
Commercial seed market, 60
Common bunt (CB), 235
Consultative Group on International Agricultural Research (CGIAR), 39
Conventional/modern breeding techniques, 9
Copper, 83, 84
Crassulacean acid metabolism (CAM), 171

D
D-genome, 95
Dietary fiber, 156
arabinoxylans, 108
AX components, 107
β-glucans, 108
carbohydrates, 107
genotype and environment effects, 107
oat and barley, 108
WE-AX, 108
WU-AX contents, 108
Dietary Zn deficiency, 88
DNA markers, 74
DNA structure variations
bread wheat, 76
durum wheat, 75, 76
einkorn wheat, 73, 74
emmer wheat, 74, 75
Domesticated wheat varieties, 59
Domestication process, 59
Drift, 1
Drought stress, 197
Durum wheat, 15, 17, 18, 20, 22, 24
canonical factors, 64
commercial cultivated, 63
diversity and distribution, 64
genotypes, 64
germplasm, 63
grain yield, 63
groups, 64
monomorphism, 63
morphological characteristics, 62
optimal altitude ranges, 64
phenotypic divergence, 64
tetraploid, 63
variation, 63, 64
Durum wheat cultivars, 106
Dwarf bunt (DB), 236
Dynamic evolutionary history, 68

E
Ecogeographic groups, 75
Einkorn germplasm, 73

Einkorn wheat (*T. monococcum* ssp. *monococcum*) landraces, 7, 24, 61, 86
Emmer wheat (*Triticum turgidum* ssp. *dicoccon Schrank Tell.*) landraces, 62
Environmental conditions, 60, 76
Environmental protection policies, 51
Enzymes, 153
Esterase/peroxidase isozymes, 72
Ethiopian tetraploid wheat landraces, 75
Ex situ conservation
  botanic gardens, 45
  collection program, 38
  erosion, plant genetic resources, 38
  evolution process, 37
  gene banks, 38
  genetic resources, 37
  Millennium Seed Bank, 39
  organizations, 37
  seed collections, 39
  seed gene banks (*see* Seed gene banks)
  Turkey, 40
  wheat, 40
Eyespot, 234

F

Fiber, 159, 160
Flavin adenine dinucleotide (FAD), 149
Flavin adenine mononucleotide (FMN), 149
Flavonoids, 157
Food safety, 89
Foreign-origin wheat materials, 36
Formal breeding programmes, 59
Formic acid, 150, 160
Functional genes, 27
Fusarium, 72
Fusarium head blight (FHB), 233

G

GEF/SGP Projects, 53
Gene banks, 38, 46
Gene Conservation and Management Areas (GEKYA), 52
Genetic diversity, 128, 216, 217
Genetic diversity analyses, 74
Genetic diversity characterization
  bread wheat landraces, 65–66
  climate change, 60
  durum wheat, 62–64
  einkorn, 61
  emmer wheat, 62
  morphological and agronomical traits, 60, 61
  phenotypic plasticity, 60
Genetic Diversity Project, 52
Genetic engineering techniques, 128
Genetic erosion, 54, 55
Genetic resources, 124, 127, 128, 137, 140
Genitors, 85
Genomes, 1
Genotypes, 8, 60
Geographical indication (GI), 138
  economic activity, 139
  economic factors, 140
  economy, 137
  environmental pollution, 137
  European Union, 139
  food chain, 138
  gastronomy, 138
  genetic resources, 137
  industrialization process, 137
  local development dynamics, 137
  local economic development, 139
  natural and human resources, 138
  negative effects, 137
  plant species, 138
  region, 138
  registration structures, 138
  wheat-based products, 139
Germplasm collections, 60
Gliadin band patterns, 68
Global warming (GW), 169
Glutenin proteins, 68
Glutenins, 97
Grain protein content (GPC), 228
Grain yield per plant (GYP), 60–61
Green revolution, 5, 218, 238

H

Harvest index (HI), 61
Heat shock factors (HSFs), 199
Heat stress
  chromosomes, 200
  dry-hot wind and high humidity temperatures, 200
  environmental factors, 199
  QTL, 200
Heat-tolerant genotype selection, 225
Heterozygosity, 68
Hexaploid wheat, 76
Hidden hunger, 83
High molecular weight (HMW), 68, 69
High-yielding varieties (HYVs), 216
HMW-glutenin proteins, 68
HMW-glutenin subunits, 70
Hybridization, 1

Hydroxybenzoic acid, 157
Hydroxycinnamic acid, 157

I
IEF polyacrylamide gel electrophoresis, 71
In situ conservation
advantages, 50
biodiversity, 48, 50
civil society movements, 46
continuity, 50
crop plants, 46
farmer conditions, landraces, 48, 49
gene banks, 46
genetic diversity, 47, 48
genetic resources, 46
grains, 47
herbaceous plants, 46
human beings, 46
landraces, 46, 48, 50
markets, 49
medicinal plants, 46
modern agricultural systems, 47
modern breeding techniques, 49
natural habitats, 46
on-site conservation practices and policies, 49
seed, 47
socioeconomic structure, farmer enterprises, 50
traditional agricultural systems, 51
traditional farming methods, 50
Turkey, 49, 51–54
wheat varieties, Turkey, 49
In Situ Conservation Protection and Management of Endangered Plant Species, 52
Inflammation process, 155
Information function value
aesthetic value, 132
artistic and spiritual value, 132
cultural and historical value, 130–132
recreation and tourism, 132
scientific and educational use value, 129
International Center for Agricultural Research in the Dry Areas (ICARDA), 4
International Winter Wheat Improvement Program (IWWIP), 4
Iranian bread wheat (*Triticum aestivum* L.), 70
Iranian durum wheat (*Triticum turgidum*), 69
Iron, 83–85
Isoelectric focusing (IEF), 71
Isoenzymes variations
bread wheat landraces, 72, 73
clustering, 71
drum wheat landraces, 72
emmer wheat landraces, 71, 72
genetic divergence, 71
populations, 71
Isozymes, 72
Isozymic variation, 73

J
Jordanian durum wheat landraces, 21
Jordanian wheat landraces, 21

K
Kamut wheat, 86

L
Landrace formation
abiotic factors, 184
agronomic and botanic responses, 175–176
agronomic traits, 176, 181
atmospheric CO_2 concentration, 180
CC and GW, 186
enzyme RuBisCo, 180
evapotranspiration and drops, 179
food crops, 174
genotypes, 182
and GW, 179
self-pollinated crop, 174
soil moisture, 180
in Turkey, 185
wheat landraces, 179
WUE, 184
Landrace genotypes, 64
Landraces, 7, 48
Leaf rust (LR), 231, 232
Lipid peroxidation, 157
Lipoxygenase (LOX), 152
Low molecular weight (LMW), 68

M
Macro-microelements, 83, 84
Macro-minerals, 83, 84, 153
Magnesium, 83, 84
Marker-assisted selection (MAS), 198, 219
Mediterranean wheat landraces, 25
Mesopotamia, wheat landrace
agricultural biodiversity, 22
allele mining, 27
bread wheat, 23–25, 27
climate change, 23, 26
durum wheat, 23, 24, 27
durum wheat breeding, Turkey, 25

einkorn wheat, 24
emmer wheat, 24
evolution, 16
Fertile Crescent, 17, 22, 23, 25, 26
folk varieties, 16
functional genes, 27
genetic diversity, 27
genetic structure, 16
genetically heterogeneous populations, 16
green revolution, 22
historical origin, 16
molecular genetic studies, 25
multiple genotypes, 25
natural and human selection, 16
nitrogen uptake, 25
nonbreeding approaches, 26, 27
nutritional value, 25
on-farm landrace conservation, 26
origin, 24
pre-anthesis uptake, 16
soluble carbohydrates, 16
sustainable conservation, 26
Turkey, 23
yield penalty, 16
Metalliferous Mountains, 61
Mexican bread wheat landraces, 77
Micro minerals, 153
Microelement contents, 89
Micronutrients, 6
Microsatellite markers (SSRs), 74
Migration, 1
Millennium Seed Bank, 39
Millennium Seed Bank Partnership (MSBP), 39
Mineral deficiencies, 85, 153
Minerals, 154
Mixograph, 110
Modern breeding techniques, 49
Modern wheat cultivars, 2, 9
Molecular analysis, 73, 75
Molecular marker-assisted selection (MAS), 127
Molecular markers, 13, 73
Monomorphism, 63
Morphological field evaluations, 62
Multidimensional scaling (MDS), 27
Mycosphaerella graminicola, 234

N

NAD phosphate (NADP), 149
National Biodiversity Strategies and Action Plans, 52
National Cool Season Cereals Research and Training Project, 4
National Plan for In Situ Conservation, 52
National Plant Genetic Resources Program, 54
National Seed Gene Bank, 41, 42
Natural selection, 1
New breeding techniques (NBTs), 196
Next-generation sequencing (NGS), 77
Niacin, 151, 160
Nicotinamide adenine dinucleotide (NAD), 149
Nonbreeding approaches, 26, 27
Non-starch polysaccharide (NSP), 159
Nutrients, 7

O

Omani durum wheat, 77
Omani wheat landraces, 65
On-farm conservation, *see* In situ conservation
Organizations, 37
Osman Tosun Gene Bank, 44, 45
Oxidative and hydrolytic enzymes, 157

P

Payne scoring method, 70
Pearson's correlation analysis, 67
Phenols, 156, 157
Phenotypic differentiation, 64
Phenotypic plasticity, 60
Phosphorus, 84
Photosensitive allele, 77
Phylogenetic analysis, 14
Phytase activity, 103
Phytochemicals and antioxidants
alkylresorcinols, 103
Carotenoid concentration, 106
carotenoids, 105
ferulic acid, 103
phytosterols, 105
polyphenols, 104
tocochromanols, 103
tocols, 104
tocopherols and tocotrienols, 104
tocotrienols, 104
volatile compound, 106
Plant breeding programmes, 62, 88
Plant genetic resources
ex situ conservation, 37 (*see also* Ex situ conservation)
in situ conservation (*see* In situ conservation)
Polymorphism, 63, 64, 74, 75
Polyphenols, 156, 159
Polysaccharides, 159
Potassium, 84

Powdery mildew (PM), 233
Principal component analysis (PCA), 27
Production function value
  biotechnology, 127, 128
  commercial use value, 124, 125
  food source, 125, 126
  genetic resources value, 129
  human and animal nutrition, 125, 126
  medical resource value, 127
  raw material value, 126
  traditional medicine, 127
Progressive adaptation, 13
Protected Area, 51
Protected Designation of Origin (PDO), 138
Protected Geographical Indication (PGI), 138
Proteins, 156, 158, 159
Pyridoxine, 150, 160

Q
Quality traits
  agronomic/genetic biofortification, 230
  allele, 228
  biofortification, 227
  encoding, 229
  endosperm texture, 228
  Fe and Zn concentrations, 227
  germplasm, 229
  gluten, 229
  grains, 228
  QTL, 228
  variants/alleles, 230
  wheat cultivars, 229
Quantitative trait loci (QTL) tool, 198, 221

R
Rapid Visco Analyser (RVA), 110
Resistant starch (RS), 157
Riboflavin, 151, 160
Root lesion nematodes, 236
Russian wheat aphid (RWA), 236
Rust diseases, 231

S
Salinity stress
  genetic opportunities, 200
  osmotic tolerance, 201
  salinity tolerance, 201
Salinity tolerance
  biomarkers, 224
  genotypes, 223, 225
  goatgrass, 224
Seed exchange system, 60
Seed gene banks
  CIMMYT, 39
  ex situ conservation, 38
  GENESYS, 39
  genetic diversity, 37, 46
  global-scale research, 38
  IBPGR, 38
  MSBP, 39
  Osman Tosun Gene Bank, 44, 45
  properties, 39
  research institutions, 37
  Seed Vault, 39
  Turkey
    Field Crops Central Research Institute, Ankara, 42
    National Seed Gene Bank, 41, 42
    renovation and reproduction activities, 41
    samples, 41
    Turkish Seed Gene Bank, 42–44
Seed management, 26
Seed storage/endosperm proteins variation
  bread wheat, 70
  durum wheat, 68, 70
  einkorn wheat, 66, 67
  emmer wheat, 67, 68
  interspecies, 66
  prolamins, 66
Selenium, 84
Septoria tritici blotch (STB), 234
Single-nucleotide polymorphism markers, 76
SNP-based diversity map, 77
Soil-borne wheat mosaic virus (SBWMV), 205
Southeastern Turkey, wheat landraces characteristics
  American wheat history, 18
  botanical varieties, 18
  bread wheat landrace, 18
  durum wheat, 17, 18
  frequency, 19
  geographical distribution, durum wheat, 19
  high-yielding Mexican wheats, 17
  micronutrient contents, 20
  morphotype approach, 19
  region, 18
  southeastern Anatolia, 20
  twentieth century, 17
  wheat program, 17
Spot blotch (SB), 235
Stagonospora nodorum blotch (SNB), 235
Stem rust, 232
Stress, 171

Syrian Arab Republic, 63
Syrian durum wheat landraces, 64

T
Tan spot, 234
T-cell-mediated disease, 159
Tetraploid wheats, 14
Thiamine, 150, 151
Tocols, 154, 156
Total economic value approach
  economic valuation, 124
  economy, 123
  ecosystem, 124
  information function value (*see* Information function value)
  labor-value theory, 123
  market-oriented change, 123
  monetary value, 123
  production function value (*see* Production function value)
  social values, 123
  total economic value, 124
  utility-value theory, 123
Traditional agricultural systems, 8, 59
Traditionally Specially Guarantee (TSG), 138
Triticum aestivum, 147, 148, 150, 154
Triticum dicoccum, 148
Triticum durum, 148, 150
Triticum monococcum, 147, 148
Triticum species, 88
Turkey
  agriculture, 35
  ex situ conservation studies, genetic resources, 40
  in situ conservation, 51–54
  seed gene banks
    Field Crops Central Research Institute, Ankara, 42
    National Seed Gene Bank, 41, 42
    renovation and reproduction activities, 41
    samples, 41
    Turkish Seed Gene Bank, 42–44
  wheat, 35, 36
  wheat breeding activities, 36
Turkish-cultivated einkorn wheat, 66
Turkish-cultivated emmer wheat, 67
Turkish Seed Gene Bank, 40, 42
  biotechnology laboratory, 43
  collection programs, 43
  conservation and sustainable use, biodiversity, 42
  drying chamber, 43
  genetic material, 43
  herbarium, 43, 44
  organization, 42
  plant breeding, 43
  seed conservation rooms, 44
  seed physiology laboratory, 43
Turkish wheat landraces, 186–187
Turkish Wheats, 36

V
Vernalization, 219

W
Water-extractable arabinoxylan (WE-AX), 108
Waterlogging, 201
Water-unextractable arabinoxylan (WU-AX), 108
Wernicke-Korsakoff syndrome (WKS), 149, 150
Wheat, 59
  climate change, 5
  evolution, 1, 2, 15
  human nutrition, 5
  income, rural areas, 5
  nutritional properties, 84
  nutritional quality, 83
  origin, 1, 2
  Southwest Asia, 2
  Turkey, 2–5
  wheat-producing countries, 5
Wheat and Maize Improvement Center (CIMMYT), 4
Wheat blast (WB), 236
Wheat breeding programs, 5, 85
Wheat grain, 24
Wheat landraces, 3, 182
  abiotic and biotic stresses, 223
  adaptability, 218
  adaptive traits, 218
  advantages, 7–9
  AFL, 13
  alleles, 220, 222, 226
  amylose and amylopectin, 99
  animal feed, 7
  antioxidant contents, 160
  bioavailability, 102
  biofortification, 102
  biological diversity, 121
  biotic and abiotic stresses, 93
  bran fraction, 149
  bred cultural varieties, 5
  breeders, 147

Wheat landraces (*cont.*)
breeding, 217
breeding methods, 93
breeding periods, 109
breeding varieties, 36
Bulgur, 111
celiac disease, 160
cereal products, 101
character, 94, 96
chemical contents
B vitamins, 149–151
carotenes, 151–153
fiber, 159, 160
macro- and microelements, 153, 154
phenols, 156, 157
proteins, 158, 159
starch, 157
tocols, 154, 156
chronic diseases, 149, 160
CIMMYT, 221
communities, 122
crop genetic resources, 36
CS, 221
cultivation, 217
definition, 6, 59
disadvantages, 7–9
diversification, 147
domestication, 216
drought and heat stress, 225, 226
durum wheat, 20
dwarfing genes, 220
economic value, 122
economy, 140
einkorn kernels, 95
endosperm, 99
environments, 93
Fertile Crescent, 13, 14
food crops, 216
food safety, 121
food scarcity problem, 237
food security, 160
food source, 6
functional and nutritional properties, 101
GA-sensitive, 220
gene pools, 36
genetic diversity, 238
genetic uniformity, 5
genotypes, 97, 217
geographical indication system (*see* Geographical indication system)
germplasms, 147
gluten, 97, 98, 160
glutenin subunits, 111
grain hardness, 95
grain quality, 96
grain size, 94
healthy foods, 94
hemoglobin, 160
hexaploid, 96
high-quality sequences, 223
human and natural selection, 217
human life, 121
human resources, 122
immune system, 160
indirect use value, 133
Iraq, 21, 22
Jordan, 21
Karacadağ, 14, 15
kernel weights, 95
lipids, 100
local products, 122
macro-microelements, 83, 85
manufacturers, 122
medicinal effects, diseases, 6
micronutrients, 6, 101
milling process, 102
mineral elements, 101
minerals, 160
molecular markers, 13
multidisciplinary, 122
nitrogen fertilization, 98
nutritional and economic value, 121
nutritional values, 94
organic farming, 148
out-of-use value, 133, 134
phenolic acids, 148
phenolics, 160
phytochemicals and antioxidants, 103
progressive adaptation, 13
protection, 122
protein content, 111
protein quality, 97
quality traits, 228
rheological analyses, 109
role, 93
rural development, 122
rural development tool
agricultural production sector, 134
agriculture, 134
agriculture-based industry, 134
animal feed, 136
economic values, 136
family farms, 135
food need and health factor, 136
geographical indication system, 136
historical process, 135

local areas, 135
local dynamics, 136
local economic development, 136
local issues, 135
local landraces, 135
orientation, 134
physical change, 134
policy approaches, 135
political revenue-oriented policies, 134
producers, 136
sectors, 134
social and cultural values, 135
socioeconomic and sociocultural structure, 135
sustainability, 136
rust diseases, 231
RVA, 110
salinity, 223, 225
seeds, 148
sensational varieties, 218
social and economic values, 140
sociocultural value, 122
Southeastern Turkey, 13
starch, 99
sustainability, 121, 122, 140
Syria, 20, 21
thiamine, 100
total economic value, 122
traditional, 93
Turkey, 6, 7, 35, 36
type 2 diabetes, 149
use, 220
USSR breeding program, 218
vernalization, 219
vitamin, 100
vs. resistance
bread and durum, 195
cultivation, 193
cultivation and uses, 194
definition, 193
distribution, 194
environmental stresses, 193
genetic characteristics, 195
genetic diversity, 196
research, 194
wheat quality, 98
wheat varieties, 147
yield-related traits, 222
Zn and Fe accumulation, 102
Wheat stem sawfly (WSS), 237
Wild einkorn, 15
Wild emmer wheat, 14
Winter wheat breeding program, 85
Work input to peak (WIP), 110

X
Xanthophylls, 151

Y
Yellow rust, 231

Z
Zinc, 83, 84
Zymogram, 71

The manufacturer's authorised representative in the EU is Springer Nature Customer Service Centre GmbH, Europaplatz 3, 69115 Heidelberg, Germany. If you have any concerns regarding our products, please contact ProductSafety@springernature.com

Printed and bound by CPI Group (UK) Ltd, Croydon, CR0 4YY
30/09/2026
02225462-0006